Advances in

MODELING *the* MANAGEMENT *of* STORMWATER IMPACTS

***Volume* 5**

edited by William James

Also in this series:

[1]*Advances in Modeling the Management of Stormwater Impacts,* Proceedings of the 1995 Conference on Stormwater and Water Quality Management Modeling.
[1]*Modern Methods for Modeling the Management of Stormwater Impacts,* Proceedings of the 1994 Conference on Stormwater and Water Quality Management Modeling.
[2]*Current Practices in Modelling the Management of Stormwater Impacts* Proceedings of the 1993 Conference on Stormwater and Water Quality Management Modeling.
[2]*New Techniques for Modelling the Management of Stormwater Quality Impacts,* Proceedings of the 1992 Conference on Stormwater and Water Quality Management Modeling.

[1]may be ordered from CHI (fax: 519-767-2770) or through Ann Arbor Press (fax: 313-475-8852).
[2] may be ordered from Lewis Publishers/CRC Press (fax: 407-998-9114; or, in continental U.S. only, 800-374-3401)

Advances in

MODELING *the* MANAGEMENT *of* STORMWATER IMPACTS

***Volume* 5**

edited by William James

Advances in

MODELING *the* MANAGEMENT *of* STORMWATER IMPACTS

Volume 5

edited by William James

CRC Press
Taylor & Francis Group
Boca Raton London New York

CRC Press is an imprint of the
Taylor & Francis Group, an **informa** business

First published 1997 by Computational Hydraulics International

Published 2019 by CRC Press
Taylor & Francis Group
6000 Broken Sound Parkway NW, Suite 300
Boca Raton, FL 33487-2742

First issued in paperback 2019

ISBN-13: 978-0-367-44835-6 (pbk)
ISBN-13: 978-1-57504-227-5 (hbk)

Visit the Taylor & Francis Web site at
http://www.taylorandfrancis.com

and the CRC Press Web site at
http://www.crcpress.com

Library of Congress Cataloguing-in-Publication Data

Catalog record is available from the Library of Congress.

CHI Catalog Number: R195

Reference Data:

Advances in Modeling the Management of Stormwater Impacts Volume 5, Proceedings of the Stormwater and Water Quality Management Modeling Conference, Toronto, Ontario, February 22-23, 1996. xv + 520 pp.
James, William, 1937- , Editor
Compiled and published by Computational Hydraulics International, Guelph, Ontario, Canada.
Cover design by Denise Eva Hands.

Acknowledgements

Special thanks once again to Dr. Lyn James for organizing the conference, collecting the papers, liaising with authors and editors, and doing the desk-top publishing; to Rob James, for his help with the graphics; and to Denise Hands for the cover design.

Thanks to all who helped; they made the publication of this book possible, and the experience fun: the reviewers, who again contributed valuable comments and judgement; the authors, who gave invigorating presentations, and responded cheerfully and quickly to the editorial reviews; and all the participants in the *Stormwater and Water Quality Management Modeling Conference,* held in Toronto, February 22-23, 1996.

Thanks also to the following: the American Society of Civil Engineers Urban Water Resources Research Council, the American Water Resources Association, the US Environmental Protection Agency, the Canadian Society for Civil Engineering, the Ontario Ministry of Environment and Energy, and the Association of Conservation Authorities of Ontario for sanctioning the conference; and my graduate students Chris Kresin and Yiwen Wang for helping at the conference.

Again, this book and the associated conference continue to be self-supporting; so no acknowledgements for financial support are necessary.

Abstracts

The chapters of this book have been abstracted by the following agencies:

- Congressional Information Services, Environment Abstracts
- Engineering Information Services
- GeoArchive

Preface

Very much like its predecessors, *Advances in Modeling the Management of Stormwater Impacts - Volume 5,* fifth in this series of monographs, depends upon contributions from authors who present their findings at the international conference on stormwater and water quality modeling. Similar meetings have been held elsewhere and at other times, but, in this present series, which started in 1992, meetings are held every winter in Toronto.

Each monograph is produced to the same high standards. Every article is reviewed by two experts, who are listed at the back, and you will agree that we have been blessed by their quality and willingness to do the work. Content is reviewed for *accuracy, relevance, merit and readability.* Every article then undergoes an extensive system of editing: the language, notation, presentation and structure is rendered consistent, and we check all reviewers suggestions. Individual authors are involved in making significant changes, where required. Sometimes a great deal of patience is called for. As a result, only half to three-quarters of the presentations make it into final print. Finally we have a monograph of textbook and journal quality.

Also, we have been blessed by some extraordinarily interesting and valuable contributions - many readers find the set a valuable source of information on emerging techniques and applications, and they are on the shelves of many North American Universities. Unusually, it seems, the monographs are distributed to the attendees at the *ensuing* conference - if you have missed any, details of the sources are provided in the end papers of this one.

Readers are cautioned, however, not to depend on the review procedures catching all errors and omissions; moreover the editors, publishers and reviewers are clearly not responsible for any opinions, or recommendations in the publication.

Subjects covered herein may be subsumed under: continuous models, urban pavement, structures and metering, combined sewer overflows, water quality management, ponds and wetlands, and planning, and the monograph ends with a chapter and CDROM on references in the non-serial literature. For readers who wish a little more detail, the chapters cover:

- models for computing pollutant removal by stormwater ponds,
- time series data management for very long term continuous modeling,
- review of statistical methods and continuous modeling,
- continuous HSPF modeling for baseflow and control of erosion,
- continuous SWMM modeling for conservation of groundwater,
- a water balance model for recharge of groundwater,
- the attributes of programs derived from and for SWMM,
- thermal effect on stormwater of urban pavement,
- infiltration through clogged porous concrete block pavers,
- contaminants from four different pavements in a parking lot,
- improvements in the performance of modern street sweepers,
- a tangential helicoidal ramp for stormwater inlets,
- errors in metering stormwater flows,
- a large CSO storage tunnel in Toronto,
- impacts and control of CSOs on large rivers,
- cost-effectiveness of various stormwater and CSO controls,
- a simplified CSO model for CSO pollution loadings,
- modeling fecal coliforms,
- small-scale stormwater treatment devices,
- water quality modeling and groundwater pollution,
- setting total maximium daily pollutant loads,
- water information management for flow modeling and sewer permits,
- modeling phosphorous dynamics in stormwater wetlands,
- retrofitting stormwater ponds for water quality enhancement,
- urban watershed planning,
- municipal stormwater guidelines,
- SCS runoff curve numbers for small watersheds, and
- a bibliographic CDROM, which provides limited access to a number of otherwise inaccessible conference papers (abstracts not included).

Once more, thanks to all mentioned above, and enjoy!

William James, Guelph.

Contents

Chapter 1

Models for Water Quality Control by Stormwater Ponds

Fabian Papa, Barry J. Adams and Graham J. Bryant

The design of stormwater management quality control ponds is largely based on empirical practices which have been developed over time by water resources engineers. Of much more value to practitioners, and more importantly the environment they protect, is sound engineering design founded on a scientific basis. At present, the modeling of stormwater management facilities for both quantity as well as quality control is often accomplished using event and/or continuous simulation analysis methods. Analytical probabilistic models, based on long-term rainfall statistics, have been developed as an alternative approach and have proven to be reliable when compared to continuous simulation models for the estimation of stormwater quantity control performance. As more emphasis has been placed on environmental impacts, and how non-point source pollution, namely urban runoff, can be mitigated, these analytical models have been extended for the prediction of water quality control performance of urban drainage systems as well.

In this chapter, these analytical model results are compared to those of continuous simulation models for the quality control performance of stormwater management ponds, namely extended detention dry and wet ponds. The estimates of pollution control performance of these facilities are compared with respect to storage volume for various detention times, catchment areas and pond depths. The objective of this exercise is to illustrate the applicability of the analytical models for use in lieu of, or in conjunction with, continuous simulation

© *Advances in Modeling the Management of Stormwater Impacts - Vol.5.* W. James, Ed.
Pub. by CHI, Guelph, Canada 1997. ISBN 0-9697422-7-4. Fax: +519 767-2770

models. The results of the analytical models compare favourably with those of continuous simulation models for planning level analyses. For extended detention dry ponds servicing catchment areas of 10 to 100 ha with depths of 1.0 to 1.5 m and detention times of 24 to 48 hr, the difference in predicted suspended solids removal is generally less than 5 to 10% for all practical storage volumes. For wet ponds servicing catchment areas of 10 to 100 ha, with permanent pool depths of 1.0 m and active storage depths of 0.5 m, and active storage detention times of 6 to 48 hr, the difference in predicted suspended solids removal is between 10 and 30% for all practical storage volumes.

The pollution control performance of extended detention wet ponds is estimated using a constant ratio of active storage volume to permanent pool volume. This ratio is demonstrated to substantially impact the long term pollution control performance of the facility since the different storage zones are subjected to different pollutant removal mechanisms. An investigation is undertaken to determine the optimal combination of active storage volume and permanent pool storage volume for a given set of conditions. These results are generated using the analytical models and can be a valuable guide to the engineer in the preliminary design of urban runoff quality control ponds.

1.1 Introduction

Stormwater management ponds can be one of the few practical and effective means of controlling both the quantity and quality of stormwater runoff from urban catchments. The quantity control aspects of analysis and design are hydraulically based and have been well established over time. Of more recent concern, however, is the non-point source pollution of receiving waters due to runoff from urban catchments. As a result, stormwater ponds have also been used to enhance the quality of runoff entering receiving waters.

The principal mechanism for pollution control from quality control ponds is the sedimentation of suspended solids in the water column. In general, the pollution control provided by such a pond increases as the detention time of the runoff increases according to traditionally used settling equations. Therefore, it is typical to design quality control ponds to detain water in storage for longer periods of time when compared to stormwater quantity control ponds, which are primarily used as a means of attenuating peak flows from the drainage system.

The primary objective of this work is to model the performance of stormwater quality control ponds, including both extended detention dry and wet ponds, with varying detention times using continuous simulation models as well as analytical probabilistic models. Furthermore, the comparison of results between the two modeling approaches (i.e. simulation vs. analytical) is undertaken to evaluate the use of analytical models as an alternative to continuous simulation models for planning level analyses. In addition, analytical models can be effectively used in studies to determine trends that may be anticipated prior to performing full scale simulations. An illustration of this application is also given herein where the analytical models are used to estimate the performance of extended detention wet ponds for varying ratios of active storage volume to permanent pool storage volume.

1.2 Continuous Simulation Model

Results from continuous simulation modeling experiments (Adams, 1996) were obtained in order to allow the comparison of analytical probabilistic models with continuous simulation models for stormwater quality control ponds. The data used was generated using a series of models; that is, the EPA Stormwater Management Model (SWMM) version 4.30 was employed to generate runoff from various catchments and MTOPOND used the output from the SWMM model to predict total suspended solids (TSS) removal efficiencies for various pond configurations.

The RAIN block of SWMM was used to read hourly rainfall data for the period of 1960 to 1992 (Toronto Pearson International Airport Meteorological Data, Station 6158733). This data represented the rainfall period (March 1 to November 30) of each year and was subsequently used in the RUNOFF block to estimate hourly runoff flow values.

Table 1.1 gives the catchment characteristics which were modeled in the simulations. Values of infiltration rates relate to Horton's equation for infiltration. In addition, SWMM requires a value for the width of the catchment in order to compute time of concentration. A catchment width of 400 m was used for the 10 ha catchment while a catchment width of 2000 m was used for the 100 ha catchment. Input to the SWMM model also included evaporation data from the Canadian Climate Normals (Hamilton, Ontario).

Table 1.1 Catchment characteristics for SWMM simulations.

Area (ha)	10 - 100
Imperviousness (%)	50
Slope (%)	1
Impervious Manning's n	0.013
Pervious Manning's n	0.25
Impervious Depression Storage (mm)	1.5
Pervious Depression Storage (mm)	4.5
Minimum Infiltration Rate (mm/hr)	7.5
Maximum Infiltration Rate (mm/hr)	50
Infiltration Decay Rate (s^{-1})	0.00055

The runoff series generated by the RUNOFF block of SWMM was then used as input to the MTOPOND model. The MTOPOND model was based on the POND model (MOEE, 1994). This model calculates particle removal from the quiescent and dynamic settling of suspended solids and tracks spills and the TSS discharge from both the controlled outflow from the pond and the spills. This model was run in the first flush mode whereby flow bypasses the pond when the pond is full or the inflow rate is higher than an input threshold value. All flow which bypasses the pond is considered a spill. For further details of the implementation of the MTOPOND model, see Adams (1996).

The time series of runoff from the 10 ha and 100 ha drainage areas were simulated through both extended detention dry and wet ponds. Extended detention dry ponds contain only an active storage zone in which detained water is treated. Extended detention wet ponds contain both a permanent pool and an active storage zone. For purposes of the present work, a constant ratio of active storage volume to permanent pool volume of 1:2 was used. For each pond configuration, simulations were run for ten different active storage volumes and four detention times (6, 12, 24 and 48 hr). In addition, for extended detention dry ponds, two active storage depths (1 m and 1.5 m) were simulated. For extended detention wet ponds, one permanent pool depth (1 m) was simulated.

1.3 Analytical Modeling of Stormwater Management Ponds

The SUDS (Statistical Urban Drainage Simulator) models are a family of analytical probabilistic models developed at the University of Toronto and follow from the original probabilistic models developed by Howard (1976), Smith (1980) and Adams and co-workers (e.g. 1984). These models are closed-form mathematical expressions requiring relatively few input parameters and, as a result, performance characteristics can be computed very efficiently.

The SUDS models are developed analytically using derived probability distribution theory assuming exponential probability density functions for various rainfall characteristics, namely the rainfall duration, rainfall volume and interevent time. The exponential distribution requires a single parameter which is equivalent to the inverse of the mean characteristic value observed as follows:

$$f_x(x) = \begin{cases} \alpha e^{-\alpha x} & \text{x} > 0 \\ 0 & \text{otherwise} \end{cases} \tag{1.1}$$

Long term meteorological records have been analyzed for various locations across Canada for varying interevent time definitions (IETDs). The IETD is the minimum temporal spacing required between rainfall events to consider the events as being separate. From the analyses of these meteorological records, parameters for the exponential distributions discussed above have been estimated. Furthermore, studies conducted at the University of Toronto (e.g. Kauffman, 1987) have shown that results from the SUDS models for Toronto meteorological statistics most closely agree with those from the STORM continuous simulation model for quantity control when an IETD of 2 hours is used. Therefore, for Lester B. Pearson (Toronto) International Airport Data, an IETD of 2 hours is used in this study. Table 1.2 gives the resulting parameter values.

Table 1.2 Parameter values for exponential probability distributions.

Interevent Time Definition, IETD	2 hr
Parameter for Exponential PDF of Rainfall Duration, λ	0.282/hr
Parameter for Exponential PDF of Rainfall Interevent Time, ψ	0.0230/hr
Parameter for Exponential PDF of Rainfall Volume, ζ	0.200/mm
Average Annual Number of Rainfall Events, θ	104.0

1.3.1 Extended Detention Dry Ponds

The extended detention dry pond has an active storage zone only and exhibits continuous flow-through conditions during the periods in which it has contents and/or receives runoff. Therefore, it is assumed that dynamic settling is the principal mechanism of TSS removal. The long term pollution control performance of extended detention dry ponds using the SUDS extension models is characterized by the following expression (Adams, 1996; derived from Adams and Bontje, 1984):

$$C_P = E_d \left\{ 1 - \left[\frac{\frac{\lambda}{\Omega}}{\frac{\lambda}{\Omega} + \frac{\zeta}{\phi}} \right] \left[\frac{\frac{\psi}{\Omega} + \frac{\zeta}{\phi} \exp\left[-\left(\frac{\psi}{\Omega} + \frac{\zeta}{\phi} \right) S_A \right]}{\frac{\psi}{\Omega} + \frac{\zeta}{\phi}} \right] \right\} \tag{1.2}$$

where:

C_P = the long-term TSS removal expressed as a fraction,
E_d = the overall TSS removal efficiency
S_A = the active storage volume (mm),
h_A = the depth of the pond (m),
Ω = the controlled constant release rate from the pond (mm/hr),
t_d = the detention (drawdown) time of a full pond (hr).
λ, ψ and ζ = meteorological parameters describing rainfall statistics (Table 1.2), and
ϕ = the runoff coefficient.

The volume measures expressed as depth (mm) are implied as depth of water across the total catchment area. The term E_d is derived using the Camp-Dobbins model for dynamic settling, viz. (Fair and Geyer, 1954):

$$\eta_d = 1 - \left[1 + \frac{1}{n} \cdot \frac{V_s}{Q/A}\right]^{-n} \tag{1.3}$$

where:

η_d = the dynamic settling removal efficiency,
n = the pond settling performance factor (or turbulence factor),
V_s = the settling velocity of the particle size of concern (m/hr), and
Q/A = the surface loading rate (m/hr) where Q is the steady-state flow through rate of the pond (m^3/hr) and A is the average surface area of the pond (m^2).

This model for TSS removal (i.e. dynamic settling) is used since, during the period of particle settling, there is flow into and/or out of the pond and, hence, there is fluid turbulence. Equation 1.2 is essentially the long-term fraction of runoff processed through the pond multiplied by the overall removal efficiency of the pond thus yielding the long-term fraction of pollution, measured as TSS, removed.

The surface loading rate can also be expressed as follows:

$$\frac{Q}{A} = \frac{h_A}{t_s} \tag{1.4}$$

where:

t_s = the average steady-state detention time of the active storage zone (hr).

This quantity (t_s) is taken to be one-half the detention, or drawdown, time of the active storage zone as follows:

$$t_s = \frac{1}{2} \cdot t_d = \frac{1}{2} \cdot \frac{S_A}{\Omega} \tag{1.5}$$

where:

t_d = the drawdown time (time required to drain a full pond with no further inflow) and is defined as S_A/Ω (h).

Thus, Equation 1.4 can be rewritten as:

$$\frac{Q}{A} = \frac{2h_A\Omega}{S_A} \tag{1.6}$$

Substituting Equation 1.6 into Equation 1.3 yields:

$$\eta_d = 1 - \left[1 + \frac{V_s}{n \cdot h_A} \cdot \frac{S_A}{2 \cdot \Omega}\right]^{-n} \tag{1.7}$$

Equation 1.7 applies only to a single particle size with a known settling velocity. A more representative measure of pollutant removal efficiency would consider the range of particle sizes found in stormwater. Using a settling velocity distribution, the overall fractional TSS removal efficiency is given by (Adams, 1996):

$$E_d = \sum_{i=1}^{n} F_i \left\{1 - \left[1 + \frac{V_{si}}{n \cdot h_A} \cdot \frac{S_A}{2 \cdot \Omega}\right]^{-n}\right\} \tag{1.8}$$

where:

V_{si} = the average settling velocity (m/hr), and
F_i = the fraction of the total mass, contained in the ith size fraction.

The turbulence or short-circuiting factor used herein is n=3 representing *good performance* (Fair and Geyer, 1954). The settling velocity distribution used herein was developed from results of the Nationwide Urban Runoff Program conducted by the U.S. EPA as well as some Canadian research efforts and is provided in Table 1.3.

Table 1.3 Settling velocity distribution of particles in stormwater (MOEE, 1994).

Size Fraction i	% of Particle Mass F_i	Avg. Settling Velocity V_{si} (m/hr)
1	20	0.00914
2	10	0.0468
3	10	0.0914
4	20	0.457
5	20	2.13
6	20	19.8

1.3.2 Extended Detention Wet Ponds

For the case of the extended detention wet pond, incoming runoff is assumed to first displace the contents of the permanent pool until such a time when the water level exceeds the permanent pool level, at which point water begins filling the active storage zone and is discharged from the active storage zone of the pond at a controlled release rate. For events where both the permanent pool and active storage volumes are exceeded, runoff is bypassed upstream of the pond thus constituting a spill. This provision results in no treatment of a spill. Treatment of stormwater in the permanent pool is assumed to occur via quiescent settling between runoff events whereas, similar to the dry pond, dynamic settling is assumed to occur for runoff passing through the active storage zone. The long term pollution control performance of extended detention wet ponds is characterized by the following expression whose derivation is described in detail by Adams (1996):

$$C_P = E_q\left[1-\exp\left(-\zeta\frac{S_P}{\phi}\right)\right]$$

$$+\,E_d\left\{\left[1-\left[\frac{\frac{\lambda}{\Omega}}{\frac{\lambda}{\Omega}+\frac{\zeta}{\phi}}\right]\left[\frac{\frac{\psi}{\Omega}+\frac{\zeta}{\phi}\exp\left[-\left(\frac{\psi}{\Omega}+\frac{\zeta}{\phi}\right)S_A\right]}{\frac{\psi}{\Omega}+\frac{\zeta}{\phi}}\right]\right]\exp\left(-\zeta\frac{S_P}{\phi}\right)\right\} \qquad (1.9)$$

where:

S_p = the volume of the permanent pool (mm),
h_p = the depth of the permanent pool (m), and
S_d = the depression storage (mm).
E_d = the TSS removal efficiency of the active storage zone (equivalent to Equation 1.8) and
E_q = the overall TSS removal efficiency of the permanent pool and is derived below.

The removal of TSS within the permanent pool of an extended detention wet pond is assumed to occur via the mechanism of quiescent settling between runoff events; that is, during periods of zero inflow. The analytical probabilistic models used herein approximate this period of zero runoff as the interevent period (i.e. period between two successive rainfalls), denoted as b. The quiescent settling model is:

$$\eta_q = \frac{V_s}{v_o} \tag{1.10}$$

where:

η_q = the quiescent settling removal efficiency for a single particle with a settling velocity V_s (m/hr) and
v_o = the overflow rate (m/hr) defined as

$$v_o = \frac{Q}{A} = \frac{h_p}{b} \tag{1.11}$$

From sedimentation theory, the removal efficiency of a single particle size by quiescent settling can be expressed:

$$\eta_q = \begin{cases} V_s \dfrac{b}{h_p} & b < \dfrac{h_p}{V_s} \\ 1 & b \geq \dfrac{h_p}{V_s} \end{cases} \tag{1.12}$$

In order to estimate the long-term pollution control performance afforded by the permanent pool zone of an extended detention wet pond, the mean, or expected value, of the removal efficiency is required. From Equation 1.12, it is evident that removal efficiency (η_q) is a function of the interevent time (b) which is a random variable that can be described probabilistically using an exponentially distributed probability density function (PDF) as follows (Eagleson, 1972; Howard, 1976; Adams and Bontje, 1984; Adams et al., 1986):

$$f_B(b) = \psi e^{-\psi b} \tag{1.13}$$

Using derived probability distribution theory (Benjamin and Cornell, 1970), the PDF of removal efficiency, η_q, can be derived from the PDF of interevent time, b (Equation 1.13), through the functional relationship described by Equation 1.12. This derivation is done in two parts (Adams, 1996):

Part 1: $b \geq \dfrac{h_p}{V_S}$

$$Prob[\eta_q = 1] = Prob\left[b \geq \frac{h_p}{V_s}\right] = \int_{b=\frac{h_p}{V_s}}^{\infty} \psi e^{-\psi b}\, db = e^{-\psi \frac{h_p}{V_s}} \tag{1.14}$$

Part 2: $b < \dfrac{h_p}{V_S}$

$$F_{\eta_q}(\eta_o) = Prob[\eta_q \leq \eta_o]$$

$$= Prob\left[b < \eta_o \frac{h_p}{V_s}\right] = \int_{b=0}^{\eta_o \frac{h_p}{V_s}} \psi e^{-\psi b}\, db = 1 - e^{-\psi\left(\eta_o \frac{h_p}{V_s}\right)} \tag{1.15}$$

which is the cumulative distribution function (CDF) of η_q over the range where the settling period (interevent time) is less than the time required to remove all the particles with the settling velocity Vs (i.e. $b < h_p/V_s$). The PDF is the derivative of Equation 1.15 as follows:

$$f_{\eta_q}(\eta_o) = \frac{d\, F_{\eta_q}(\eta_o)}{d\eta_o} = \psi \frac{h_p}{V_s} \exp\left(-\eta_o \psi \frac{h_p}{V_s}\right) \tag{1.16}$$

The expected value of η_q is then

$$\overline{\eta_q} = \int_0^1 \eta_o \cdot \left[\psi \frac{h_p}{V_s} \exp\left(-\eta_o \psi \frac{h_p}{V_s}\right)\right] d\eta_o + 1 \cdot \exp\left(-\psi \frac{h_p}{V_s}\right)$$

$$= \frac{V_s}{\psi h_p}\left[1 - \exp\left(-\psi \frac{h_p}{V_s}\right)\right] \tag{1.17}$$

Equation 1.16 can be further generalized into an expression for the overall fractional TSS removal efficiency employing a settling velocity distribution; viz:

$$E_q = \sum_{i=1}^{n} F_i \left\{ \frac{V_{Si}}{\psi \cdot h_P} \left[1 - \exp\left(\frac{-\psi \cdot h_P}{V_{Si}} \right) \right] \right\} \tag{1.18}$$

1.3.3 Experiments

The above analytical models were used to estimate pollutant removal efficiencies for comparison with the continuous simulation model estimates. The input parameters used in the SUDS models included the runoff coefficient, ϕ, which is taken to be 0.5. This value is based upon results generated using the RAIN and RUNOFF modules of SWMM.

Total suspended solids removal efficiencies are calculated for both the extended detention dry and wet ponds for varying detention times (i.e. 6, 12, 24 and 48 hr) as well as for varying storage volumes. In addition, two pond depths (1 m and 1.5 m) are analyzed for the dry pond case and only one configuration is modeled for the wet pond case (1 m permanent pool depth and 0.5 m active storage depth). The analytical models for pollution control are based on unit catchment area calculations and are thus insensitive to the catchment area used. As with the continuous simulation modeling runs, a constant ratio of active storage volume to permanent pool volume of 1:2 is assumed. The results from the analytical models are compared to the results obtained from the continuous simulation models discussed above.

As it is inappropriate to base the design of a wet pond on an arbitrary ratio of active storage volume to permanent pool volume, the effect of this ratio on pollution control performance is also analyzed. It is anticipated that these results will contribute to the scientifically-based design of wet ponds with respect to how much storage should be allocated to both the permanent pool and active storage zones.

1.4 Discussion of Results

1.4.1 Extended Detention Dry Ponds

Figures 1.1 to 1.6 plot the percent total suspended solids (TSS) removal against active storage volume in the range of 0 to 20 mm normalized over the catchment areas. The quality control performance results were generated for dry ponds with detention times of 6, 12, 24 and 48 hours. Of these detention times, the 24 and 48 hour detentions are probably the more realistic detention times for most applications.

Figure 1.1 shows the pollution control performance of an extended detention dry pond, servicing a 10 ha catchment with a pond depth of 1 m for detention times of 6 to 12 hours, with respect to storage volume. In general, the analytical models over-estimate the TSS removal, when compared to the simulation results, by less than about 10%. The departure between the results decreases for the longer detention times. These detention times (i.e. 6 and 12 hours) are typically shorter than what would be expected in practice. Nonetheless, the model results tend to agree favourably with each other. Figure 1.2 presents the results for the 24 and 48 hour detention times for the same catchment size and pond depth. The correlation between the models for these longer detention times is quite remarkable, exhibiting a maximum departure of approximately 5%. For the 24 hour detention time, the analytical models consistently over-estimate the TSS removal when compared to the continuous simulation model results. For the 48 hour detention times, the analytical model results are virtually coincident with the continuous simulation model results for storage volumes up to 500 m^3 (or 5 mm normalized over the catchment area). Thereafter, the analytical models marginally under-estimate the pollution control performance when compared to the simulation results. For the sake of brevity, not all

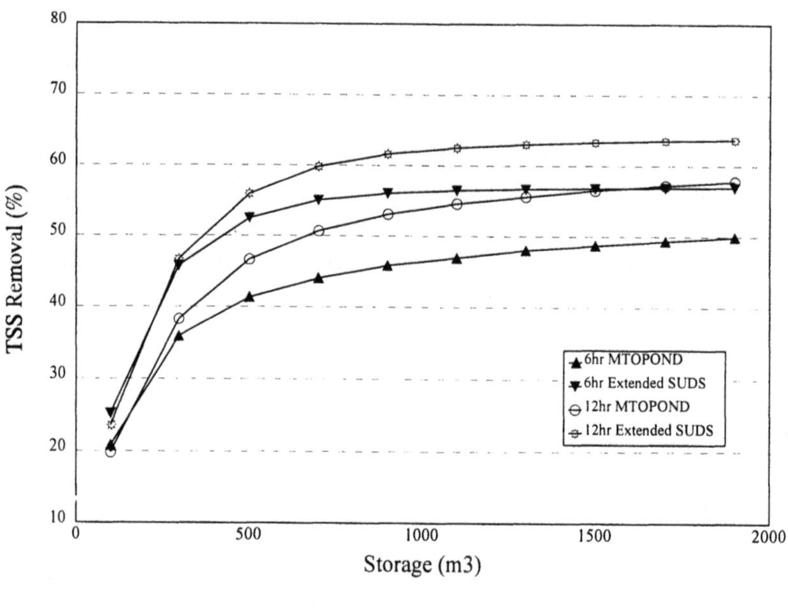

Figure 1.1 TSS removals of extended detention dry ponds, 10ha catchment, 1m pond depth, 6 & 12 hr cases.

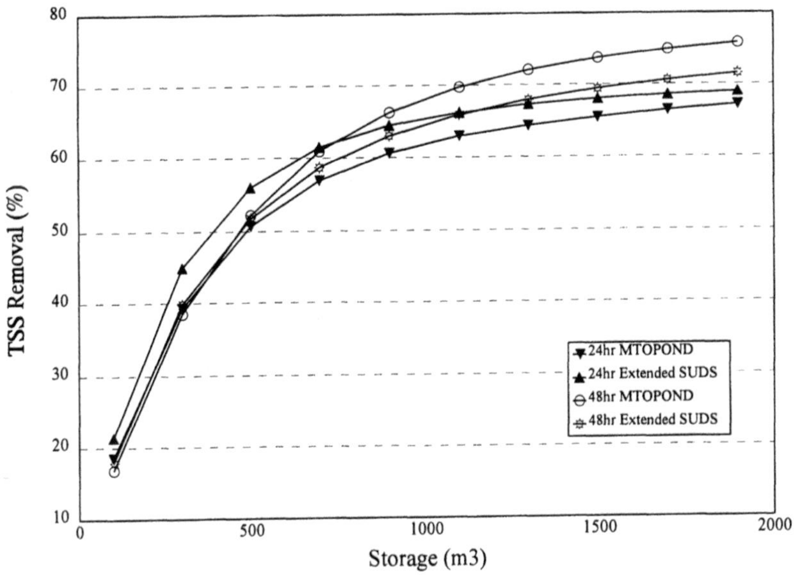

Figure 1.2 TSS removals of extended detention dry ponds, 10ha catchment, 1m pond depth, 24 & 48 hr cases.

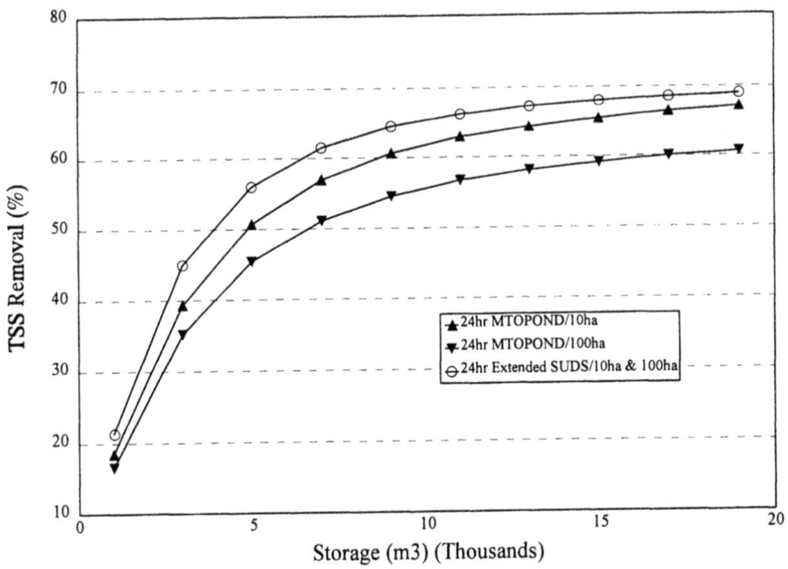

Figure 1.3 TSS removals of extended detention dry ponds, 1 m pond depth, 24 hr case.

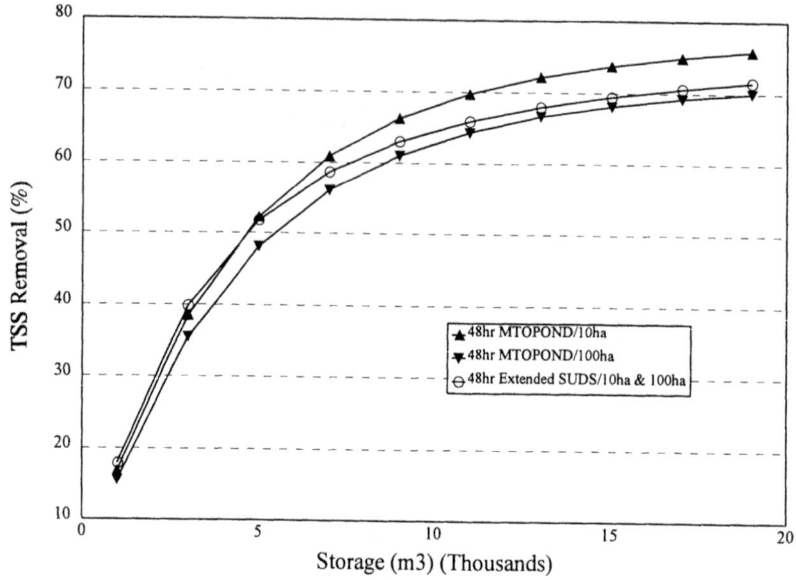

Figure 1.4 TSS removals of extended detention dry ponds, 1 m pond depth, 48 hr case.

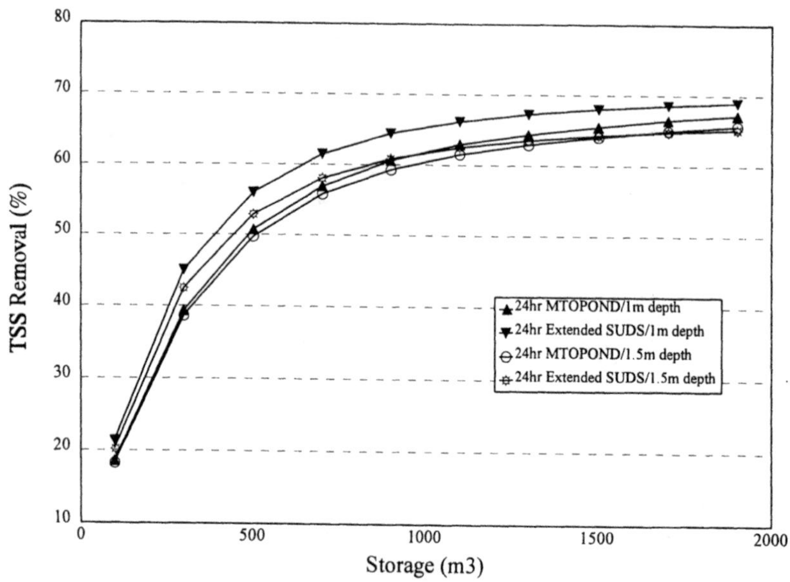

Figure 1.5 TSS removals of extended detention dry ponds, 10 ha catchment, 24 hr case.

the comparisons are shown for the 100 ha catchment areas nor for the 1.5 m ponding depth. Figure 1.3 shows the comparison of model results for a detention time of 24 hours and for different catchment areas (i.e. 10 and 100 ha). It is important to note that the analytical models are derived using unit catchment areas and are thus insensitive to variations in catchment area. Figure 1.4 presents this comparison for a detention time of 48 hours. In general, the analytical model results are effectively the same results as the continuous simulation models. Figures 1.5 and 1.6 show the model comparisons for ponding depths of 1 m and 1.5 m where the catchment area is 10 ha.

The results shown in Figure 1.5 are those for a 24 hour detention time. For a pond depth of 1 m, the departure between modeling results is typically less than 5% whereas for a pond depth of 1.5 m, this figure is consistently less than about 3%. Figure 1.6 gives the comparison for a detention time of 48 hours. In general, the analytical model results compare quite favourably to the continuous simulation results within the practical range of pond depths.

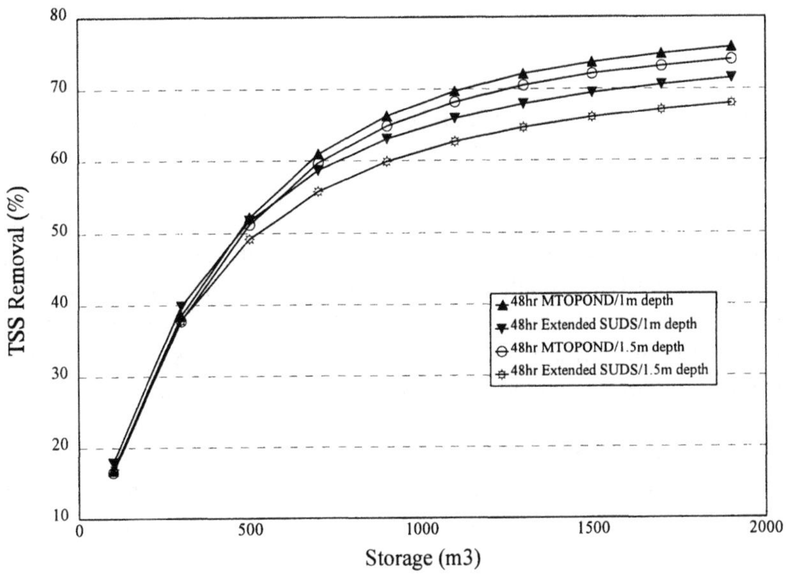

Figure 1.6 TSS removals of extended detention dry ponds, 10 ha catchment, 48 hr case.

1.4.2 Extended Detention Wet Ponds

The comparison of the analytical modeling results with the continuous simulation modeling results for extended detention wet ponds is given in Figure 1.7. The analytical models exhibit a sharp rise in the level of pollution control performance as storage volumes increase whereas the curves for the continuous

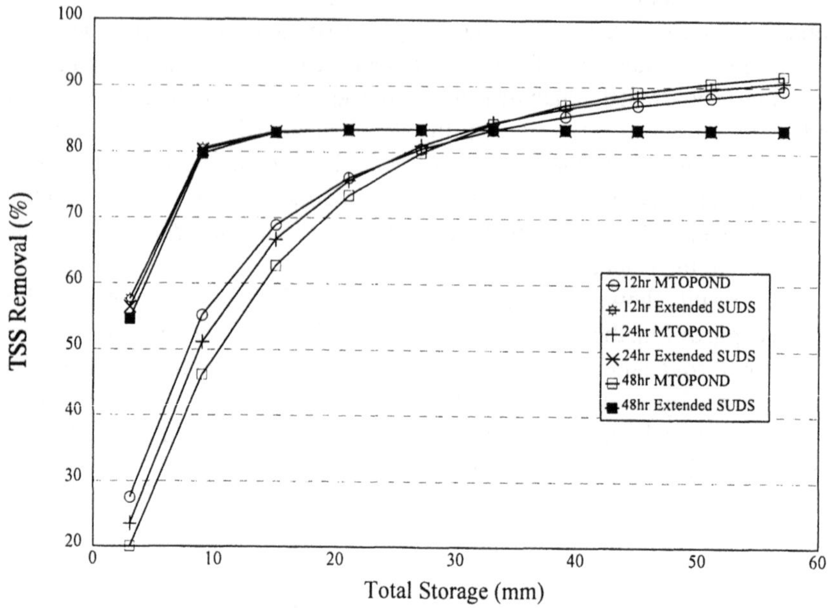

Figure 1.7 TSS removals of extended detention wet ponds, 10 ha catchment, 1 m permanent pool, 0.5 m active storage depth.

simulation model are more gradual. Both sets of curves reveal that the pollution control performance of wet ponds is relatively insensitive to the active storage detention time used. The quality control performance results from the analytical models show negligible sensitivity to the detention times analyzed and in fact the curves representing different active storage detention times converge onto a single curve as storage volumes increase. This is a result of the active storage zone not contributing significantly to the overall pollution control performance. This, in turn, is directly related to the arbitrarily-set ratio of active storage volume to permanent pool volume, in this case 1:2. Therefore, as the storage volume increases, the permanent pool storage increases by twice the amount that the active storage volume increases and, thus, fewer rainfall events will utilize the active storage zone, thereby reducing its importance in long-term pollution control. Therefore, it is considered useful to investigate the quality control performance of extended detention wet ponds for varying ratios of active storage volume to permanent pool volume, which is the topic of the subsequent section.

In general, the analytical models do not seem to estimate the relative quality control performance of extended detention wet ponds when compared to the continuous simulation model. Therefore, further development of these models is warranted to improve their reliability such that they may be used with reasonable confidence by researchers and practitioners alike.

1.5 Investigation of Active Storage Volume to Permanent Pool Volume Ratio

This section investigates the effect of changing the ratio of the active storage volume to the permanent pool volume (hereafter referred to as the ratio) of extended detention wet ponds in order to illustrate its effect on the pollution control performance of such ponds. The analyses presented in Figures 1.8 and 1.9 are performed using a constant pond depth for consistency, whereas Figure 1.10 illustrates the effect of varying pond depths on the performance characteristics. In each of Figures 1.8 to 1.10, the thickest curve with the square markers represents the same set of conditions and is, hence, the same curve. This is done in an attempt to aid readers visualize the results.

Figure 1.8 indicates that, depending on the detention time of the active storage zone, there exists an optimal ratio at which TSS removal is maximized. The figure also indicates that there is a *starting level* of pollution control at a ratio

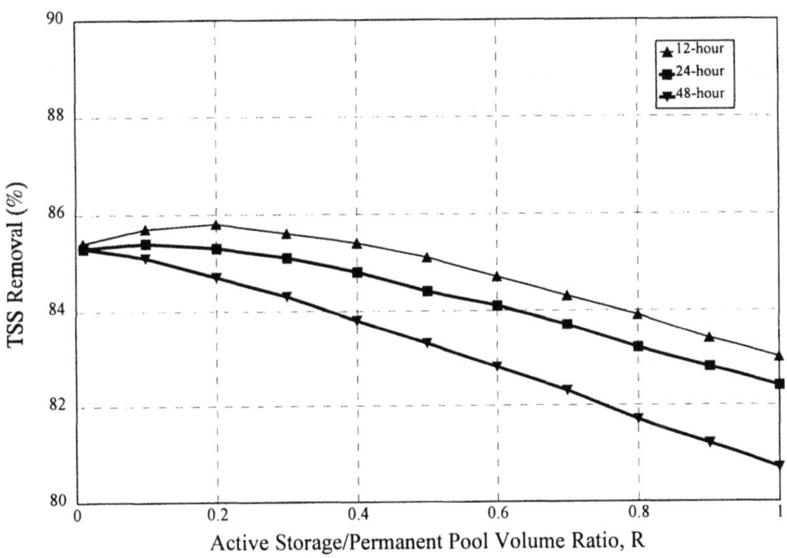

Figure 1.8 Extended detention wet pond analysis using analytical models, 10 ha catchment, 1.5 m total pond depth, 6 mm total storage.

approaching zero which is due solely to the permanent pool. By increasing the proportion of active storage there is an improvement, albeit marginal, in TSS removal to a point and then a decrease in performance thereafter. This effect is especially serious for longer detention times (i.e. 48 hours) where the quality

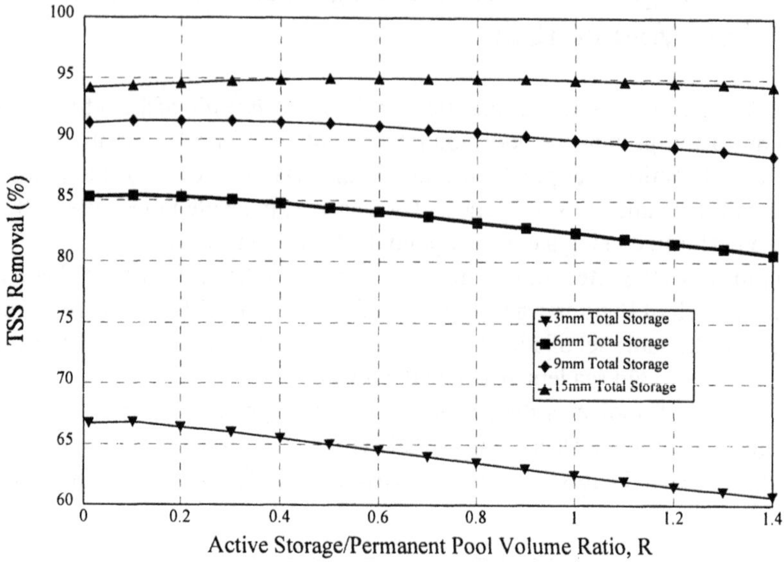

Figure 1.9 Extended detention wet pond analysis using analytical models, 10 ha catchment, 1.5 m total pond depth, 24 hr detention time.

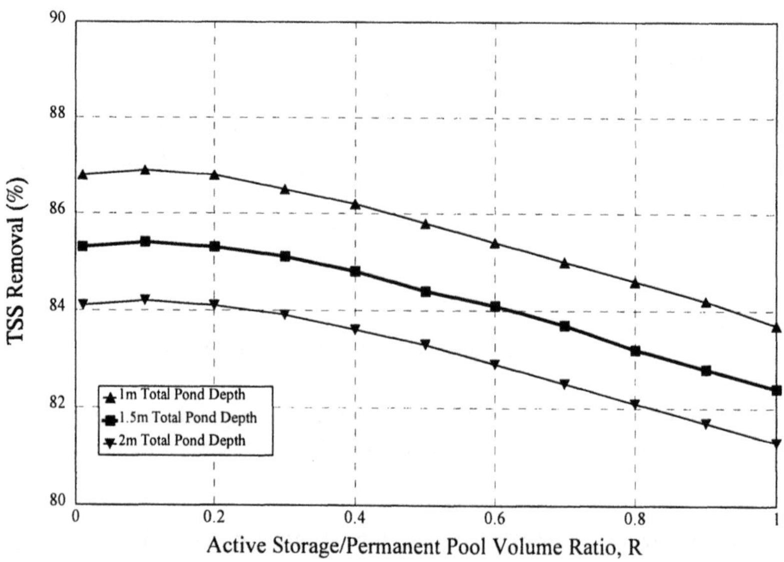

Figure 1.10 Extended detention wet pond analysis using analytical models, 10 ha catchment, 24 hr detention time, 6 mm total storage.

control provided by the permanent pool alone is greater than any combination of active storage and permanent pool storage for the set of conditions analyzed herein.

Figure 1.9 illustrates the impact of the ratio on TSS removal for various total storage volumes (active and permanent pool combined). The results indicate that there is a ratio, depending on the total storage volume, which optimizes pollution control. In general, as the storage volume increases, the value of the optimal ratio increases or, simply put, the active storage zone can contribute relatively more to long term TSS removal as the storage volumes increase. However, the increased benefit attainable by including an active storage zone is slight at best, and is not considered of great significance.

Figure 1.10 shows the general impact of total pond depth (active and permanent pool combined). The results are quite similar and encourage the use of shallow ponds for stormwater quality control. There exist practical limits, however, to the extent to which shallow pond depths can be constructed. All three curves show essentially the same trend and optimal combination of storage zones with the exception that the levels of control are shifted vertically.

1.6 Summary and Conclusions

The comparison of the analytical model results for long term pollution control with those from continuous simulation model for extended detention dry ponds is very favourable, especially considering the level of effort required to produce each set of results. A practical implication of this finding is that, at the planning stage of urban drainage systems, the analytical models can be employed with confidence to estimate system performance for pollution control for significantly less cost than their more comprehensive counterparts. This is especially useful where the time and funding required to perform full scale simulations may not be available. It is important to note that the employment of such models does not preclude the subsequent use of more comprehensive continuous simulation models for more detailed analysis and design.

For the extended detention wet pond, the agreement between the analytical models with the continuous simulation models is much less favourable. Further investigation of these models is therefore recommended, namely the investigation of the underlying causes of high-early TSS removals.

The ratio of active storage volume to permanent pool volume is assumed to be constant for all analyses conducted earlier in this work, in order to allow comparison of continuous simulation model results with analytical model results. This ratio, however, can affect the quality control performance of the stormwater management facility differently for different design characteristics. Although the agreement between modeling results for the wet pond is not good, it is useful to illustrate the application of the analytical models in performing less than routine investigations. The analytical models are easily used to investigate various ratios of active storage volume to permanent pool volume in terms of quality control performance. Furthermore, these results are anticipated to be similar to those that would be obtained using a continuous simulation, since both the analytical models and the continuous simulation models exhibited similar characteristics with respect to the lack of importance of the active storage zone, especially for large storage volumes. These analyses yield an optimal combination of storage components which maximizes the level of pollution control attainable by the facility. In general, wet ponds favour no active storage zone for improving the long term pollution control performance.

The results presented herein indicate that the active storage zone contributes relatively little to the quality-control performance of extended detention wet ponds particularly as detention times increase. This would encourage the use of wet ponds of this sort with an active storage zone to be used primarily for runoff quantity control purposes. This will eliminate the need for dual-cell stormwater management facilities which, in most cases are eyesores in an urban landscape. It will also provide perhaps a recreational facility with social benefits. The net result, in terms of stormwater management, will be peak flow attenuation by the active storage zone and quality control by the permanent pool.

Acknowledgements

The authors would like to extend their appreciation to Mr. Wan M. Wong, P.Eng. of the Ontario Ministry of Transportation for supporting this project.

Notation

A	Average Surface Area of the Pond (m^2)
b	Rainfall Interevent Time (hr)
C_P	Long Term Pollution Control Performance
E_d	Overall Removal Efficiency of Suspended Solids by Dynamic Settling
E_q	Overall Removal Efficiency of Suspended Solids by Quiescent Settling
F_i	Fraction of Total Mass Contained in i^{th} Size Fraction
h_A	Depth of Active Storage Zone (m)
h_P	Depth of Permanent Pool (m)
IETD	Interevent Time Definition
n	Turbulence or Short-Circuiting Constant in Camp-Dobbins Equation

PDF	Probability Density Function
Q	Steady-State Flow Through Rate of the Pond (m^3/hr)
S_A	Active Storage Volume (mm)
S_d	Depression Storage (mm)
S_P	Permanent Pool Storage Volume (mm)
t_d	Average Detention Time of Active Storage Zone (hr)
t_s	Average Steady-State Detention Time of Pond (hr)
TSS	Total Suspended Solids
V_{Si}	Average Settling Velocity of ith Size Fraction (m/hr)
ϕ	Runoff Coefficient
η_d	TSS Removal Efficiency for a Single Particle Size by Dynamic Settling
η_q	TSS Removal Efficiency for a Single Particle Size by Quiescent Settling
λ	Parameter for Exponential PDF of Rainfall Duration (hr^{-1})
Ω	Controlled Release Rate from Pond (mm/hr)
ψ	Parameter for Exponential PDF of Rainfall Interevent Time (hr^{-1})
ζ	Parameter for Exponential PDF of Rainfall Volume (mm^{-1})

References

Adams, B.J. 1996. Development of Analysis Methods for Stormwater Management with Ponds. Report to the Ontario Ministry of Transportation, Toronto, Canada.

Adams, B.J. and J.B. Bontje. 1984. Microcomputer Applications of Analytical Models for Urban Drainage Design. Proceedings of the Conference on Emerging Computer Techniques in Stormwater and Flood Management, American Society of Civil Engineers. pp. 138-162.

Adams, B.J., H.G. Fraser, C.D.D. Howard and M.S. Hanafy. 1986. Meteorologic Data Analysis for Urban Drainage System Design. Journal of Environmental Engineering, ASCE, Vol. 112, No. 5, October. pp. 827-848.

Benjamin, J.R. and C.A. Cornell. 1970. Probability, Statistics, and Decision for Civil Engineers. McGraw-Hill, New York. pp. 100-134.

Eagleson, P.S. 1972. Dynamics of Flood Frequency. Water Resources Research, Vol. 8, No. 4. pp. 878-897.

Fair, G.M. and J.C. Geyer. 1954. Water Supply and Waste-Water Disposal, John Wiley & Sons, New York. p. 596.

Howard, C.D.D. 1976. Theory of Storage and Treatment Plant Overflow. Journal of Environmental Engineering, American Society of Civil Engineers, 102(EE4). pp. 709-722.

Kauffman, G. 1987. A Comparison of Analytical and Simulation Models for Drainage System Design: SUDS vs. STORM. M.A.Sc. Thesis, Department of Civil Engineering, University of Toronto, Canada.

MOEE, Ontario Ministry of the Environment and Energy. 1994. Stormwater Management Practices Planning and Design Manual, Toronto, Canada.

Smith, D.I. 1980. Probability of Storage Overflow for Stormwater Management. M.A.Sc. Thesis, Department of Civil Engineering, University of Toronto, Canada.

Chapter 2

Integration of US Army Corps of Engineers' Time-Series Data Management System with Continuous SWMM Modeling.

Yiwen Wang and William James

Of the computer applications currently widely used in civil engineering, hydrological modeling has been mostly restricted to discrete events. However, with improved affordability of powerful computing, long-term continuous modeling has now become much more attractive. At the same time, long-term urban impacts on water quality have also become an important design concern in recent years. Continuous simulation is known to be a fundamental and reliable method for modeling nonpoint source pollution and its long-term effects on aquatic habitat. In the past, the main argument against using continuous modeling has been the difficulty of managing large amounts of input and output data. A very large portion of this data is time-series (TS) in nature and requires special time series management (TSM) software. Different data formats and software also cause difficulties for continuous modeling of stormwater management. Our purpose in this effort is to provide code that mitigates these arguments against continuous modeling.

A survey of modelers was conducted to confirm the necessity and importance of developing easy-to-use software that assists long-term continuous modeling. A graphical user-interface called CASCADE2 was designed and developed to integrate the Hydrologic Engineering Center Data Storage System (HECDSS) with the Storm Water Management Model (SWMM). Our purpose at this stage is not to integrate HEC application programs with SWMM, but merely to facilitate the easy export and import of datasets from HECDSS to SWMM.

Pub. by CHI, Guelph, Canada 1997. ISBN 0-9697422-7-4. Fax: +519 767-2770

CASCADE2 automatically manages and processes hourly and 15-minutely TS data for use with continuous SWMM. It allows the user to export or import any amount data from HECDSS by creating a dynamic macro, which overcomes the limitation in HECDSS of export or import of only monthly data. Storm-event statistics can also be computed by CASCADE2 automatically, and a summary file which includes station information, the total depth of rain, and the top ranked storm events is given at the end. Both SWMM and HECDSS program are run within CASCADE2, which thus acts like a shell, designed for application in typical, busy design offices.

A performance assessment of CASCADE2 in terms of computational speed and storage space is provided. A panel of professional engineers tested CASCADE2 and the results are summarized. Long-term TS data can now be collected, stored, and retrieved in minutes not days, as compared to conventional efforts.

2.1 Introduction

In this section, the result of survey of experts in stormwater management is presented: a literature review related to current data management systems is given; and the HECDSS data management system is introduced.

2.1.1 Survey of Experts in Stormwater Management

A survey of experts in stormwater management for research purposes is important and useful, because it helps identify current needs in urban stormwater management.

Eight survey questions were designed for the following purpose:

1. to obtain opinions on using continuous or event modeling in stormwater management;
2. to obtain a practical perspective on how many professionals are familiar with the existing TS data management systems and have real experience with the basic needs of linking models with these data management systems;
3. to ascertain the need for software tools for managing long-term TS data.

Copies of the questionnaire were distributed to 75 attendees at the workshops on the USEPA SWMM4 model with PCSWMM and other shells and the annual Toronto conference on Stormwater and Water Quality Management Modeling, February 19-23, 1996, Toronto, Canada. Attendees included private consultants, government representatives and education professionals, who were diversified in their experience and professional background, most of them having experience with computer modeling techniques.

Of these, 35 surveys were completed and returned. Three prominent categories were identified from the set of responses: environmental and water resources engineers (68%), professors and students (15%), and related professionals (29%). Related professionals include planners, software developers, and civil engineers.

The survey results showed that 60% of the respondents believed that continuous modeling is necessary in stormwater management. 100% of the respondents believed that a software tool for managing long-term TS data is needed. Also:

1. some stormwater modelers would prefer to use continuous modeling instead of event modeling if long-term TS data were easy to access;
2. the main difficulties when using continuous modeling are: obtaining relevant rainfall data, input and output files manipulation and storage, data collection and data entry, too many data formats which can't be accepted by SWMM, the high cost of calibration, and the time required;
3. sources of TS data include: Environment Canada, Atmospheric Environment Services (AES), National Weather Service (NWS), EARTHINFO, the National Oceanic and Atmospheric Administration (NOAA), and local governments; and
4. the two TS data management systems HECDSS and ANNIE are not well known by the stormwater management modelers.

2.1.2 Evolving Data Environment for Hydrology

Credible modeling requires reliable data. A computational hydrology work group often receives hydrological data from external data gathering agencies. Digital data comes in many formats, from many sources.

Canadian Information Agencies

Various data providers in the Canadian system of data acquisition are:

1. Inland Waters Directorate of Environment Canada, who recently introduced a Monitoring and Information Branch for the Ontario Region, responsible for collecting and disseminating water data from over six hundred sites across Ontario.
2. Water Survey of Canada (WSC), who handle hydrometric and sediment data. Data Management and Client Services are responsible for managing the data, providing client and administrative services, as well as ensuring quality control.

3. Atmospheric Environment Service (AES), who have a large network of observation stations in its weather and climate programs. All the collected meteorological data are managed by AES, a branch of Environment Canada, responsible for acquiring and archiving sufficient data to define the climate of Canada (Allsopp, 1987).
4. National Water Quality Data Base (NAQUADAT), who serve as a storage and retrieval system for field and laboratory measurements from 8,500 sampling sites across Canada. A variety of computer software has been developed for retrieval functions. Additionally, graphical and statistical functions were available to the user (Environment Canada, 1990).

United States Information Agencies

The US Geological Survey, the US Environmental Protection Agency, and the National Oceanic and Atmospheric Administration are the three most prominent agencies in the United States responsible for collecting, managing, or otherwise manipulating information that relates to water and earth sciences.

1. US Geological Survey produce and distribute information on a wide variety of earth-science specialties such as geology, hydrology, cartography, geography, and remote sensing, as well as information on land use and energy, mineral and water resources. This information is available in many forms, and in various data formats. It is the largest water resources research and information agency in the United States, and has taken a lead role in maintaining water databases (USGS, 1994). The USGS National Centre operated and maintained two systems: the National Water Data Exchange (NAWDEX) and the National Water Data Storage and Retrieval System (WATSTORE) (USGS, 1993). The Water Resources Division began the National Water Information System (NWIS) in 1983, which was designed to integrate and replace both WATSTORE and NAWDEX, and represented the first endeavor to construct a distributed database management system specifically for water data (Edwards et al., 1987).
2. USEPA water resources data collection programs are not as extensive as the USGS, but they have developed a large-scale database utility similar to WATSTORE, for the storage and retrieval of water quality data: Storage And Retrieval System (STORET).
3. National Oceanic and Atmospheric Administration (NOAA) has two notable data archival centers. (i) The National Weather Service (NWS) is responsible for the meteorological and hydrological services of NOAA, and operates hundreds of facilities across the

United States. (ii) The National Climatic Data Centre (NCDC) has the basic responsibility of collecting, archiving, processing, and disseminating climatological data in association with the National Weather Service (US National Weather Service, 1985).

2.1.3 Storm Water Management Model (SWMM)

The Storm Water Management Model (SWMM) was originally developed for the Environmental Protection Agency in 1969 by Metcalf and Eddy Inc., Water Resources Engineers Inc., and the University of Florida. In addition to the computational modules of SWMM (RUNOFF, TRANSPORT, EXTRAN, and STORAGE/TREATMENT), there are several TS management modules (Huber and Dickinson, 1988): RAIN, TEMP, STATISTICS, and COMBINE. Both RAIN and STATISTICS modules are used in CASCADE2. The purpose of the RAIN block is to read long time series of precipitation records, perform an optional storm event analysis, and generate a precipitation interface file for input into RUNOFF. The STATISTICS module is used to separate the continuous hydrograph record and pollutographs (concentration as a function of time) into independent storm events; it also calculates statistics and performs frequency analysis.

2.1.4 Hydrological TS Data Management System

Continuous simulation of hydrographs for long periods at many locations in a study area requires the management of large volumes of TS data. It has been found that SWMM has inefficient data processing techniques for continuous modeling. Modeling for urban impacts on water quality often requires the linking of several hydrological and hydraulic models, and incompatible datafile formats often give difficulties to modelers for this application. Thus, it becomes increasingly important to use external TS management systems that are independent of the hydrology models (Lumb, 1995).

The two most widely-used systems developed specifically for managing hydrological TS data are ANNIE (from USGS) and HECDSS (from HEC). Since CASCCADE (Gregory, 1995) links ANNIE with SWMM, CASCADE2 was written to link HECDSS with SWMM.

HECDSS

The Hydrologic Engineering Centre (HEC, of the US Army Corps of Engineers) developed the data storage system (DSS - *note: .dss in this chapter refers to a file-extension*) to meet needs for data storage and retrieval for water resources studies. The system, which has been under development since 1979, enables efficient storage and retrieval of TS and other data-types for which storage in blocks of contiguous data elements is most appropriate.

Like ANNIE, HECDSS is nonproprietary and written in the FORTRAN language. Datasets are stored in records, identified by a unique name, called the pathname. HECDSS performs its searches based on this pathname reference, which includes the project name, station identification, TS type, starting date, timestep, and an additional user-defined descriptor (US Army Corps of Engineers, 1990). Four data "types" are recognized by HECDSS: flow, incremental precipitation, stages, and precipitation mass curve. Data blocks are labeled with a six-part pathname. Several DSS utility programs will generate a list of the pathnames in a DSS file and store that list in a "catalog" file. Figure 2.1 shows a sample HECDSS catalog file that describes an hourly rainfall dataset for Oregon.

HECDSS Complete Catalog of Record Pathnames in File OREGONP.DSS
Catalog Created on May 29, 1995 at 16:26 File Created on May 9, 1995
Number of Records: 252 DSS Version 6-FO
Sort Order: ABCFED

Ref. Number	Tag	Record Pathname
1	T277	/OREGON/MISSOURI/PRECIP-INC/01JAN1974/1HOUR/OBS/
2	T278	/OREGON/MISSOURI/PRECIP-INC/01FEB1974/1HOUR/OBS/
3	T279	/OREGON/MISSOURI/PRECIP-INC/01MAR1974/1HOUR/OBS/
4	T280	/OREGON/MISSOURI/PRECIP-INC/01APR1974/1HOUR/OBS/
5	T281	/OREGON/MISSOURI/PRECIP-INC/01MAY1974/1HOUR/OBS/
6	T282	/OREGON/MISSOURI/PRECIP-INC/01JUN1974/1HOUR/OBS/
7	T283	/OREGON/MISSOURI/PRECIP-INC/01JUL1974/1HOUR/OBS/
8	T284	/OREGON/MISSOURI/PRECIP-INC/01AUG1974/1HOUR/OBS/
9	T285	/OREGON/MISSOURI/PRECIP-INC/01SEP1974/1HOUR/OBS/
10	T286	/OREGON/MISSOURI/PRECIP-INC/01OCT1974/1HOUR/OBS/
11	T287	/OREGON/MISSOURI/PRECIP-INC/01NOV1974/1HOUR/OBS/
12	T288	/OREGON/MISSOURI/PRECIP-INC/01DEC1974/1HOUR/OBS/
13	T289	/OREGON/MISSOURI/PRECIP-INC/01JAN1975/1HOUR/OBS/
14	T290	/OREGON/MISSOURI/PRECIP-INC/01FEB1975/1HOUR/OBS/
15	T291	/OREGON/MISSOURI/PRECIP-INC/01MAR1975/1HOUR/OBS/
16	T292	/OREGON/MISSOURI/PRECIP-INC/01APR1975/1HOUR/OBS/
17	T293	/OREGON/MISSOURI/PRECIP-INC/01MAY1975/1HOUR/OBS/
18	T294	/OREGON/MISSOURI/PRECIP-INC/01JUN1975/1HOUR/OBS/
19	T295	/OREGON/MISSOURI/PRECIP-INC/01JUL1975/1HOUR/OBS/
20	T296	/OREGON/MISSOURI/PRECIP-INC/01AUG1975/1HOUR/OBS/

Figure 2.1 Sample HECDSS catalog file.

Utility programs have been developed to manipulate or display data stored in a DSS file. Several of these programs are described briefly below (US Army Corps of Engineers, 1990). HECDSS is depicted in Figure 2.2.

- DSSUTL - General Utility Program
  - allows users to manipulate or edit data in HECDSS files. Like ANNIE, datasets may be exported or imported with sequential ASCII text files. Records may be compressed, and a summary table of attributes can easily be generated.
- DSPLAY - Graphical Display Program
  - enables users to graphically display or tabulate TS data. Users can enlarge or reduce plots, and edit data directly on-screen.
- MATHPK - Mathematical Manipulation of DSS Data
  - provides utilities for mathematically manipulating TS data, or for computing statistics.
- Data Entry Programs

A variety of utility programs are available for entering data into a DSS database file. Some are designed to enter data from another data base, WATDSS which reads data from a file retrieved from the USGS WATSTORE system, or a WATSTORE format file from a Compact Disk, and enters it into a DSS file. Other programs are designed to enter data "manually" or in a generic form. DSSTS is a prompt-driven program for entering regular-interval TS data.

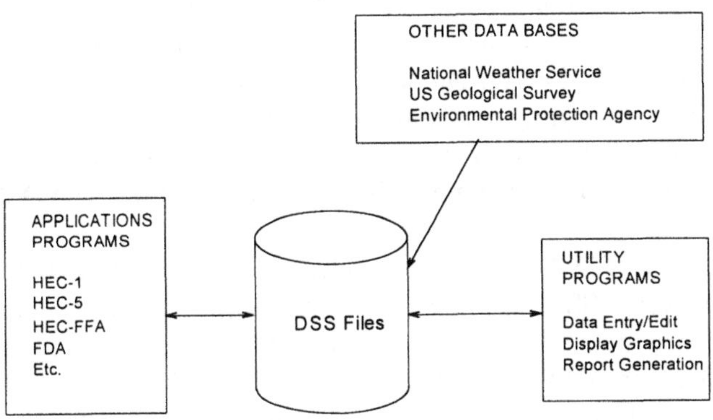

Figure 2.2 HECDSS data storage system.

2.2 Time-Series Manager Development

In this section, the data storage system utility program (DSSUTL) in HECDSS is introduced; the development of TS manager CASCADE2 (co-evolving assistant software for changing computational and data environments) is described; and the graphical user-interface is presented.

2.2.1 Software Structure

In this study, three utilities were considered important to meet the continuous modeling needs of typical engineering design offices:

1. retrieving TS data;
2. archiving TS data; and
3. data analysis operations.

Microsoft Windows was used to develop CASCADE2 and Visual Basic was selected as the programming language. The total program requires 9MB of disk space. CASCADE2 will run on any system capable of running Windows, requiring a personal computer using a 386SX or higher processor and running Microsoft Windows version 3.1 and MS-DOS version 5.0 or later. Problems may occur with Windows NT. Processing time can be much shorter if CASCADE2 is installed on a 486DX2-66 or higher processor. At least 4MB of RAM is required, 12MB of RAM or more is recommended. Hard disk with 50MB should be available, 100MB is recommended. Approximately 570K of free memory is required for running SWMM.

2.2.2 Data Storage System Utility Program (DSSUTL) in HECDSS

The previous section describes several utility programs in HECDSS, of which DSSUTL is the only one used in CASCADE2, as it provides a means of performing utility functions on data stored in HECDSS. These functions include tabulating, editing, copying, renaming, and deleting data. The program can also format and copy data into an ASCII sequential file for transfer to another computer, or for use by a program without DSS capabilities. The DSS file is a binary file.

A capability of DSSUTL is the exportation or importation of TS data for exchange between DSS and an MS-DOS spreadsheet. This is accomplished by the user defining data sets to be exchanged and an exchange format. When exporting data from DSS, the data is written to an ASCII (text) file with the defined format. The user then exits DSSUTL and executes the PC program and imports that ASCII file. Importing data to DSS essentially follows the reverse procedure. Exporting or importing data with DSSUTL requires the use of several commands to perform one function. The "exchange format" command is inherently more complex than any other command. Because of this, users typically write these commands in an input file or in the PREAD macro file.

Once the exchange format and exchange variables have been entered, data may be exported with the EXPORT command or imported with the IMPORT command. A time window must be used when exporting data, and may be used when importing (if the import file does not contain the date and time of the data). The file name to export to, or import from, follows the command name. Figure 2.3 shows an example of export and import macro.

```
Example 1. Export one month 1h interval time series data
MACRO EXPORT
TIME 01JAN1992 0100 31JAN1992 2400
EV PRECIP=/OREGON/MISSOURI/PRECIP-INC/01JAN1992/1HOUR/OBS/
EF 61533009        [DATE:YY] [DATE:MM] [DATE:DD] [TIME:HH] [TIME:MM]
//[PRECIP]
EX EXDATA.DSX
ENDMACRO

Example 2. Import one month 1h interval time series data
MACRO IMPORT
TIME 01JAN1992 0100 31JAN1992 2400
EV PRECIP=/GREEN RIVER/GLENFIR/PRECIP-INC/01JAN1992/1HOUR/OBS///
U=INCHES T=PRE-CUM
EF [SKIP] [SKIP] [SKIP] [SKIP] [SKIP] [SKIP] [PRECIP]
IMP IMPORT.IMP
ENDMACRO
```

Figure 2.3 Example export and import macro.

The PREAD macro file can contain many sets of commands that are identified by a macro name. DSSUTL will automatically connect to the macro file in the same directory. To execute those commands, the user enters "!RUN macro-name" at the DSSUTL prompt (where macro-name is the name of the macro to execute).

The export and import macros in Figure 2.3 have been used in the CASCADE2 source code with the following changes:

- DSSUTL allows export of at most five months TS data at a time, but by using repeated lines of EX and TIME step through the time window, in 5-month increments, this method can be repeated as many times as required. In the following example macro, exdata1.dsx and exdata2.dsx are two export files, each including five months data (exdata1.dsx includes Jan., 1992 -May, 1992; exdata2.dsx includes June, 1992 - Oct., 1992):

```
MACRO EXPORT
TIME 01JAN1992 0100 31MAY1992 2400
EV PRECIP=/OREGON/MISSOURI/PRECIP-INC/01JAN1992/1HOUR/OBS/
EF 61533009        [DATE:YY] [DATE:MM] [DATE:DD] [TIME:HH] //
[TIME:MM] [PRECIP]
EX EXDATA1.DSX
TIME +5M +5M
EX EXDATA2.DSX
ENDMACRO
```

- Each export macro cannot be longer than 50 lines, so it is tough and boring work to export say 75 years TS data. CASCADE2 solved this problem by chaining macros as shown in Figure 2.4.

```
macro example
line 1
.
line 45
!r chain1
endmacro
macro chain1
line 46
line 47
.
line 90
!r chain2
endmacro
macro chain2
line 91
line 92
.
endmacro
```

Figure 2.4 Example chain macros.

2.2.3 Overview of the TS Manager Utilities

CASCADE2 is intended to be an easy-to-use, graphical interface that links HECDSS with SWMM. Two types of TS data are utilized by CASCADE2, hourly and 15-minutely rainfall data. Rainfall datafiles are directly exported from HECDSS, and the resulting sequential file is then converted to a format recognizable by SWMM. Once retrieved, the appropriate SWMM format can easily be determined, by noting the location (starting and ending character positions) of the station ID string or other identifiers. With CASCADE2, the sequential file may be processed by the RAIN module of SWMM, which creates an interface file to be used by the RUNOFF module, and may also be used to calculate storm event statistics using the STATS module of SWMM.

The utilities of CASCADE2 are classified into three components, namely: retrieving TS data; archiving TS data; and data analysis operations, described in detail in subsequent sections. Figure 2.5 shows the layout of CASCADE2.

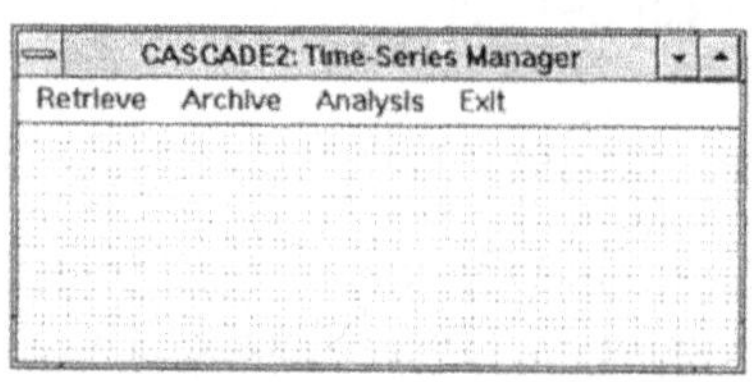

Figure 2.5 Layout of CASCADE2.

2.2.4 Retrieving Time-Series Data

After selecting the "Retrieve From HECDSS" menu option, CASCADE2 displays all the .dss files in the user's computer. Figure 2.6 shows the dialog box. If there is no catalog file of the selected .dss file, CASCADE2 will ask the user to create a catalog file for further use. This message box is shown in Figure 2.7.

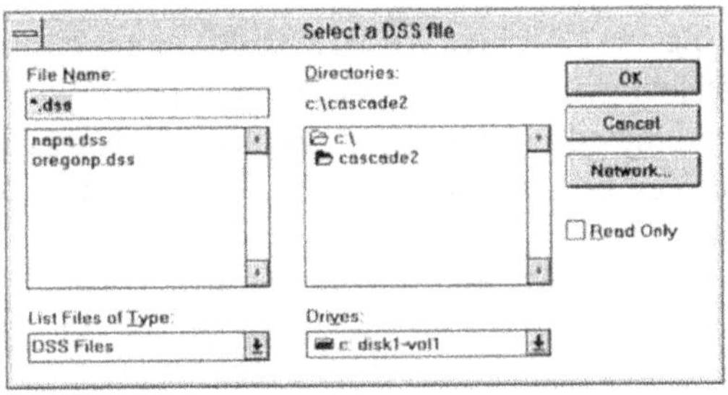

Figure 2.6 Export file selection dialog box.

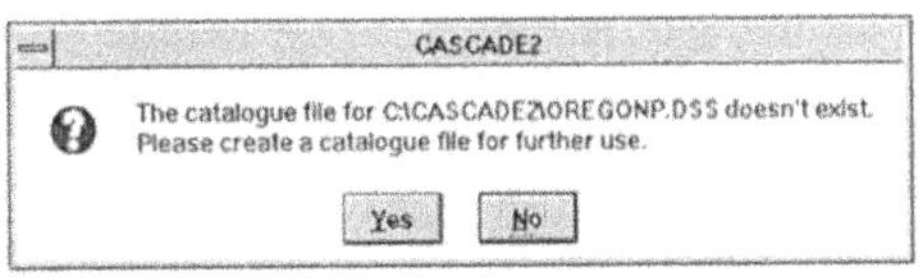

Figure 2.7 Create catalog file message box.

Then the HECDSS Retrieve user interface, which is displayed in Figure 2.8, will be shown on the screen. The user is prompted for the start and end year to be exported (CASCADE2 allows the user to export yearly data), and the name under which the export file will be saved. CASCADE2 then starts the HECDSS program and writes a dynamic macro file that automatically extracts the selected HECDSS datasets into a sequential export file. The HECDSS run will commence in a separate window, independent of CASCADE2.

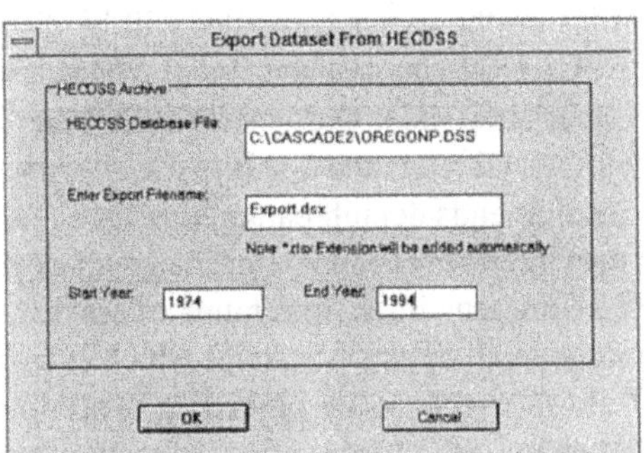

Figure 2.8 HECDSS export option in CASCADE2.

After the HECDSS run is completed, the export file will then be converted into SWMM format (IFORM 3) and compressed in order to save space, and an extension name .if3 will be added automatically. This file will be processed to run a statistics analysis by running the two SWMM modules RAIN and STATS.

2.2.5 Archiving TS Data

After selecting the "Archive To HECDSS" menu option, the HECDSS Archive user interface will be displayed on the screen. Figure 2.9 shows this form. The user is prompted to enter the import filename, and a HECDSS database filename. This file can be a new .dss file or an existing .dss file.

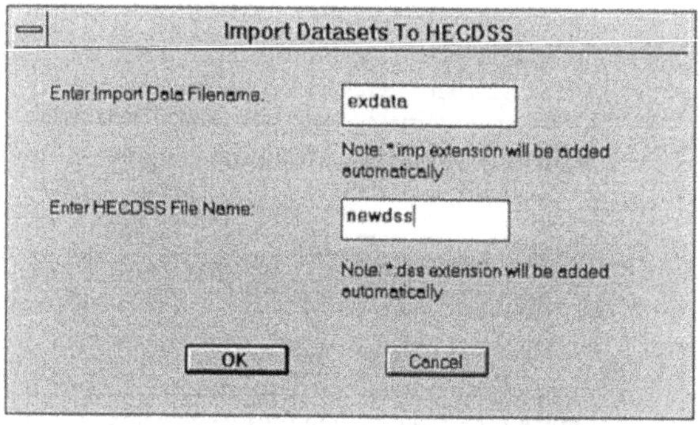

Figure 2.9 HECDSS import option in CASCADE2.

Then the DSSUTL run will commence in a separate window, independent of CASCADE2. A dynamic macro file in CASCADE2 will automatically create a catalog file that describes dataset attributes.

2.2.6 Data Analysis Operations

After selecting the "Time-series Data Analysis" menu option, all SWMM format rainfall datafiles on the user's computer will be displayed. Figure 2.10 shows the form. CASCADE2 can manage five SWMM formats so far: NWS (Fixed, .if0); NWS (Variable, .if1); Old NWS (.if2); User-Defined (.if3); Old AES Rain (.if5). CASCADE2 reads and displays file information in the top of the user interface, and recommends default parameters for a SWMM input datafile in the bottom of the user interface (the user can change these default parameters), including file name, comment lines, minimum interevent time, storm event threshold (the minimum depth of hourly rain for which events will be analyzed), and the number of storm events to be ranked in descending order of depth. SWMM interface file names are also specified in this user interface. Figure 2.11 shows the form. The SWMM run will also commence in a separate window. The SWMM engine is a separate application, running independently of CASCADE2.

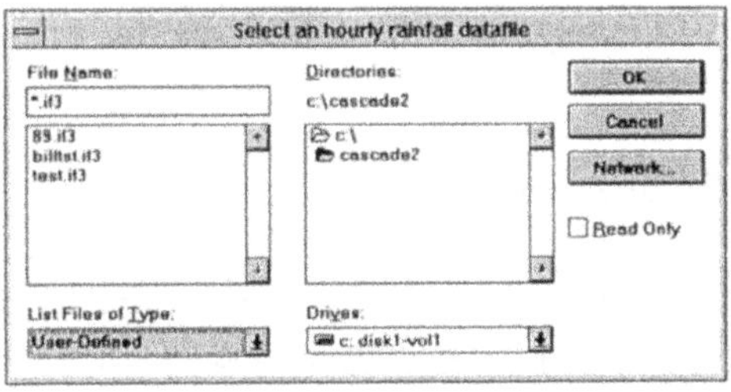

Figure 2.10 Rainfall datafile selection dialog box.

After the SWMM run is completed, it automatically goes back to the user interface. Finally, a message asks the user if a storm event summary file is desired. CASCADE2 reads the SWMM output file and creates a summary table that includes descriptive station information, the total depth of rain, and the top ranked storm events.

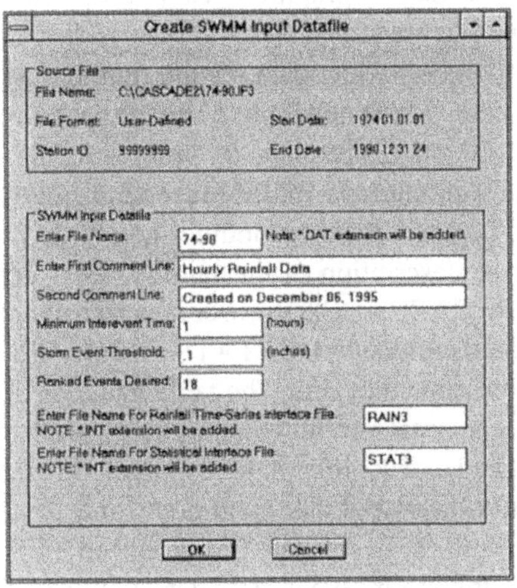

Figure 2.11 Storm event analysis option in CASCADE2.

2.3 Time-Series Data Manager Evaluation

The aim of CASCADE2 is to automatically process file transactions between SWMM and HECDSS in a Windows environment. CASCADE2 provides a visually- appealing, easy-to-use interface that 1. integrates the TS management functions of HECDSS with the continuous modeling applications of SWMM, 2. reduces the model learning time and 3. allows modelers to take more time to conduct sensitivity, calibration, error, and uncertainty analysis.

The performance and utility of CASCADE2 was assessed by the authors, using hourly and 15-minutely rainfall datafiles, and by a panel of professional engineers.

2.3.1 Overall CASCADE2 Features

Data management, storage and retrieval are essential for typical engineering design offices, especially for such long-term data. In general, CASCADE2 provides the user with the following features:

- Long-term continuous modeling requires that large amounts of TS data be manipulated for the most part externally and individually by the user. Lack of available data and inability to manage the large input datasets and output datasets are two major arguments against using continuous modeling. CASCADE2 retrieves data from a large data management system HECDSS, and then converts the data to one standard format. A graphical, easy-to-use software tool has been developed to assist this task.
- Large hydrometeorological datasets are required as input in continuous modeling, which needs large storage space in the computer. CASCADE2 solved this problem by processing the sequential file using the RAIN module of SWMM to create an interface file, which can then also be used by other modules in SWMM.
- An informative description of the rainfall TS is often needed by modelers. With CASCADE2, the user may also choose to calculate storm event statistics using the STATS module of SWMM, which gives a table of ranked events. The table includes time of day that each event began, total rainfall volumes during each event, duration of each event and a rank order table of magnitude of the event being analyzed, return period of that magnitude, and frequency. Figure 2.12 shows the storm event statistics produced by STATS. Figure 2.13 displays the storm event statistics summary file created by CASCADE2.

2.3.2 Processing Time

Managing TS data for long periods is a critical and time-consuming task. As an important performance parameter therefore, processing times were measured using eight .dss files which include 1, 2, 3, 5, 10, 13, and 21 years of hourly and 15-minutely rainfall data on a 486DX-66 computer. The average processing time to export one year hourly data is 67 seconds. The average processing time to run the statistics analysis on the exported datasets is about 60 seconds. The average processing time to export one year of 15-minutely data is 267 seconds. The average processing time to run the statistics analysis on it is about 120 seconds.

Table of Magnitude, Return period and Frequency
Constituent analyzed: Rain (inches) Hourly rainfall data 1974.out
Event parameter analyzed: Rain Volume

Date	Time (Hour)	Magnitude (inches)	Return Period (Years)	Percent < or=
740518	1.00	0.770	2.000	100.00
740518	2.00	0.600	1.000	96.55
740517	4.00	0.600	0.667	93.10
740517	23.00	0.510	0.500	89.66
740518	4.00	0.400	0.400	86.21
740517	2.00	0.370	0.333	82.76
740517	3.00	0.320	0.200	79.31
740516	23.00	0.320	0.182	75.86
740304	0.00	0.320	0.167	72.41
740517	0.00	0.300	0.105	68.97
740420	13.00	0.220	0.100	65.52
740928	4.00	0.200	0.095	62.07
740420	15.00	0.190	0.065	58.62
740420	16.00	0.180	0.063	55.17
740420	12.00	0.180	0.061	51.72
740421	0.00	0.180	0.043	48.28
740518	6.00	0.170	0.043	44.83
740510	23.00	0.160	0.042	41.38

Figure 2.12 Storm Event Statistics File.

Hourly Rainfall Data (1974-1990) Created on December 11, 1995
Minimum Interevent Time : 1 hours Rainfall Depth Threshold Value : .1 inches
Total depth of rainfall : 245.69 inches Number of storm events analyzed : 591.

Rank	Time of Occurrence (Year/MM/DD/Hour)		Value (inches)
1	1987 04 23	11:00	6.43
2	1987 05 28	7:00	5.23
3	1989 05 10	15:00	2.50
4	1986 05 17	23:00	2.50
5	1989 05 09	14:00	2.12
6	1986 05 16	22:00	2.12
7	1987 08 13	3:00	2.06
8	1987 12 13	17:00	2.05
9	1987 09 04	2:00	1.92
10	1987 06 08	19:00	1.91
11	1983 06 18	1:00	1.75
12	1983 03 26	3:00	1.58
13	1987 06 10	16:00	1.44
14	1989 07 26	5:00	1.37
15	1988 04 20	3:00	1.28
16	1985 09 01	1:00	1.27
17	1984 09 01	1:00	1.27
18	1986 04 20	8:00	1.19

Figure 2.13 Storm event statistics summary file.

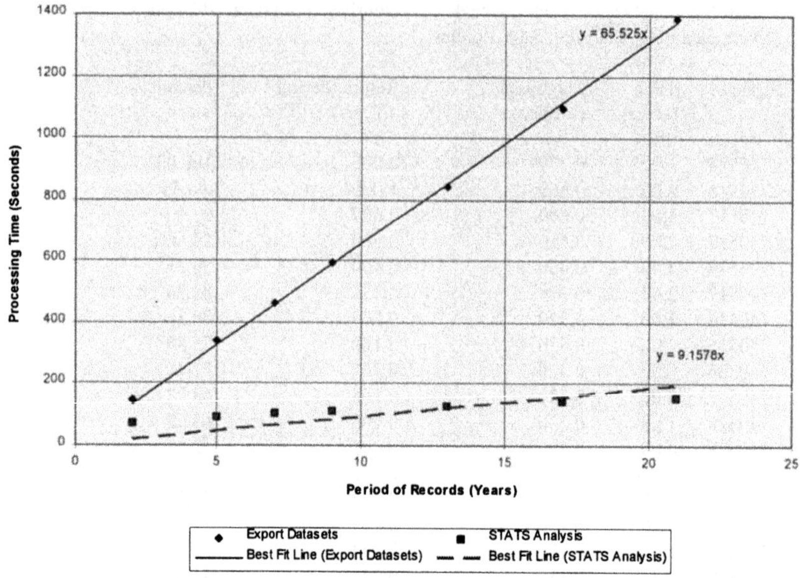

Figure 2.14 Processing time to export hourly datasets from HECDSS and STATS analysis.

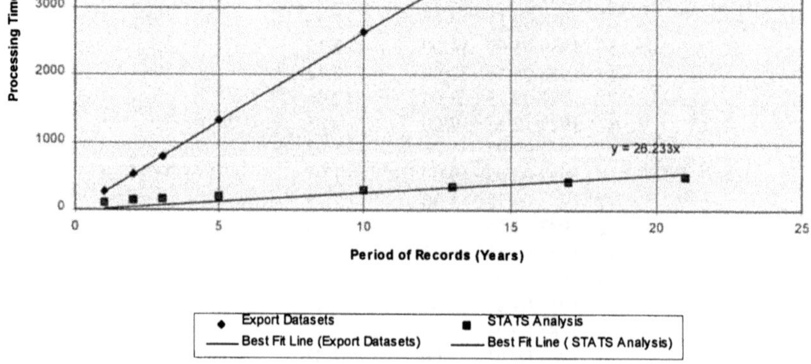

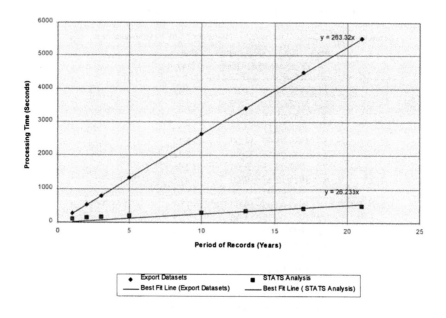

Figure 2.15 Processing time to export 15-minutely datasets from HECDSS and STATS analysis.

Figures 2.14 and 2.15 show the plots of processing times versus period of record for hourly rainfall data with best fit lines and their equations. From the equation, it can be inferred that to export 75 years hourly rainfall data needs about 80 minutes if using a 486DX-66 computer, and then needs 11 minutes to run the statistics analysis; to export 75 years of 15-minutely rainfall data needs about 300 minutes if using 486DX-66, and 33 minutes to run the statistics analysis.

Compared to the same work in a typical design office, it may take days to create the 75-year input datasets for continuous modeling.

2.3.3 File Size

As mentioned at the beginning of this section, storage of long-term TS rainfall data will occupy a large amount of computer space. The HECDSS sequential export files and various SWMM format rainfall datafiles are all ASCII text files, which are easy to visualize, but difficult to store. CASCADE2 creates binary interface files instead of the ASCII text files by using SWMM.

For the hourly rainfall data, the average size of one year .dss file is approximate 50KB; the export text file is 341KB, which is about seven times the size of the .dss file; but the binary SWMM interface file of one-year datasets is only about 2KB, which is 4% of the original .dss file. Figure 2.16 displays the plots of file size versus period of record with best fit lines and their equations.

For the 15-minutely rainfall data, the average size of one-year .dss file is approximate 155KB; the export text file is 1,360KB, which is about eight times

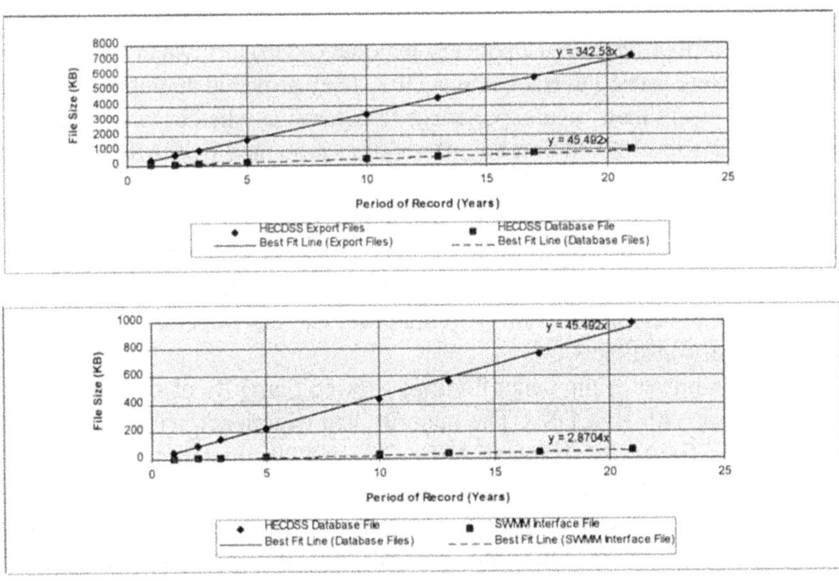

Figure 2.16 File size comparison (hourly data).

the size of the .dss file; but the SWMM binary interface file of one-year datasets is only about 60KB, which is about 40% of the original .dss file. Figure 2.17 displays the plots of file size versus period of record for 15-minute data with best fit lines and their equations.

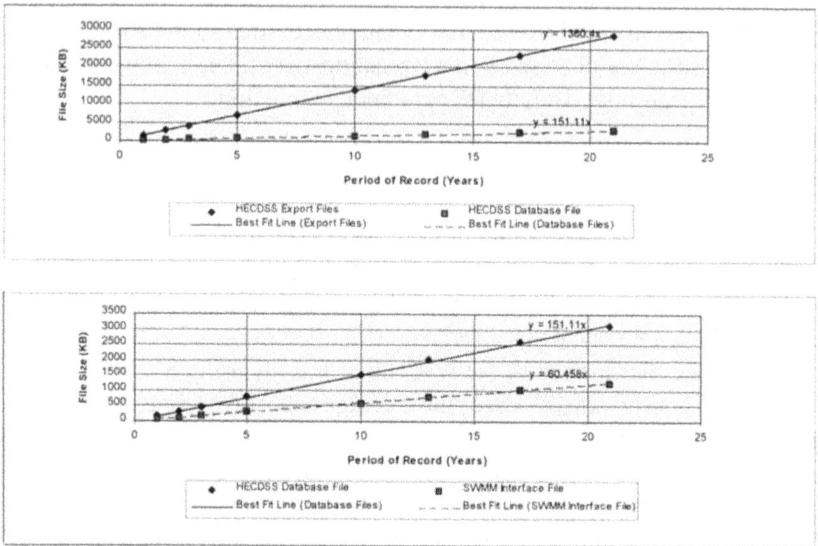

Figure 2.17 File size comparison (15-minutely data).

2.3.4 Test by Panel of Engineers

A panel of professional engineers, experienced in stormwater management modeling, were invited to test CASCADE2. They provided an unbiased review of the tool, using their own experiences to determine whether CASCADE2 is helpful, and also checked that the information presented by CASCADE2 was correct.

The User's Guide for the test version of CASCADE2 and the evaluation questionnaire were provided. Eleven evaluation questionnaires were completed and returned, a 70% response rate. Three prominent categories were identified from the set of responses, namely: researchers and developers; professors and students; and consultants.

The summary of the panel findings showed that 50% of the respondents were familiar with HECDSS. The majority had experience with SWMM and HEC models. Generally, it took a CASCADE2 user 20 minutes to read the User's Guide, 20 minutes to learn how to use the program, and 40 minutes to test CASCADE2, about one hour and 20 minutes in total. Nearly 100% had no difficulty using CASCADE2 and were all basically satisfied with the user interface, nearly 80% of the respondents found that two utilities in CASCADE2 (data retrieval and data archival) were most important in their practice. All the tests showed that the results of the statistics analysis were correct, and the summary table created by CASCADE2 is simpler and clearer than the SWMM output datafile.

The following is a summary of improvements recommended by the panel:

1. Include another utility program in HECDSS called DSPLAY as an option to view DSS data and create a catalog file of DSS file; include a simple text viewer so the user does not have to exit CASCADE2 to see the results.
2. Make it possible to export other data such as hydrographs stored in DSS into a SWMM readable format; import SWMM results into DSS format so that results can be plotted using DSPLAY program.
3. When running DOS applications, have them run in the 'Window' mode instead of full screen, then have a message with instructions on the file-name or whatever in the background and visible with DOS in the active window.

Based on tests with this representative panel of users, it may be concluded that:

1. CASCADE2 provides effective software to manage and process TS data;
2. a single TS manager is very useful in the organization and analysis of datasets used in stormwater management;
3. as a tool, CASCADE2 is easy to learn and use because of its user interface; and
4. CASCADE2 saves modeling time, since different agencies prefer different models for stormwater management and switching models should be easier with this program.

2.4 Conclusions

The result of the survey of experts in stormwater management has shown that the majority believe that continuous modeling is necessary in stormwater management. The main difficulties when using continuous modeling are: input and output file manipulation and storage, data collection and data entry, the different data formats and software tools. Stormwater modelers feel that a TS data manager (which can retrieve long-term continuous rain data from data management systems automatically on the user's computer) is necessary and useful. Furthermore, the two TS data management systems HECDSS and ANNIE are not well known by stormwater management modelers.

HECDSS was developed to meet needs for data storage and retrieval for water resources studies. Several utility programs have been developed to manipulate or display data stored in a DSS file. Exporting or importing datasets requires the use of several commands to perform one function. These commands are quite complex, often taking the user a few days to learn, and also, in HECDSS, the datasets are stored monthly, each dataset comprising one month data - at most five months data can be exported at a time. It is time-consuming to export several years data. CASCADE2 overcomes this limitation by writing a dynamic macro, allowing any amount of data to be exported at a time.

Our purpose at this stage is not to integrate HEC application programs with SWMM, but merely to facilitate the easy export and import of datasets from HECDSS to SWMM. At present, CASCADE2 is limited to manipulating both hourly and 15-minutely raindata. It includes a user-interface integrating HECDSS with SWMM and its TS data routines (STATS).

Performance of CASCADE2 was tested on both 486-DX33 and 486-DX2 personal computer using 70 years of rainfall data. The performance was measured in terms of both file size and processing time. The average processing time to export one year hourly data is around 1 minute when using 486-DX2, and another 1 minute is needed to run a statistical analysis. To export 1 year of 15-minutely data needs 4.5 minutes, and 2 minutes to run the statistical analysis.

The original and sequential file formats are not storage-space-efficient, and therefore not practical for storage or transmission of very long-term datasets. Binary files are more efficient. The performance assessment results show that SWMM can store hourly rainfall data in interface files about 6% the size of HECDSS, and 15-minutely data about 40%.

A panel of professional engineers, experienced in stormwater management modeling, were invited to test CASCADE2. The purpose of these tests was simply to gather feedback on the simplicity of the interface: how the interface could be made simpler in order to be made more efficient. Findings of the test panel indicated that CASCADE2 is robust and suitable for professional applications.

This study provided code that mitigates arguments against continuous modeling, and also revealed both the feasibility and practicality of linking continuous stormwater modeling with a TS-data management system. CASCADE2 automatically manages and processes TS data for use with a continuous stormwater model. It was shown that long-term TS data can now be collected, stored, and retrieved in minutes rather than days.

CASCADE2's utility could benefit from further enhancements: it should be elaborated to manage other types of TS data, such as streamflow data, water quality data; and other time-step data, for example 5-minutely rainfall data, which is widely used in forecasting flooding caused by individual cells within a storm. Creating a package that includes several widely-used databases, able to exchange data information, and combine the functions and features of all databases, fills a real need in stormwater modeling.

Acknowledgements

The authors gratefully acknowledge those who supported and guided this work: Troy Nicolini at the USACE Hydrologic Engineering Centre in Davis, CA; Wayne Huber at Oregon State; Julia Biedermann and Deborah Stacey at the University of Guelph; Rob James at CHI in Guelph. We also really appreciate the people who took time to test our code, including: Robert Dickinson at XP-Software Inc.; Alan Lumb and Kate Flynn at the U.S. Geological Survey; Barry Adams, Fabian Papa and Glen Thoman at University of Toronto; Chris Kresin at the University of Guelph; and consulting engineers: Ronald Kilmartin and Michael Gregory. Examination of code developed as part of a M.Sc. thesis is an unusual departure.

References

Belk, A.F., and Heathcote, I.W. (1995). Computer-Accessible Resources For Canadian Water Resources Management. Universities Council on Water Resources, Water Resources Update, Issue No. 99. pages 26-29.

Bugliarello, G. and Gunther, F. (1974). Computer Systems and Water Resources. Developments in Water Science 1. Elsevier Scientific Publishing Company, Amsterdam, The nethlands. 202 pages.

Carvalho, L. (1992). Computer-Integrated Crop-Management System (CICMS) With On-Line Weather Station. M.Sc. Thesis, University of Guelph. 120 pages.

Chatfield, C. (1989). The Analysis Of Time-Series: An Introduction (4th Edition). Chapman and Hall. ISBN 0-412-31820-2. 241 pages.

Chorafas, D.N. (1982). Databases for Networks and Microcomputers. Petrocelli Books Inc., Princeton, 280 pages.

Davis, D.W. (1981). Data Management Systems for Water Resources Planning. Technical Paper No. 81, Hydrologic Engineering Center, U. S. Army Corps of Engineers. 11 pages.

Donigian Jr., A.S., and Huber, W.C. (1991). Modelling Of Nonpoint Source Water Quality In Urban And Non-Urban Areas. US Environmental Protection Agency, Report No. EPA-600/3-91-039. 72 pages.

Edwards, M.D., Putnam, A.L., and Hutchison, N.E. (1987). Conceptual Design For The National Water Information System. US Geological Survey, Bulletin 1792. 22 pages.

Gregory, M. (1995). Management of Time-Series Data for Long-Term, Continuous Stormwater Modelling. M.Sc. Thesis, University of Guelph. 124 pages.

Huber, W.C., and Dickinson, R.E. (1988). StormWater Management Model, Version 4: User's Manual. US Environmental Protection Agency, Report No. EPA-600/3-88-001a. 569 pages.

James, W. (1995). On Reasons Why Traditional Single-Valued, Single-Event Hydrology (Typical Design Storm Methodology) Has Become Simple-Minded, Dishonest, and Unethical. US Army Corps of Engineers, Workshop on Urban Hydrology and Hydraulics, Davis, CA.

James, W., and James, R.C. (1994). PCSWMM Getting Started (Version 4.3). Computational Hydraulics International. ISBN 0-9697422-3-1. 47 pages.

James, W., and Robinson, M.A. (1981). Standards for Computer-Based Design Studies. Journal of the Hydraulics Division, ASCE, Vol. 107, No. HY7. Pages 919-930.

James, W., and Unal, A. (1984). CHGTSM - A Combined Hydrological Time Series and Topographic Database Manager. Proceedings of the Stormwater and Water Quality Modelling Meeting, USEPA, Detroit, Michigan. EPA-600/9-85-003. pages 217-232.

Nicolini, T. (1995). Personal Communication. US Army Corps of Engineers, Hydrologic Engineering Centre, Davis, CA.

Pabst, A.F. (1983). Data Management for Hydrologic Engineering. US Army Corps of Engineers, Hydrologic Engineering Center. 9 pages.

Salas, J.D. (1993). Analysis And Modelling Of Hydrologic Time-series. In Handbook of Hydrology, D.R. Maidment (Editor), McGraw-Hill Inc.. ISBN 0-07-039732-5. pages 19.1-19.72.

Unal, A. (1984). Computational Hydraulics Group Time Series Management System - User's Manual. CHI Report R125, Hamilton. 50 pages.

Unal, A. (1986). Centralized Time-Series Management For Continuous Hydrology On Personal Microcomputer Networks. Ph.D. Thesis, McMaster University. 170 pages.

US Army Corps of Engineers (1990). HECDSS User's Guide And Utility Program Manuals. US Army Corps of Engineers, Hydrologic Engineering Centre, Report No. CPD-45. 302 pages.

US Army Corps of Engineers (1991). HECLIB Volume 2: HECDSS Subroutines Programmer's Manual. US Army Corps of Engineers, Hydrologic Engineering Centre, Report No. CPD-57. 273 pages.

US Army Corps of Engineers (1993). Hydrologic Engineering Centre: 1993 Annual Report. US Army Corps of Engineers, Hydrologic Engineering Centre. 30 pages.

Chapter 3

On Integrating Continuous Simulation* and Statistical Methods for Evaluating Urban Stormwater Systems

James P. Heaney and Leonard T. Wright

The purpose of this chapter is to discuss various approaches for estimating the pollutant removal by urban stormwater detention systems. After a brief description of detention basins and their components, the characteristics of the inflow and its quality are described, including the effect of covariance between flow and concentration. Solids removal processes are discussed briefly.

Various modeling approaches for estimating pollutant removal effectiveness are compared, including single storm event simulation, continuous simulation of multiple storm events, statistical methods for evaluating pollutant removal effectiveness, and spreadsheet-based approaches which include Monte Carlo simulation. The pros and cons of the various approaches are described, and a simple example is used to illustrate the potential integration of these approaches. This chapter *reviews* the topic rather than develops a hard-and-fast methodology.

3.1 Performance of Detention Systems

A wide variety of methods exist for evaluating the effectiveness of detention systems for removing pollutants in urban runoff. These methods are described in contemporary textbooks, e.g. Wanielista and Yousef (1992), Debo and Reese (1995). Virtually all methods view the detention pond as a settling basin wherein the removal rate depends on the residence time in the basin and the reaction rate.

© *Advances in Modeling the Management of Stormwater Impacts - Vol. 5.* W. James, Ed. Pub. by CHI, Guelph, Canada 1997. ISBN 0-9697422-7-4. Fax: +519 767-2770

* *see Editor's note at end of Chapter*

These processes occur in any storage device be it a small pond, large lake, primary clarifier, or stormwater detention pond. Detention systems can be classified based on detention time:

Detention Time	*Descriptor*
Few hours to few days	Detention
Weeks or months	Retention
Intermittent	Dry
Intermittent	Wetland

The theory for estimating the performance of detention basins is based on Stokes' law for discrete particle settling of suspended solids, or by reaction kinetics. These theories are described in Nix, Heaney, and Huber (1981), Nix (1982), Goforth, Heaney, and Huber (1983), Nix and Heaney (1984), Nix (1985), Nix and Heaney (1988), and Nix, Heaney, and Huber (1988). The mechanisms affecting removal in detention facilities are shown in Figure 3.1. For stormwater detention systems, the basins fill and empty at relatively short intervals such as every few hours, days or weeks. Thus, it is vital to properly characterize the process dynamics. Major components are discussed below.

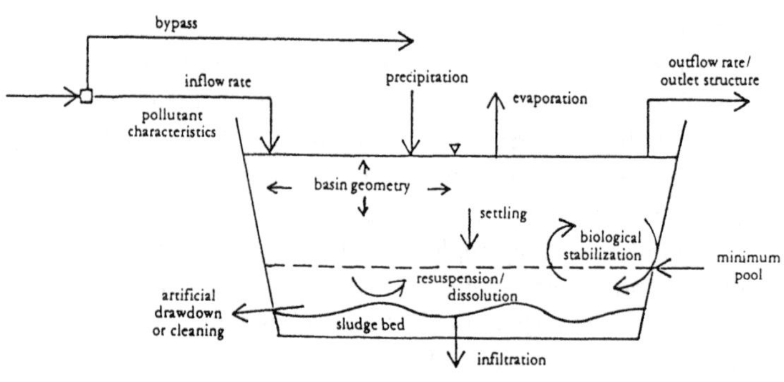

Figure 3.1 Mechanisms affecting pollutant removal in detention facilities (Nix et al. 1981).

3.2 Inflow Characteristics

3.2.1 Inflow Quantity

The mass balance for the flow into and out of a detention system is:

$$\frac{dS}{dt} = \sum I - \sum O \tag{3.1}$$

where:

dS/dt = rate of change of volume in the detention system,
I = inflow, and
O = outflow.

Many of the results in the literature are for this simple steady-state case. For steady-state operation of detention systems, such as a typical primary clarifier in a wastewater treatment plant (WWTP), dS/dt=0. However, stormwater systems are much more complex. The volume in the detention system and the inflows and outflows typically vary over time. Thus, it is essential to track the dynamics of the operation of stormwater detention systems. The accuracy of the depiction of the performance of the detention system depends upon the selected time step in Equation 3.1. Some of the criteria for selecting the appropriate time step are:

- Availability of data. Precipitation data may be available in time steps as short as five minutes; however, fifteen-minute to one-hour data are widely available.
- Average detention time of the basin. A simple definition of detention time is:

$$t = \frac{S}{Q} \tag{3.2}$$

  where:

  t = detention time,
  S = volume of the basin, and
  Q = average outflow rate.
- As mentioned above, the average detention time may vary from a few minutes to several months. Thus, methods of analysis vary accordingly. This is discussed in more detail later.
- Pollutant uptake rate. If the pollutant removal rate is very rapid, then a relatively small pond will suffice and vice versa.
- Purpose of the analysis. Very short time steps would be appropriate for real-time control and longer time steps may be used for preliminary planning.
- Computational considerations, e.g. need for numerical stability in solving the differential equations and the ease of computing using various time steps.

For urban stormwater analyses, the most popular time steps have been (i) five to fifteen minutes, (ii) one hour, or (iii) storm event. The five to fifteen minute time step is often used in detailed single-event simulations for evaluating the effect of surcharging in the pipe system. The one-hour time step is used in STORM and similar models to evaluate various storage-release strategies. Lastly, the hourly data may be aggregated into a "storm event" which is defined as ending when a specified number of dry hours has elapsed. Storm events are used in statistical models of stormwater systems. The critical question in defining a storm event is to specify the appropriate number of dry hours which will terminate an event. Various event definitions have been used but there is no correct definition. It depends on the nature of the problem. Hourly precipitation data for Boulder, Colorado for the period from August 1948 to December 1993 were analyzed to determine the number of wet-weather events as a function of the definition of an event. The results are shown in Figure 3.2. If an event is assumed to end when no precipitation has occurred during the previous hour, then over 11,000 events would result. A two-hour event definition reduces this total to over 5,000 events. The so-called knee of the curve occurs at a three-hour definition with about 4,400 events. The number of events continues to decrease as the event definition increases to 24 hours when the number of events is about 2,500. Thus, the assumed event definition has a major impact on the results.

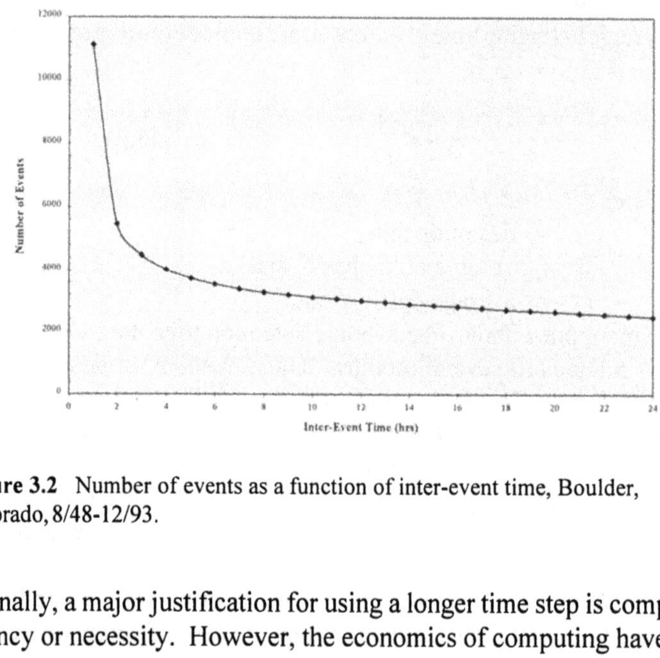

Figure 3.2 Number of events as a function of inter-event time, Boulder, Colorado, 8/48-12/93.

Traditionally, a major justification for using a longer time step is computational expediency or necessity. However, the economics of computing have changed radically since the introduction of PCs.

3.2.2 Inflow Quality

While the quantity of stormwater varies widely during a storm event, the quality varies even more widely. Depending on the nature of the storm and the sources of pollutants, the concentration of pollutants may exhibit a first flush, i.e.

concentration decreases as the flow duration increases. However, a first flush may not occur in all cases. Process-oriented models such as EPA SWMM provide various buildup-washoff relationships for estimating pollutant concentrations and loadings. However, these methods have only had limited success in depicting concentration variability.

A popular way to describe input quality to a detention system is to use the event mean concentration (EMC), the flow-weighted mean concentration:

$$EMC = \frac{L}{V} \tag{3.3}$$

where:

L = total load during the storm event, and
V = total runoff during the storm event.

The introduction of flow-weighted composite sampling devices made it easier to measure the EMC. However, the EMC is limited by the ambiguities of defining an "event" as described above. "Typical" stormwater pollutant EMCs may be found in textbooks, e.g. Debo and Reese (1995) and planning manuals, e.g. Schueler et al. (1992). The variance in EMCs has also been tabulated based on the results of the NURP studies (US EPA, 1983). Driscoll et al. (1990) summarize this data for highway runoff. A more accurate way to estimate pollutant input to a detention system is to use observed monitoring data. Unfortunately, such data tend to be scarce.

In order to gain a process-level understanding of urban runoff quality and to explain its variability and treatability, it is important to evaluate seasonal water quality changes. For example, Thomson et al. (1994) examined Minneapolis highway runoff consisting of 211 events, of which 47 are snow events. They divided the storm events into snow, rain, and mixed rain and snow. They show how the probability density functions (PDFs) of these events differ. One would expect strong seasonal variations in urban runoff quality, especially in areas like Minneapolis with strong influence of sanding and salting during the winter and spring, and the influence of fallen leaves in the fall.

Thomson et al. (1994) evaluated the impact of first flush for the Minneapolis database. Their results indicate a strong first flush effect for most suspended solids and chlorides. If a first flush exists, it can greatly benefit the design of detention basins because the design and operation can focus on the first flush.

A critical component of the inflow characterization is to accurately describe the particle size distribution of the influent. Pisano and Brombach (1996) summarize suspended solids characterizations in North America and Germany. Pisano and Brombach (1996) make several important points regarding the evaluation of "sewer solids":

- The test method is critical and only recently have appropriate testing procedures been used.

- It is important to characterize the entire range of solids from very coarse to very fine material. If rapidly settling particles are excluded, then the results are skewed since the "removal efficiencies" are based on the final vs. the initial concentration. This has been a chronic problem in comparing the effectiveness of urban stormwater detention systems.
- The median settling velocities vary widely even for similar waste streams. For example, the median settling velocity for CSO solids ranges from about 8 cm/sec to 0.08 cm/sec., a difference of two orders of magnitude.
- The median settling velocities vary widely depending on the type of waste. A summary of the North American and German studies is shown in Table 3.1. The German data indicate higher settling velocities than the North American. However, Pisano and Brombach (1996) attribute much of this difference to different evaluation methods.

Table 3.1 Reported settling velocities in North America and Germany.

	Settling Velocities (cm/sec)					
	North America			Germany		
	Geometric Mean	Medians		Geometric Mean	Medians	
		Low	High		Low	High
Dry weather wastewater	0.045	0.03	0.066	0.32	0.1	0.8
Stormwater	0.011	0.0015	0.15			
CSO	0.217	0.01	5.45	0.42	0.12	1.4
Sediment	3.23	0.8	6.75	14	0.8	0.23
Slime				0.8	0.2	6

Reference: Pisano and Brombach (1996)

3.3 Covariance between Flow and Concentration

Figure 3.3 shows the daily influent biological oxygen demand (BOD) concentration to the Boulder WWTP during a wet period that extended over three months in Spring 1995. For normal flows into the WWTP, the average BOD concentration is about 250 mg/l. As the inflow increased to 45 mgd (2.0 m^3/s), the corresponding concentration dropped to about 50 mg/l, 20% of its normal value. The correlation coefficient for this data set is -0.82, indicating a strong negative correlation: as flow increased, concentration decreased. Accounting for covariance, when it exists, is critical. The load of BOD entering the WWTP is found using:

$$L = cQ \tag{3.4}$$

where:

L = load,
c = concentration, and
Q = flow rate.

Using the Boulder WWTP data, the total daily influent load, shown in

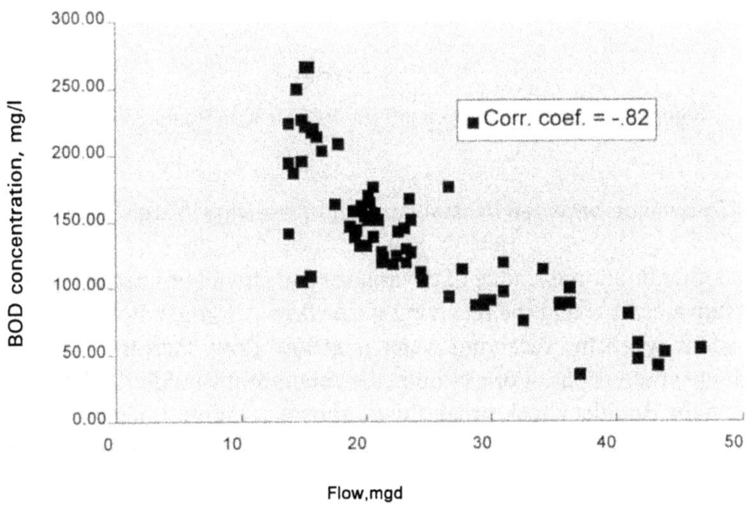

Figure 3.3 Effect of flow on influent BOD concentrations - Boulder, CO. WWTP.

Figure 3.4, indicates a constant or slightly decreasing total load as flow increases. In this case, the added infiltration to the sewer system is "clean water" which does not cause any increase in pollutant load.

Driscoll et al. (1990) evaluated the covariance between flow and concentration of highway runoff. They examined data for a given site and the aggregate database for all sites. This approach ignores the seasonal effects described above, e.g. the flow-concentration relationship would be expected to be different during early spring as opposed to midsummer. They concluded that significant correlation between flow and concentration occurs only about 25% to 35% of the time and that this covariance can be neglected. They also ignore the covariance among pollutants. By neglecting these covariances, the statistical approach proposed by Driscoll et al. (1990) is simplified.

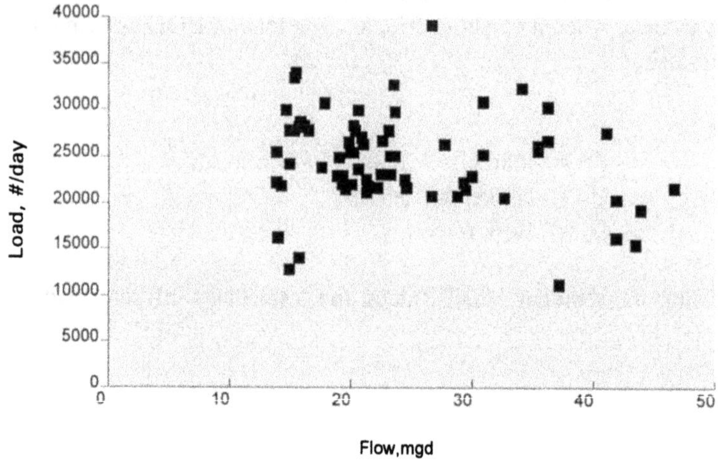

Figure 3.4 Effect of flow on influent BOD load to Boulder WWTP.

3.3.1 Covariance between Wastewater and Receiving Water Flows

Another important source of covariance that should be checked is between the wastewater flow and the receiving water flow. If larger stormwater runoff flows occur when the receiving water is at low flow, then the "worst case" conditions would occur. For example, the concurrent Boulder WWTP influent flows and the Boulder Creek streamflows, shown in Figure 3.5, indicate a strong positive correlation (r = +0.81). There were at least 23 days during a five year

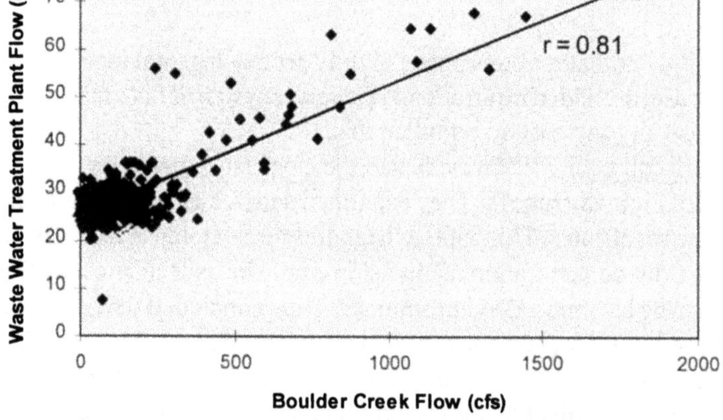

Figure 3.5 Boulder WWTP flow vs. Boulder Creek flow.

period when the flow in the WWTP was at least 40 cfs (1.1 m^3/s). Because of the strong positive correlation, when the flow in the WWTP is 40 cfs, the flow in Boulder Creek would be expected to be over 500 cfs (14.1 m^3/s), or a dilution ratio of over 14:1. At a WWTP influent of 70 cfs (2.0 m^3/s), the expected flow in Boulder Creek would be over 1,600 cfs (45.2 m^3/s), a dilution ratio of over 23:1. If the WWTP was forced to bypass during these high flow conditions, the bypass would be the flow in excess of the WWTP's capacity, or about 10 to 20 cfs (0.28 to 0.56 m^3/s) . Thus, the joint probability of CSOs or sanitary sewer overflows (SSOs) and large flows in the receiving water is very high in this example. Accordingly, the overflows are less serious, due to the high dilution.

3.4 Basin Characteristics

3.4.1 Provision for Bypass

Referring to Figure 3.1, an important design consideration regarding the inflow is whether to include a bypass for high flows. If no bypass is included, then the deposited solids can be washed out at higher flows causing a major load to leave the detention basin in a short period of time. Because of the flashiness of urban runoff, high peak flows can occur, even for light or medium storms. It may be desirable to route as much flow through the detention system as possible because this water will receive some pollutant removal. There is also a regulatory issue: all bypasses may be counted as system "failures" if the regulations limit the number of overflows per year. A critical issue with regard to bypasses is determining the "capacity" of the detention system.

3.4.2 Basin Configuration and Volume

The size and geometry of the basin depend on many factors including: whether it is a single or multipurpose facility (e.g. stormwater quantity and quality control, wetland); its relation to the groundwater table; the estimated rate of solids accumulation and removal; the evaporation rate; ownership of the stormwater; efficient geometry and baffles to encourage settling; and efficient geometry for cleaning.

3.4.3 Removal Efficiencies of Detention Basins

Virtually all popular methods for evaluating the expected removal of pollutants from a detention basin are based on simplified representation of settling processes. For example, Brune's trap efficiency curves, shown in Figure 3.6, express removal efficiency as a function of the capacity-to-inflow ratio (a measure of detention time) and the type of suspended solids. While the reservoir capacity is known, the inflow is an average for the year and may vary widely. Similarly, the nature of the suspended solids can vary widely. Brune's curves show the importance of properly estimating the nature of the suspended solids.

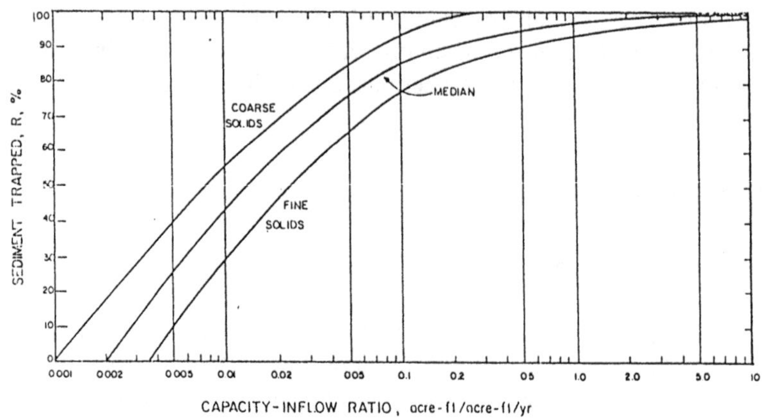

Figure 3.6 Brune's trap efficiency curves (Nix, Heaney, Huber 1981).

3.4.4 Reaction Rates

A popular approach to estimating removal efficiency is to parameterize the settling process as a first order reaction, i.e.

$$\frac{dc}{dt} = -kc$$

$$\frac{c}{c_0} = e^{-kt} \tag{3.5}$$

Equation 3.5 assumes that removal rates are *independent* of the initial concentration. Thus, we get the same removal rate independent of c_0.

Taking the log of Equation 3.5 yields:

$$\ln\left(\frac{c}{c_0}\right) = -kt \tag{3.6}$$

A semi-log plot of the data should yield a straight line with a slope of k. Whipple and Hunter (1980) have analyzed the settleability of urban runoff. The settleability of hydrocarbons is shown in Figure 3.7. A semi-log plot of this data shows that k is not constant but can be divided into three stages: an initial relatively rapid removal rate for the first several hours, and subsequent lower reaction rates for longer detention times as shown in Figure 3.7. Nix and Heaney (1984) provide a more general solution wherein the reaction order is included directly in the formulation, as shown in Equation 3.7. This improvement allows the dependency of k on concentration to be included. This is very important. If an "average" value of k is used, then the effectiveness of the detention system is underestimated for short detention times and overestimated for longer detention times (Goforth, Heaney, and Huber 1983).

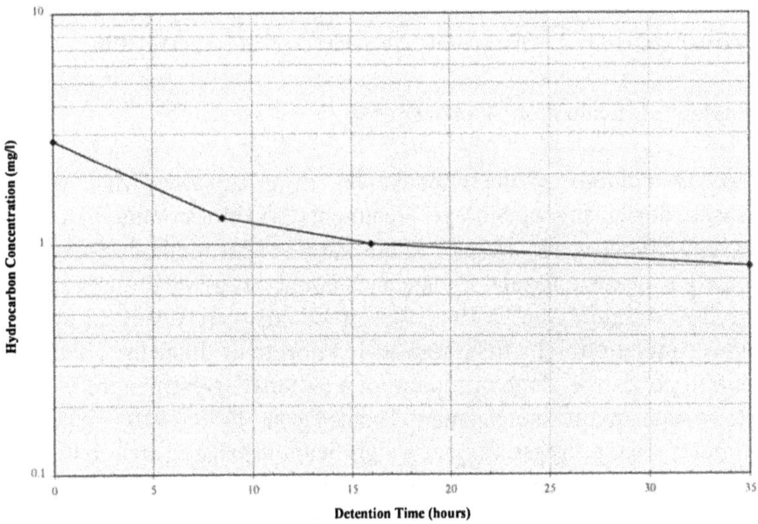

Figure 3.7 Hydrocarbon data on semi-log scale (Whipple and Hunter 1980).

The equation expressing the change in concentration may better be described by a higher order reaction, i.e.

$$\frac{dc}{dt} = -kc^n$$

$$\frac{c}{c_0} = \left[\left(\frac{n-1}{c_0^{1-n}}\right)kt + 1\right]^{\frac{1}{1-n}}, n \neq 1 \tag{3.7}$$

where:

n = reaction order
k = reaction rate constant, and
c, c_0 = concentration

The reaction order, n, and rate constant, k, may be determined by transforming the differential equation by the natural logarithm; i.e.

$$\ln\left(-\frac{dc}{dt}\right) = n\ln(c) + \ln(k) \tag{3.8}$$

(Nix and Heaney 1984).

This method of analysis greatly improves the characterization of settling dynamics in a detention basin by using a higher order reaction equation to describe mass removal as a function of initial concentration.

3.5 Evaluation of Detention Systems Performance

3.5.1 Review of Evaluation Approaches

Early evaluations of the effectiveness of urban stormwater detention systems were done using the Storage/Treatment (S/T) block of the EPA SWMM (Heaney and Huber et al., 1975). These single-event methods proved to be inadequate because the regulatory agencies could not agree on the criteria for selecting the "design event". Thus, these methods were extended to include continuous simulation. The first large-scale effort to evaluate the effectiveness of detention ponds was done in support of a national assessment of the cost of urban stormwater quality management (Heaney et al., 1977, 1979). In this study, cost estimates were prepared for every significant urbanized area in the United States. Detailed studies were done in five cities: Washington, D.C., Atlanta, Minneapolis, Denver, and San Francisco. The optimal mix of storage and treatment or release rate was found by running continuous simulations of hourly rainfall using the STORM model. The resulting storage-treatment isoquants and associated costs of storage and treatment were used to find the least costly mix of storage and treatment for a given level of pollution control. This assessment methodology is still used today.

The so-called statistical method was first introduced to the stormwater field by DiToro (1975), Howard (1976), and DiToro and Small (1979). Adams and his students at the University of Toronto have led the extension of this earlier work (Adams and Bontje 1984). Their work includes software to make these methods easier to use. The statistical method was used extensively to summarize the results of the EPA-sponsored Nationwide Urban Runoff Program (NURP) (US EPA 1983). Driscoll et al. (1990) have shown how these methods can be used for preliminary planning of highway stormwater detention systems. Loganathan and his students have also made recent contributions to this area, e.g. Seggara-Garcia and Loganathan (1994).

The early efforts to evaluate the effectiveness of storage-treatment systems revealed that the problem is more complex than meets the eye. Medina (1976), and Medina, Huber, and Heaney (1981a, 1981b) showed how to model the detailed process dynamics associated with storage-treatment systems. An improved storage-treatment block for the EPA SWMM model was developed that attempted to incorporate the dynamics of detention ponds to estimate their removal efficiencies (Nix 1994). The reaction rate is typically not constant. For example, the settling of suspended solids proceeds at different rates as a function of time. Thus, a more general formulation of removal rates is needed to capture these dynamics. A detailed description of these methods is given in Nix (1982, 1985), Nix, Heaney and Huber (1988), and Nix and Heaney (1988).

As part of the above work on stormwater detention ponds, the available data on pond removal efficiencies were reviewed. Virtually none of these studies measured pond removal efficiencies correctly: typically, no account was taken of the change in storage (all that was measured was the influent and effluent quality during a storm). The resulting estimates revealed wide ranges in performance.

3.5.2 Very Simple Approaches

Debo and Reese (1995) show that the expected effectiveness of wet detention systems is "high". Similar descriptors are found in stormwater design manuals, e.g. for Washington, D.C. (Schueler et al. 1992) and Camp, Dresser and McKee (1993). Debo and Reese (1995) summarize the estimated annual pollutant removals of wet ponds as shown in Table 3.2.

In order to use Table 3.2, one only has to input the ratio between the basin volume and the mean storm event volume. However, the definition of mean storm volume depends on how storm events are defined. *The mean volume per event changes drastically as the definition of a storm event is varied from ending after one, two, or more dry hours.* For example, using precipitation statistics for Boulder, Colorado, the mean volume per event ranges from 0.17 inches (4.3 mm) for a two dry hour definition to 0.391 inches (9.93 mm) for a 24-hour dry hour definition. The proper definition of a storm event depends on the relative size of the detention pond. In the statistical method, the event is defined so that the coefficient of variation is about one, which is important for simplifying the statistical analysis. However, it is not better than other storm event definitions. Thus, the pond designer will find confusion using even this very simple ratio.

The expected performance of wet ponds based on interim Federal Highway Administration (FHWA) guidance for designing wet detention systems for highway runoff is shown in Figure 3.8 (Dorman et al., 1988). The expected TSS removal is plotted as a function of the basin surface area expressed as a percentage of the contributing catchment area. The database for this curve is

Table 3.2 Average annual pollutant removal capability of wet retention ponds (Debo and Reese 1995).

Pollutant	Wet Pond Design Type		
	0.5 in./ acre	Vb/Vr = 2.5	Vb/Vr = 4.0 Two week det.
TSS	60-80%	60-80%	80-100%
TP	40-60%	50-70%	60-80%
TN	20-40%	20-40%	40-60%
BOD	20-40%	20-40%	40-60%
Metals	20-40%	60-80%	60-80%

Data from Schueler (1987 1993).
Vb = volume of basin, and Vr = volume of runoff from the mean storm.

the results of eleven NURP studies in Michigan (6), New York (1), Washington, D.C. (2), and Illinois (2). Thus, extensive transposing was required to extend these results to the entire United States.

Lastly, Wanielista and Yousef (1992) present a performance curve for removal of suspended solids and total phosphorus as a function of the detention basin volume/runoff volume from the mean storm as shown in Figure 3.9. The database for these curves is from Denmark and the United States (the eight NURP sites mentioned above for the FHWA curves). The wide variance in the estimates is apparent. Also, the measure of performance, while related to detention time, does not accurately reflect components of detention time such as interevent time for storms.

Nix, Heaney, and Huber (1981) compared various methods of evaluating the effectiveness of detention systems. The simple methods were borrowed from earlier results on sediment control systems. For example, Brune's trap efficiency curves are based on the pond volume, the annual inflow to the pond, and the physical characteristics of the suspended solids as shown in Figure 3.2. The database consists of 44 normally ponded and semi-dry reservoirs located in twenty different states.

Interestingly, Brune's curves, developed many years ago, are more refined than contemporary performance curves for estimating the effectiveness of stormwater detention ponds. Brune's curves at least vary according to both hydraulic and suspended solids characteristics.

3.5.3 Statistical Methods

Statistical approaches emerged in the 1970's by which probability density functions (PDFs) were fitted to storm event data in order to predict the performance of detention systems based on the characteristics of the precipitation events, the

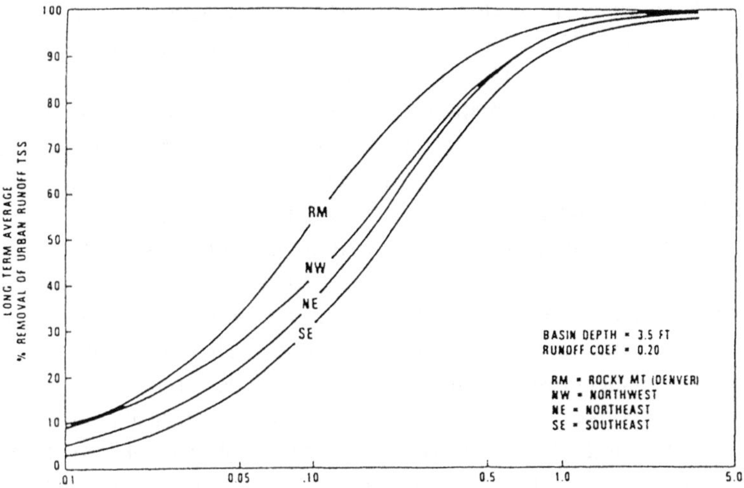

Figure 3.8 Basin surface area as a percent of contributing catchment area (Dorman et al.,1988).

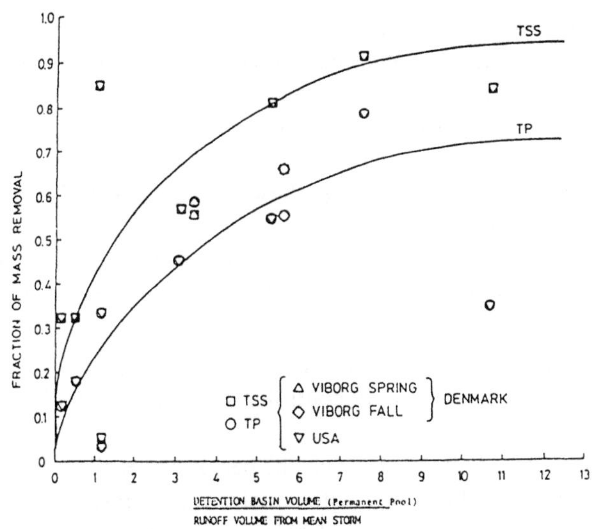

Figure 3.9 Removal of suspended solids and total phosphorus in detention ponds (Wanielista and Yousef, 1992).

drainage area, and the detention system. This work was pioneered by Howard (1976), DiToro and Small (1979) and Small and DiToro (1979). The statistical method has been used as a preliminary screening tool in presenting the results of the NURP studies (US EPA 1983) and has been used to evaluate highway-related runoff (Driscoll et al. 1990). This method allows the PDF of the output to be derived directly. The same information can be obtained by running a continuous simulation model such as SWMM, STORM, or HSPF. Numerous simplifying assumptions must be made to use the statistical method including: restrictions on the type of PDF, ignoring the covariance between flow and concentration, a simple runoff coefficient, and simplistic treatment removal kinetics. The statistical method was originally designed to be a preliminary screening tool for planning purposes, and to be used as a way of efficiently summarizing the results of process-based continuous simulation models, e.g. Howard et al. (1979) compared their statistical results to the output from STORM runs.

Statistical methods are an improvement over simple empirical approaches because they incorporate the storm event statistics (Small and DiToro, 1979; DiToro and Small, 1979). They also incorporate the "average" removal rate coefficient. The resulting equation for estimating basin performance is:

$$R = a\{1/[1+(bQ_r)/kA]\}^{(k+1)} \tag{3.9}$$

where:

R = percent removal,
a = removal efficiency coefficient,
b = kinetic coefficient,
Q_r = average runoff rate,
k = inverse of the coefficient of variation of the runoff rate, and
A = surface area of sedimentation tank.

A key assumption is that removal follows first order kinetics. As discussed earlier, this may be inaccurate.

Li and Adams (1994) cite two advantages of the statistical method for estimating the long-term pollution control performance of storage-treatment systems:

1. closed-form equations describing the relationships between input rainfall statistics and output control system performance statistics are available; and
2. preliminary cost optimization can be performed easily with knowledge of the cost functions of the control measures.

Segarra-Garcia and Loganathan (1994) also used the statistical approach for deriving estimates of the performance of stormwater systems. The assumptions and results of the Li and Adams (1994) approach are:

1. Rainfall event characteristics such as rainfall event volume (v), duration (t), and interevent time (b), can all be described by the single parameter exponential probability distribution:

$$f(w) = ze^{-\xi w:} \tag{3.10}$$

where:

z = 1/E(w), and the reciprocal of the average magnitude of v,b,t (ξ, λ, ψ),

$f(w)$ = the probability density function of w, and

w = the rainfall event characteristic (v,b,t);

2. Runoff can be estimated from rainfall using a simple coefficient:
3. Rainfall pulses can be divided into events using an event definition which provides the best fit for the exponential distribution. Thus, the criteria for the event definition is one of analytical convenience, rather than based on the expected response time of the control system.
4. Covariance between concentration and flow is ignored.
5. Treatment efficiencies are assumed to be constant and do not depend on residence time and reaction rates.
6. A constant pollutant concentration is assumed for all events.

Given these assumptions, the final equation for estimating the expected performance of the storage treatment system is a complex function of eight variables which describe the rainfall, catchment and drainage system characteristics and treatment efficiency.

3.5.4 Simulation Models

Early research on evaluating the effectiveness of detention systems for urban stormwater quality evaluations was done using the Storage-Treatment block of early versions of the EPA SWMM (Heaney et al., 1975). This initial effort estimated the removal efficiency of a storage-release system for a single design event. It was soon recognized that continuous simulation was needed to meaningfully characterize the overall performance of these systems. Concurrently, there was interest in getting planning level estimates of the performance of these systems. This interest was stimulated by the large 208 planning effort and our nationwide assessment of the cost of controlling stormwater pollution. Thus, the STORM model, with hourly rainfall data as input, was used to define the trade-offs between storage volume and treatment or release rate. These results are summarized in Heaney and Nix (1977), Heaney et al. (1977), Heaney et al. (1978), and Heaney et al. (1979). The continuing theme of this research was to obtain a *process-level* understanding of how real detention systems work, so that improved designs could be developed.

3.6 Comparison of Methods

The above methods range from very simple planning level estimates , e.g. Brune's curve, to detailed process-oriented approaches, e.g. EPA SWMM S/T block. Nix, Heaney, and Huber (1981) compare all of the above approaches using a common example. Also, Goforth, Heaney, and Huber (1983) compare the statistical method to using continuous simulation using the Storage/Treatment block of SWMM. The results, shown in Figure 3.10, indicate that the statistical method does not perform well. The statistical method estimated the required basin size to be 2.28 times bigger than the simulator estimate. The associated cost for the statistical design was 1.87 times the cost of the S/T design. Li and Adams (1994) compare the results of their statistical approach to continuous simulation and feel that the results are close.

As seen from this comparative analysis, the statistical method is a very simplified characterization of the problem. In our opinion, it should only be used as a companion to process-oriented, continuous simulation models as a way of summarizing the findings from these studies. *It should not be used as the primary*

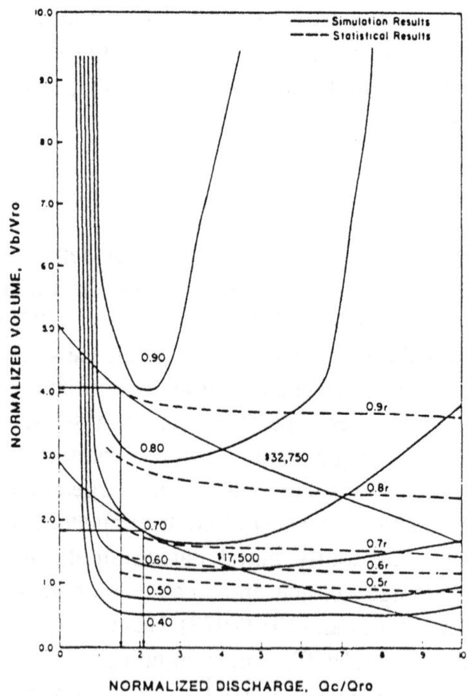

Figure 3.10 Determination of least cost combination of basin volume and drawdown rate (Goforth et al 1983).

tool for design-level evaluations. Driscoll and Strecker (1993) report wide variability in the performance of detention basins. It is essential to explain this variability using process-based approaches. Using a process-oriented simulator like the S/T block of SWMM, the engineer can explicitly incorporate:

1. a wide variety of detention facility geometries and outlet structures;
2. sludge accounting;
3. the capability for dry-weather drawdown;
4. the effect of various assumptions about buildup and washoff rates;
5. a variety of particle size/specific gravity distributions;
6. a wide variety of pollutant removal equations; and
7. multiple pollutants

3.7 Statistical Method

The statistical method described below is a form of risk or reliability analysis. As part of this risk analysis, continuous simulation and scenario analysis are used to evaluate how the proposed design performs for a variety of forcing functions. The U.S. Army Corps of Engineers has already moved strongly in this direction by developing guidelines and mandates to incorporate formal risk analysis into their evaluations (Greeley-Polhemus Group, 1992). The availability of @Risk, a spreadsheet add-in for Monte Carlo simulation, was critical in helping engineers understand and accept the risk-based approach. The US EPA has also embraced risk reduction as a priority-setting procedure (Finkel and Golding, 1994). The U.S. water supply industry is interested in developing and implementing formal approaches to evaluate risk and reliability.

Even though the background theory has been available for nearly half a century, systematic applications of risk and reliability have a very short history of the past few years. The main reason for this long gestation period is that the analytical methods are difficult to understand and require advanced knowledge of probability, statistics, and calculus. Even with the ability to use these advanced methods, closed-form solutions are available only for very simple, well-behaved systems. A major breakthrough in developing easy-to-understand-and-use methods for estimating risk and reliability was the introduction of Monte Carlo simulation software as an add-in for spreadsheets (Palisade Corp., 1994). Other risk analysis software is also available. However, state-of-the-art versions of spreadsheets allow simple risk and reliability analysis even without these add-ins.

Reliability is defined as: 1-risk. A large engineering literature exists on this subject, e.g. Pierucshka (1962), Hahn and Shapiro (1967), and Kapur and Lamberson (1977). The analytical techniques are very similar. Reliability engineering deals with "failures" of any type whereas the recent interest in risk analysis has been prompted by "failures" that cause public health problems.

Mays and Tung (1993) present a summary of risk analysis applications in water resources. Another major source of information on the use of risk analysis in water resources is the set of seven proceedings of Engineering Foundation Conferences on this subject which summarize developments in the field through the end of 1995, e.g. Haimes et al. (1994).

3.8 Computational Methods for Risk Analysis

3.8.1 Simple Sensitivity Analysis

Simple sensitivity or what-if analysis uses the output from a design event simulation and systematically varies the values of selected parameters to test the sensitivity of the solution to the assumed values of the parameters. This method can be done on spreadsheets using one, two, or three-way data tables which automate the sensitivity analysis process. Scenario analysis wherein a discrete number of alternatives are evaluated can also be done (a scenario is a vector of assumed values of the key parameters). Single or design event evaluation is very restrictive since a design can rarely be reduced to evaluating performance for a single future scenario.

3.8.2 Continuous Simulation

Continuous simulation models can track the status of the detention basin at all times. The summary output could include the cumulative density function (CDF) of how often the reservoir was at any given stage or spilling. Sensitivity analysis entails varying the values of key parameters and rerunning the simulation to evaluate the impact on the solution. In addition to older Fortran models, continuous simulation can be done on spreadsheets.

3.8.3 Monte Carlo Simulation

Monte Carlo simulation can be an efficient alternative to continuous simulation (Law and Kelton 1991). Probability density functions are fitted to the data and the distributions are sampled to estimate the variability of the solution. Monte Carlo simulation is now easier using spreadsheet add-ins. The CDF is determined directly. Using Monte Carlo simulation removes virtually all of the major objections against the current statistical method for stormwater analysis, e.g. the theory and derivation are hard to understand; only well-behaved distributions can be used; covariance is ignored; seasonality is ignored. Similar limitations have impeded the adoption of these methods in other areas of water resources engineering.

3.8.4 Advanced Continuous Simulation

A process-based continuous simulation is essential for understanding and properly designing detention basins. The EPA SWMM with the Storage/ Treatment block simulates conventional treatment processes. However, the S/ T block was written fifteen years ago and is not as user-friendly as engineers now expect from software. An excellent dynamic wastewater treatment simulator called GPS-X is available for Work Stations. It is being used to evaluate various wet-weather control options. [*Editor's note: Several shells for SWMM also render SWMM S/T user-friendly.*]

3.9 Towards More Robust Evaluation Methods

After more than twenty years experience in evaluating the effectiveness of stormwater detention systems using single event and continuous simulation and analytical statistical methods, there appears to be agreement on the following:

1. Evaluation of at least one year of precipitation data is essential in order to estimate the overall performance of stormwater detention systems. This requires the use of continuous simulation and/or statistical approaches.
2. The advent of the PC has made computing much easier than anticipated even a decade ago.
3. Widespread availability of databases for precipitation, water quality, etc. is permitting us to go from an *analytical* to an *information* based approach.

Because of the historical difficulty in dealing with information, we have been conditioned to replacing data by an approximating equation. Given such equations, we can conduct sophisticated analytical evaluations of the equations to find "optimal" solutions. A fundamental problem with replacing data by equations is that the equations may not accurately characterize the response surface. For example, Nix and Heaney (1988) could not find an accurate equation to describe the relatively simple production function of pollution control (y) as a function of storage volume (S) and release rate (T), or

$$y = f(S,T) \tag{3.11}$$

Thus, the production function was found by fitting cubic splines to the database and outputting the numerical result to a data table. Similarly, only a few probability distributions such as the exponential and log-normal have been used in the statistical method because they are "well-behaved" in the mathematical sense. Thus, the analyst's choice of functional relationships is heavily influenced by analytical tractability. Fortunately, we are no longer bound by these restrictive approaches.

There is active discussion in the literature on the relative merits of continuous simulation vs. analytical statistical methods. The statistical methods purport to decrease computational time by reducing the need to run continuous simulations using models such as STORM. However, most of these debates predate the advent of personal computers. Our more recent excursions into evaluating these problems indicates that the economics of computing and database acquisition have resulted in a significant shift in strategy. One can now get hourly precipitation data directly off CD-ROMs. The statistics of this hourly data can be easily determined for an assumed precipitation event definition. With PCs, spreadsheet-based models can easily process the precipitation data and perform continuous simulations. With the introduction of risk analysis software such as @Risk, it is now possible to replace the relatively complex analytical statistical analysis with a much more robust Monte Carlo simulation. Given these improved methodologies, how do the continuous simulation and statistical methods compare? The following simple example, developed using Boulder, Colorado data, provides a preliminary evaluation of these two approaches.

3.10 Case Study of Boulder, Colorado

Given that we can analyze stormwater problems using spreadsheets with Monte Carlo simulations, which method or combination of methods seems to be most appropriate? First, consider event-based methods, be they continuous simulation or statistical.

3.10.1 Continuous Simulation using Precipitation Events

The required steps to perform a continuous simulation using precipitation events are:

1. Adopt a definition of precipitation events based on a maximum number of dry hours allowed within an event.
2. Aggregate the hourly precipitation data into storm event data based on the above definition.
3. Set up a storm event water budget for the period of investigation keeping track of the status of the detention pond during and between storms.
4. Perform the simulation and record the summary statistics on the performance of the storage-release system. Repeat this process by varying the assumed storage volume (S) and release rate (T).

Sensitivity analysis can be expedited using a spreadsheet feature called two-way data tables (the analyst creates a two dimensional matrix of assumed values of S and T). The two-way data table feature is invoked and the model runs for each assumed pair of S and T values. If more than three parameters are to be varied for each simulation, then the scenario analysis spreadsheet tool can be used. For example, we may wish to vary depression storage and the runoff coefficient in addition to varying S and T.

A sample calculation using this method is shown in Table 3.3. For an assumed event definition, 10 hours in this example, the precipitation events for June 1987 are listed. Knowing the ending hour of the event, its duration, the total precipitation, and the storage in the reservoir at the beginning of the simulation, it is straightforward to calculate the performance of the system as shown in Table 3.3.

Table 3.3 Event modeling: sample calculation for Boulder, Colorado.

Event Rule = 10 hrs
Depression Storage (in) = 0.05
Drawdown Rate (in/hr) = 0.004
Runoff Coefficient = 0.5
Total Storage (in) = 0.6

Date	Time	Rain Event Totals	Event Duration	Inter-event time	Runoff (in)	Storage at start of event (in)	Storage at end of event (in)	Volume stored (in)	By-pass Volume (in)
6/7/87	2200	0.2	2		0.05	0.60	0.558	0.05	0.00
6/8/87	2400	2.9	5	26	1.4	0.60	0.000	0.60	0.80
6/10/87	1600	0.1	1	40	0	0.16	0.164	0.00	0.00
6/24/87	1800	0.1	1	338	0	0.60	0.604	0.00	0.00
6/29/87	2400	2.4	14	126	1.15	0.60	0.000	0.60	0.55
TOTAL		5.7			2.6			1.25	1.35
MEAN		1.14			0.52			0.25	0.27
% Stored = 48%									

3.10.2 Continuous Simulation using Hourly Precipitation Data

The procedure for continuous simulation using hourly precipitation data is a simplification of the storm event procedure described above. The major difference is that it is unnecessary to define a storm event since the hourly precipitation data are being used directly. The simpler procedure is described below.

1. Set up an hourly water budget for the period of investigation keeping track of the status of the detention pond during and between storms.
2. Perform the simulation and record the summary statistics on the performance of the storage-release system. Repeat this process by varying the assumed storage volume (S) and release rate (T).

An example of this simulation is shown in Table 3.4 for the same period of June 1987. The general calculations are the same as for the event simulation except that hourly accounting and intrastorm variability is incorporated. For this example, the number of calculations increases from five for the event analysis to 24 for the hourly analysis. Thus, the event analysis does save some computations. On the other hand, the event analysis aggregates the 6/29/87 event which actually goes from 7 am to midnight into a single event. Thus, the entire dynamics of the behavior of this storm are lost when the event simulation is done. This may introduce significant errors in calculating the performance of the detention pond. Major advantages of the hourly accounting include not having to select a storm event definition and the ability to calculate detention times more accurately. This avoids significant sources of error that can result from aggregation. With the current economics of computing, it is very easy to simply use hourly data. Thus, we recommend using the hourly data directly for continuous simulation and avoiding the use of storm events.

The recommended procedure for generating the final storage-treatment production function and the cost minimization is to follow the procedure described by Nix and Heaney (1988) wherein approximating splines are used to find the isoquants and the output is in the form of a tabular production function. This avoids the errors introduced by trying to fit equations to the data. Then the final cost analysis is done by simply multiplying the various S, T pairs by their respective unit costs in order to derive the optimal expansion path.

3.10.3 Statistical Method

An alternative to continuous simulation is to use the statistical method whereby PDFs are fitted to the hourly precipitation data. The monthly results for Boulder, Colorado are shown in Figure 3.11 for the number of events per month, the hours per event, the volume per event, and the volume per month. Similar statistics can be generated for other factors including the interevent time. Using the statistical method, one selects a period or periods for which PDFs will be developed. One option is to take all of the precipitation events in the period of record and develop the PDFs for all of the data. This approach can be inaccurate due to seasonal differences that clearly exist. Another option is to select a period of interest, e.g. the summer months, and generate the PDFs for the summer period only. Finally, some analysts derive PDFs for each month.

Table 3.4 Hourly simulation: example calculation for Boulder, Colorado.

Runoff Coefficient = 0.5
Total Storage (in) = 0.6
Depression Storage (in) = 0.05
Drawdown Rate (in/hr) = 0.004

Date	Time	Rain	"Dry" Time (hrs)	Runoff (in)	Storage at Start of Event (in)	Storage at End of Event (in)	Volume Stored (in)	By-pass Volume (in)
6/1/87	100	0.00	182	0	0.60	0.60	0.00	0.00
6/7/87	2000	0.10	163	0	0.60	0.60	0.00	0.00
6/7/87	2200	0.10	0	0.05	0.60	0.55	0.05	0.00
6/8/87	2000	0.60	22	0.25	0.60	0.35	0.25	0.00
6/8/87	2100	0.50	0	0.25	0.35	0.11	0.25	0.00
6/8/87	2200	0.90	0	0.45	0.11	0.00	0.11	0.34
6/8/87	2300	0.60	0	0.3	0.00	0.00	0.00	0.30
6/8/87	2400	0.30	0	0.15	0.00	0.00	0.00	0.15
6/10/87	1600	0.10	40	0	0.16	0.16	0.00	0.00
6/24/87	1800	0.10	338	0	0.60	0.60	0.00	0.00
6/29/87	700	0.10	109	0	0.60	0.60	0.00	0.00
6/29/87	800	0.30	0	0.15	0.60	0.45	0.15	0.00
6/29/87	900	0.20	0	0.1	0.45	0.36	0.10	0.00
6/29/87	1000	0.10	0	0.05	0.36	0.31	0.05	0.00
6/29/87	1100	0.20	0	0.1	0.31	0.22	0.10	0.00
6/29/87	1200	0.10	0	0.05	0.22	0.17	0.05	0.00
6/29/87	1300	0.20	0	0.1	0.17	0.07	0.10	0.00
6/29/87	1500	0.20	0	0.1	0.07	0.00	0.07	0.03
6/29/87	1600	0.20	0	0.1	0.00	0.00	0.00	0.10
6/29/87	1700	0.10	0	0.05	0.00	0.00	0.00	0.05
6/29/87	1800	0.30	0	0.15	0.00	0.00	0.00	0.15
6/29/87	1900	0.20	0	0.1	0.00	0.00	0.00	0.10
6/29/87	2100	0.10	0	0.05	0.00	0.00	0.00	0.05
6/29/87	2400	0.10	0	0.05	0.00	0.00	0.00	0.05
TOTAL		5.70		2.60			1.2820	1.3180
% stored = 49%								

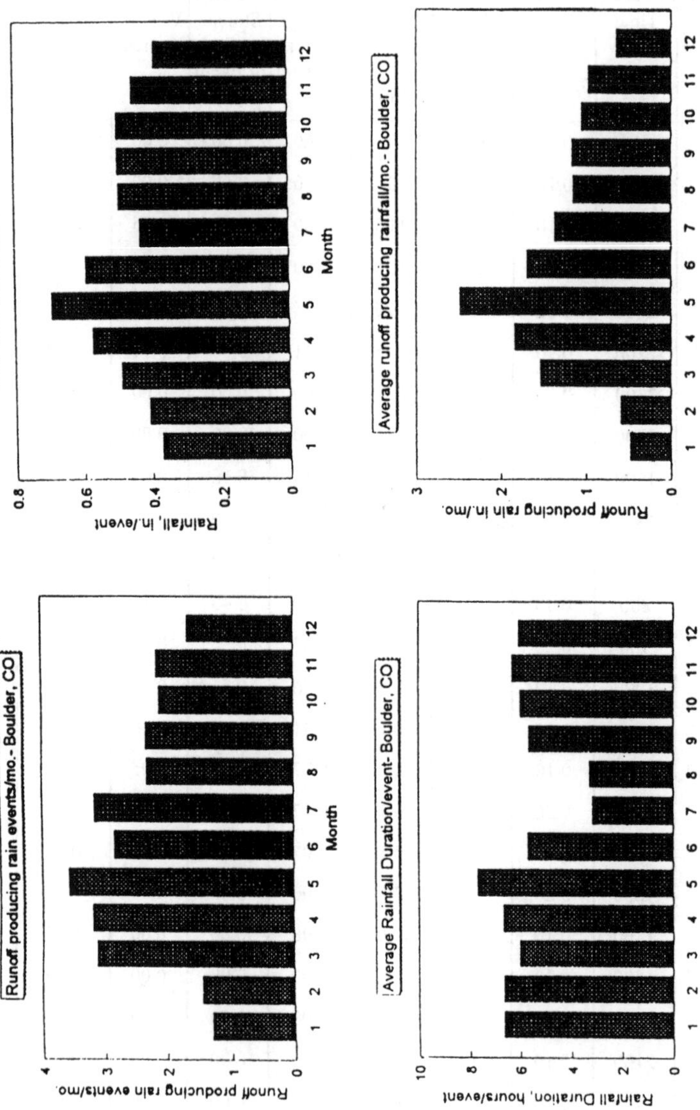

Figure 3.11 Seasonal variability of rainfall for Boulder, Colorado.

3.10.4 Monte Carlo Simulation

A more robust approach is to use Monte Carlo simulation which is much less restrictive than analytical methods (types of distributions can be selected and the covariance among these distributions can easily be included in the simulation). The steps in the analysis are:

1. Define a precipitation event in terms of the maximum number of dry hours allowed within an event.
2. Given this event definition, generate the rainfall statistics for event volume, duration, and interevent time.
3. Find the best PDFs for the above distributions, e.g. using BestFit (Palisade Corp. 1994).
4. Upon completion of the Step 3, specify the three input distributions for volume, duration, and interevent time.
5. The covariance among these three distributions is calculated by the spreadsheet.
6. These three PDFs and their correlation matrix are entered into the spreadsheet, where storage and detention time characteristics are used to calculate performance.
7. The Monte Carlo software samples the input distributions, and after the spreadsheet calculates the basin performance, output distributions are developed, e.g. using @Risk (Palisade Corp. 1994).
8. The output CDF is used to estimate the statistics of the long-term performance of the system.

An example of the Monte Carlo approach is shown in Table 3.5. The spreadsheet calculation is very simple because the simulation proceeds from event to event as new estimates are generated by the Monte Carlo method. The major limitation of this approach is the inaccuracies introduced by converting the original hourly precipitation data into approximating PDFs and the associated covariance. Increased accuracy can be obtained by fitting PDFs for each month but that complicates the calculations from month to month. It is more accurate to work directly with the original hourly time series data.

3.11 Summary and Conclusions

A wide variety of methods exist for evaluating the effectiveness of detention systems for removing pollutants in urban runoff. The purpose of this chapter was to describe various approaches for estimating the pollutant removal capability of urban stormwater detention systems. Virtually all of the methods view the detention pond as a settling basin whose removal rate depends on the residence time in the basin and the reaction rate. These processes occur in any storage

Table 3.5 Example Monte-Carlo simulation for Boulder, Colorado.

Event Rule (10 hrs) = 1000
Depression Storage (in) = 0.05
Drawdown Rate (in/hr) = 0.004

Runoff Coefficient = 0.5
Total Storage (in) = 0.6

	Event Volume (in)	Event Duration (hr)	Inter-Event Time (hr)	Runoff Volume (in)	Storage at Start of Event (in)	Storage at End of Event (in)	By-Pass Volume (in)	% Stored
	0.06	1.42	136.97	0.00	0.60	0.61	0.00	
Input PDF's	Inverse Gaussian	Pearson Type VI	Truncated Lognormal					
Minimum	0.02	0.03	10.14	0.00	0.00	0.00	0.00	0.00
Mean	0.33	5.26	133.15	0.12	0.57	0.49	0.02	0.96
Maximum	2.33	89.15	2101.05	1.11	0.60	0.60	1.11	1.00

device be it a small pond, large lake, primary clarifier, or stormwater detention pond. The theory for estimating the performance of detention basins is based on either Stokes' law for discrete particle settling of suspended solids, or more complex reaction kinetics. For stormwater detention systems, the basins fill and empty at relatively short intervals such as every few hours, days or weeks. Thus it is important to properly characterize the process dynamics.

The mass balance for the flow into and out of a detention system is determined using a mass balance equation. The accuracy of the depiction of the performance of the detention system depends upon the selected time step in the mass balance equation. For urban stormwater analyses, the most popular time steps have been five to fifteen minutes, one hour or storm event. The five to fifteen minute time step is often used in detailed single event simulations such as those used to evaluate surcharging in the pipe system. The one hour time step is used in STORM and similar models to evaluate various storage-release strategies. Lastly, the hourly data may be aggregated into a "storm event" which is defined as ending when a specified number of dry hours has elapsed. Storm events are used in statistical models of stormwater systems. The critical question in defining a storm event is to specify the appropriate number of dry hours which will terminate an event. Various event definitions have been used but there is no correct definition. Traditionally, a major justification for using a longer time step is computational expediency or necessity. However, the economics of computing have changed radically since the introduction of PCs.

Virtually all popular methods for evaluating the expected removal of pollutants from a detention basin are based on simple removal kinetics. However, experimental work with urban runoff indicates that there are multiple stages of removal. Detention time is defined as the residence time of a parcel of water in a detention basin. Detention times vary widely during and immediately after a storm event due to the dynamics of the storm event. Existing simplified methods calculate an "average" detention time which may give a very inaccurate measure of real detention times because of the nonlinearities involved.

The earliest evaluations of the effectiveness of urban stormwater detention systems were done using early versions of the Storage/Treatment block of the EPA SWMM . [*Editor's note: EPA SWMM now runs S/T continuously.*] These single event methods proved to be inadequate because the regulatory agencies could not agree on the criteria for selecting the "design event". Thus, these methods were extended to include continuous simulation. The optimal mix of storage and treatment or release rate was found by running continuous simulations of hourly rainfall using the STORM model. The resulting storage-treatment isoquants and associated costs of storage and treatment were used to find the least-costly mix of storage and treatment for a given level of pollution control. The so-called statistical method was first introduced to the stormwater field in the late 1970's.

Current practice includes very simple approaches which estimate pollutant removal as a simple function of the ratio of detention basin volume and the mean storm event volume. However, the definition of mean storm volume depends on how storm events are defined. The mean volume per event changes drastically as the definition of a storm event is varied from ending after one, two, or more dry hours.

Statistical approaches for evaluating stormwater detention systems were reviewed. They are an improvement over the simple empirical approaches because they incorporate storm event statistics. Two types of statistical approaches can be used. One option is to derive the performance of the system analytically. The other approach is to use Monte Carlo simulation. The major limitation of these approaches is the need to aggregate the hourly precipitation data into storm events. The results are very sensitive to how events are defined.

Single event and continuous simulation models have been used for many years to evaluate stormwater detention systems. The EPA SWMM and STORM have been the two most popular simulation models. With the advent of the PC, it is now possible to do many of these calculations with much easier to use software or to use spreadsheets.

We recommend using the statistical procedures as a companion to process-oriented continuous simulation models as a way of summarizing the findings from these simulation studies. However, they should not be used as the primary tool for design.

[*Editor's note: In this chapter "continuous simulation" is evidently taken to mean preprocessing the rain record into a sequence of wet events which are then input to the model - as opposed to the usage in following chapters, wherein the long-term precipitation record is directly input to a model that includes all algorithms active for the complete record, independent of any definition of minimum interevent dry-weather period.*]

References

Adams, B.J. and J.B. Bontje. 1984. Microcomputer application of analytical models for urban stormwater management. in James, W.D. Ed. Emerging Techniques in Stormwater and Flood Management, ASCE New York, p. 138-162.

Camp, Dresser, and McKee. 1993. Best Management Practice Handbook-Vol. 1, Municipal. California Stormwater Task Force.

Debo, T.N. and A.J. Reese. 1995. Municipal Storm Water Management. Lewis Publishers, Boca Raton, FL

DiToro, D.M. 1975. Statistical design of equalization basins. Jour. of Env. Engg. Div., ASCE, 101, EE6, p. 917-933.

DiToro, D.M. and M.J. Small. 1979. Stormwater interception and storage. Jour. of Env. Engg. Div., ASCE, 105, EE1, p. 43-54.

Dorman, M.E., Hartigan, J., Johnson, F. and B. Maestri. 1988. Retention, Detention, and Overland Flow for Pollutant Removal from Highway Stormwater Runoff: Interim Guidelines for Management Measures. FHWA/RD-87/056, Federal Highway Administration, McLean, VA

Driscoll, E.D., Shelly, P.E. and E.W. Strecker. 1990. Pollutant loadings and impacts from highway stormwater runoff. Vol. I. Design Procedure (FHWA-RD-88006), Volume II. User's Guide for Interactive Computer Implementation of Design Procedure (FHWA-RD-88-007), Vol. III. Analytical Investigation and Research Report (FHWA-RD-99008), and Volume IV. Research Report Data Appendix (FHWA-RD-88-009). Federal Highway Administration, McLean, VA

Driscoll, E.D. and E. Strecker. 1993. Assessment of BMP's being used in the U.S. and Canada. Proc. 6th *Int. Conf. on Urban Storm Drainage. Niagara Falls, NY*

Field, R., Struzeski, E.J., Masters, H.E. and A.N. Tafuri. 1974. Water pollution and associated effects of from street salting. Jour. Env. Engg. Div., ASCE, 100 (EE2), p. 459-477.

Finkel, A. and D. Golding. 1994. Worst Things First? The Debate Over Risk-Based National Environmental Priorities. Resources for the Future, Washington, D.C.

Goforth, G.F.E., J.P. Heaney, and W.C. Huber. 1983. Comparison of basin performance modeling techniques. J. of Environmental Engg. Div., ASCE, 109, EE5, p. 1082-1098.

Greeley-Polhemus Group, Inc. 1992. Guidelines for Risk and Uncertainty Analysis in Water Resources Planning. Report prepared for U.S. Army Corps of Engineers, Institute for Water Resources, Fort Belvoir, VA.

Hahn, G.J. and S.S. Shapiro. 1967. Statistical Models in Engineering. J. Wiley and Sons, New York.

Haimes, Y.Y., D.A. Moser, and E.Z. Stakhiv, Eds. 1994. Risk-Based Decision Making in Water Resources VI, ASCE, New York.

Heaney, J.P., Huber, W.C, Sheikh, M., Medina, Jr., M.A., Doyle, J.R., Peltz, W.A., and J.E. Darling. 1975. Urban stormwater management modeling and decision making. EPA-670/2-75-022, Cincinnati, OH

Heaney, J.P., and S.J. Nix. 1977. Storm Water Management Model: Level I-Comparative evaluation of storage-treatment and other management practices. EPA-600/2-77-083, Edison, NJ

Heaney, J.P., Huber, W.C., Medina, Jr., M.A., Murphy, M.P., Nix, S.J., And S.M. Hasan. 1977. Nationwide evaluation of combined sewer overflows and urban stormwater discharges, Vol. II, Cost assessment and impacts. EPA-600/2-064b, Cincinnati, OH

Heaney, J.P.,Nix, S.J., and M.P. Murphy. 1978. Storage-treatment mixes for stormwater control. Jour. of Env. Engg. Div., Proc. ASCE, 104, EE4, p. 581-592.

Heaney, J.P., W.C. Huber, R. Field, and R.H. Sullivan. 1979. Nationwide cost of wet-weather pollution control. Jour. Water Poll. Control Fed., Vol. 51, No. 8, pp. 2043-2053.

Howard, C.D.D. 1976. Theory of storage and treatment plant overflows. Jour. Env. Engg. Div., Proc. ASCE, 102, EE4, p. 709-722.

Howard, C.D.D., Flatt, P.E. and U. Shamir. 1979. Storm and combined sewer storage treatment theory compared to computer simulation. Grant No. R-805019, US EPA, Cincinnati, OH

Kapur, K.C. and L.R. Lamberson. 1977. Reliability in Engineering Design. J. Wiley and Sons, New York.

Law, A.M. and W.D. Kelton. 1991. Simulation Modeling and Analysis. 2nd Ed. McGraw Hill, NY, NY.

Li, J.Y. and B.J. Adams. 1994. Statistical water quality modeling for urban runoff control planning. Water Science and Technology, Vol. 29, No. 1-2, p. 181-190.

Mays, L.W. and Y-K. Tung. 1993. Hydrosystems Engineering and Management. McGraw-Hill, New York.

Medina, M.A. 1976. Interaction of urban stormwater runoff, control measures, and receiving water responses. PhD Dissertation, U. of Florida, Gainesville, FL

Medina, M. A., W.C. Huber, and J.P. Heaney. 1981a. Modeling stormwater storage/ treatment transients: theory. Jour. of Environmental Engg. Div., ASCE, 107, EE4, p. 781-797.

Medina, M.., W.C. Huber, and J.P. Heaney. 1981b. Modeling stormwater storage/ treatment transients: applications. Jour.. of Environmental Engg., Div. ASCE, 107, EE4, p. 799-816.

Nix, S.J., Heaney, J.P., and W.C. Huber. 1981. Water quality benefits of detention. Chapter 12 of Urban Stormwater Management, Special Report No. 49, American Public Works Assn., Chicago, IL

Nix, S.J. 1982. Analysis of storage-release systems in urban stormwater quality management. PhD Dissertation, U. of Florida, Gainesville, FL

Nix, S.J. and J.P. Heaney. 1984. Characterization of suspended solids settling. Proc. Int. Symposium on Stormwater Management. U. of Kentucky, Lexington.

Nix, S.J. 1985. Residence time in stormwater detention basins. Jour. of Environmental Engg., 111, 1, p. 95-100.

Nix, S.J. and J.P. Heaney. 1988. Optimization of storage-release strategies. Water

Resources Research, 24, 11, p. 1831-1838.

Nix, .J., J.P. Heaney, and W.C. Huber. 1988. Suspended solids removal in detention basins. J. of Environmental Engg., ASCE, 114, 6, p. 1331-1343.

Nix, S.J. 1994. Urban Stormwater Modeling and Simulation. Lewis Publishers, Boca Raton, FL

Palisade Corp. 1994. @Risk, Newfield, NY.

Palisade Corp. 1994. BestFit!, Newfield, NY.

Pierucschka, E. 1962. Principles of Reliability. Prentice-Hall, Englewood Cliffs, NJ

Pisano, W.C. and H. Brombach. 1996. Solids settling curves. Water Environment and Technology, Vol. 8, No. 4, p. 27-33.

Schueler, T.R. 1987. Controlling Urban Runoff-A Practical Manual for Planning and Designing Urban Best Management Practices. Metropolitan Washington Council of Government, Washington, D.C., 202 p.

Schueler, T. 1993. Performance of stormwater ponds and wetland systems. in Engineering Hydrology, C. Kuo, Ed., ASCE

Schueler, T.R., Heraty, M. and P. Kumble. 1992. A current assessment of urban best management practices: Techniques for reducing nonpoint source pollution in the coastal zone. Metropolitan Washington Council of Governments, Washington, D.C.

Segarra-Garcia, R. and G.V. Loganathan. 1994. A stochastic pollutant load model for the design of stormwater detention facilities. Water Sci. Tech., Vol. 29, No. 1-2, p. 327-335.

Small, M.J. and D.M. DiToro. 1979. Stormwater treatment systems. Jour. of Environmental Eng. Div., ASCE, 105, EE3, p. 557-569.

Thomson, N.R., McBean, E.A., Mostrenko, I.B., and W.J. Snodgrass. 1994. Characterization of Stormwater Runoff from Highways. Chapter 9 in James, W. Ed. Current Practices in Modeling the Management of Stormwater Impacts, Lewis Publishers, Boca Raton, FL

Urbonas, B. and P. Stahre. 1993. Stormwater Best Management Practices and Detention. Prentice-Hall, Englewood Cliffs, NJ.

Wanielista, M.P. and Y.A. Yousef. 1992. Stormwater Management. J. Wiley and Sons, New York.

Whipple, Jr., W. and J.V. Hunter. 1981. Settleability of Urban Runoff Pollution. Jour. Water Pollution Control Federation, Vol. 53, No. 12, p. 1726-1731.

Chapter 4

Use of Continuous Simulation for Evaluation of Stormwater Management Practices to Maintain Base Flow and Control Erosion

Raymond T. Guther, Ronald B. Scheckenberger and William R. Blackport

Maintaining the natural distribution of flow within watercourse ecosystems is desirable for the support of downstream fisheries, and preservation of natural channel forming processes. Increases in impervious land coverage and efficient conveyance systems, associated with urban development, can lead to a decrease in infiltration and associated decrease in base flow to downstream watercourses, along with localized temporal and absolute increases in high flows, which ultimately impact on channel form and associated habitat.

4.1 Background

The Regional Municipality of Hamilton-Wentworth and the City of Hamilton have identified approximately 46 ha of land proposed for prestige industrial development within a headwater area of the Red Hill Creek in southern Ontario (see Figure 4.1). The total contributing subwatershed consists of approximately 133 ha with two tributary watercourses traversing the development area. The present land use within the area is primarily agricultural with a single commercial user and a number of single family residences.

© *Advances in Modeling the Management of Stormwater Impacts - Vol. 5.* W. James, Ed.
Pub. by CHI, Guelph, Canada 1997. ISBN 0-9697422-7-4. Fax: +519 767-2770

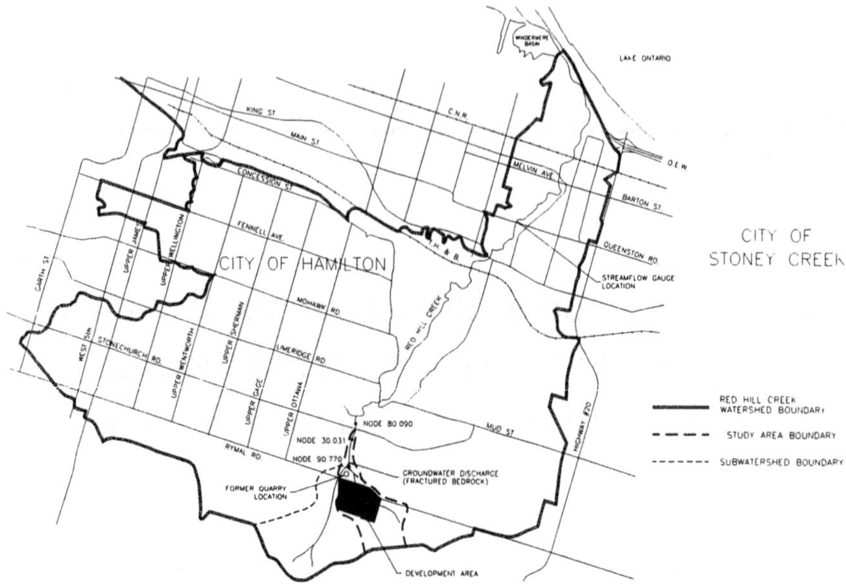

Figure 4.1 Redhill Creek watershed - location plan.

The Red Hill Creek, located almost entirely within the urban area of Hamilton and Stoney Creek, has historically been subjected to many anthropogenic factors and has more recently been subject to intensive study regarding a proposed expressway development, as well as a provincially-funded restoration project.

In 1993, the Regional Municipality of Hamilton-Wentworth commissioned a study to examine the potential to utilize an abandoned quarry located immediately north (downstream) of the development area for stormwater management purposes (Dames and Moore Canada, 1993). This study was singularly focused on the use of the quarry site and associated impacts of such use. However, concerns were raised by the Hamilton Region Conservation Authority (HRCA), owners of the former quarry site, and the Ministry of Natural Resources regarding environmental impacts of the project on baseflow and temperature within the downstream watercourse, and flooding and erosion impacts associated with increased imperviousness and reductions in channel storage capacity. Additionally the HRCA indicated a requirement that any proposed stormwater management facility would need to complement the use of the site for passive recreational uses, indicating a requirement for a comprehensive study of stormwater management throughout the development and including an assessment of on-site control measures.

In order to address these concerns the Regional Municipality of Hamilton-Wentworth (RMHW) and City of Hamilton initiated a comprehensive environmental assessment. The potential impacts of developing the 46 ha industrial land use which were considered in the assessment included: surface water quality, temperature, erosion and baseflow, as well as the effects that these potential impacts may have on the physical and social environment (i.e. fisheries and terrestrial resources, stream morphology, and socio-economic effects). The assessment was undertaken in accordance with requirements of the Ontario Municipal Engineers Association Class Environmental Assessment process (MEA, 1993).

It should be recognized that a dry stormwater management flood control facility is currently under construction approximately 1 km. downstream of the development area. Additionally, the lands located between the development area and the water quantity facility are in public ownership and are used as passive open space, and they are not considered to be subject to damage during flood events. Therefore water quantity control for flood protection has not been required as a key objective in developing a stormwater management strategy, beyond the control of localized flood and erosion impacts within, and immediately downstream of, the development area.

4.2 Baseline Conditions

4.2.1 Soils

The watershed is characterized by rolling topography with average slopes generally ranging from 1.5 to 4.0%. The soils are classified as a mixture of clay and silt loams with drainage characteristics ranging from hydrologic classes BC to D. Most soil types in the western portion of the watershed exhibit class C and D characteristics, with class BC soil types located in the eastern portion of the study area. Notwithstanding, field observations indicate that the infiltration capacity of the soils would likely exceed the infiltration predicted based on the soil class alone.

4.2.2 Fisheries

An assessment of the fish habitat downstream of the Development Area indicates that the stream flow exhibits high temperatures in the summer months and fish species diversity is very limited. Significantly cooler temperatures were however noted at the base of pools, indicating the contribution of base flow to the stream.

4.2.3 Stream Morphology

Watercourse reaches immediately downstream of the developing area are relatively steep and feature cobble substrate with local outcrops of bedrock. Further downstream the watercourse substrate becomes finer, channel slopes are reduced, and critical velocities are lower. Therefore, the downstream sections of the watercourse are considered to be more vulnerable to erosion due to changes in peak flow rates and duration of flow, than the watercourses immediately downstream of the study area.

4.2.4 Geology

The overburden throughout the study area generally consists of a clayey silt or clay till, likely associated with the Niagara Falls Moraine. The overburden thickness varies from approximately 10 m in the southern portion of the study area to bedrock outcropping particularly along portions of Red Hill Creek and its tributaries. Throughout most of the study area the overburden is less than 5 m thick and is commonly oxidized and fractured. The existence of oxidation and fractures gives rise to a more significant potential for movement of water within the overburden as compared to a massive, un-oxidized clay till.

The upper 7 to 10 m of the underlying bedrock consists of a grey brown to dark brown petroliferous dolostone. The upper bedrock appears to be highly fractured and contains numerous vuggy zones. The fractured vuggy nature of the dolostone allows for a significant pathway for groundwater movement. The bedrock topography tends to follow the surficial topography in the study area.

4.2.5 Hydrogeology

Water well records and geotechnical borehole logs indicate that the water table in the shallow bedrock is generally within 5 m below the top of the bedrock. This general water table trend, combined with the oxidized and fractured nature of the overburden, the highly fractured nature of the shallow (upper 10 m) bedrock, and the headwater nature of the study area, supports the following groundwater flow scenario.

Recharge occurs throughout most of the study area. The infiltration into the fractured, thin overburden will be higher than a more massive, thicker overburden. The higher rate of infiltration was confirmed by an observation of a distinct lack of ponded water within the watershed during an early winter snowmelt. The infiltrating water moves vertically to the bedrock and then moves, to a greater degree, horizontally within the shallow bedrock, following the bedrock topography. The horizontal groundwater flow, as it follows the bedrock topography, is directed to an area along Red Hill Creek channel just downstream of the developing area where it discharges through a relatively small area of highly fractured karstic bedrock (see Figure 4.1).

A groundwater balance for the area indicates that the infiltration rate throughout the subwatershed may be of the order of 150 to 200 mm/year, which is generally consistent with fractured clay silt tills

The fractured nature of the flow system as presented typically allows for minimal attenuation of contaminated waters entering the subsurface.

4.3 Alternative Stormwater Management Strategies

Based on consideration of baseline environmental conditions, various stormwater management techniques were developed for consideration and evaluation in order to determine the optimum plan for mitigating the impacts due to development. A watershed model was used to evaluate the effects of various stormwater management strategies on baseflow, peak flow and runoff volume. Stormwater management strategies which were considered included:

1. *Future Uncontrolled Development* This alternative involved a traditional type of development (i.e. urban form) with no stormwater management controls. It would not address potential impacts on water quality, erosion or fisheries habitat as required by Provincial and Federal policy; therefore it was considered unacceptable, however it continues to serve as a measure of maximum impact for comparative purposes.
2. *Source Controls* This alternative featured the use of mitigative measures within developing areas on a lot-by-lot basis and included techniques such as: on-site infiltration, reduced lot grading, roof leader discharge to grassed areas, and rural road standards (i.e. roadside ditches).
3. *End-of-Pipe Facility* The end-of-pipe alternative was based on traditional development drainage utilizing an urban road cross section (i.e. curb & gutter) and collection of runoff by storm sewers. Stormwater would then be conveyed to a stormwater management facility (wet pond or constructed wetland) for water quality treatment and additional storage to prevent potential increased incidence and duration of erosive flows.

   Two general end-of-pipe alternatives, plus a number of variations were considered and evaluated. The primary difference between these alternatives relates to the conveyance of external drainage through the site:

   1: no mixing of external and internal stormwater runoff; or
   2: external and internal stormwater runoff mixes and is conveyed to an end-of-pipe facility.

A potential location for such a stormwater management facility was previously identified (Dames and Moore, 1993) in a former quarry immediately north (downstream) of the development area.

4. *Combination of End-of-Pipe and Source Control Techniques*
   This alternative consisted of the use of source controls (i.e. rear yard ponding areas/soakaway pits, reduced lot grading), to promote infiltration from roof tops and landscaped areas in conjunction with an end-of-pipe facility to treat runoff from roads and parking areas, which would be constructed to a full urban standard including storm sewer conveyance.

   This option would provide the benefit of promoting the infiltration of "clean" water from rooftop areas, while providing treatment of parking lot, and roadway runoff, typically much higher in suspended sediments, nutrients and heavy metals than roof-top drainage. The combination alternative also included provisions for open watercourse systems to convey external runoff through the developing area without mixing with urban runoff from parking areas and roads.

Based on the proposed land use within the developing area, the use of water quality inlets to pretreat stormwater and provide spill protection for runoff from roads and parking areas, was also considered.

4.4 HSP-F Analysis Approach

In order to determine existing and future land use flows to assess the impact of development on the Red Hill Creek tributary watersheds, a continuous hydrologic simulation technique to determine frequency of flows for system evaluation and design was recommended. Continuous modeling uses historical rainfall data as input to generate a time-series of modeled peak flows. These time-series (i.e. at various key points in the watershed) are then further analyzed for recurrence, based on frequency analysis, thereby associating a peak flow with frequency.

The use of a continuous simulation model offers a significant benefit in terms of allowing evaluation of various water quality and quantity management strategies through a full range of flow conditions including baseflow conditions. This provides a more realistic evaluation of the overall effectiveness of management strategies in replicating the natural hydrologic characteristics of the watercourse.

The Hydrologic Simulation Program - Fortran (HSP-F), developed by Hydrocomp Inc. 1980, for the US-EPA (Bicknell et al. 1993), was selected as a Ministry of Natural Resources - approved continuous simulation technique for use in this study.

4.4.1 Calibration/Verification

The HSP-F model dataset is based on the original calibrated model of the Red Hill Creek which was developed for the *Mountain East-West and North-South Transportation Corridor Study* (Philips Planning and Engineering, 1989) as refined through subsequent design studies (Philips Planning and Engineering, 1993; Bishop and Scheckenberger, 1995). The original model was calibrated to storm flows at a gauge station located at Queenston Road (see Figure 4.1). A portion of the original model was modified and discretized to a greater level of detail as required for the study purposes. The base model was calibrated and validated based on stream flow measurements taken as part of this study. In addition to precipitation and evaporation, snowmelt processes were also simulated in the updated HSP-F model in order to model the performance of stormwater management techniques for the full year as well as assisting in determining water levels within the stormwater management facility.

4.4.2 Low Flow Calibration

Low flow measurements were undertaken on June 1, 1995. The resulting baseflow/interflow from the general study area was observed to be approximately 11 l/s (0.011 m^3/s) at a location immediately downstream of the development area. The measured flow was determined to occur on the receding limb of the hydrograph caused by a rainfall event of 18 mm on May 28, 1995. The original base model parameterization computed a base flow of 70 l/s for the same event.

An analysis of interflow and soil storage characteristics was undertaken to determine the sensitivity of these various parameters. Hydrogeological investigations which were undertaken to assess interflow conditions in the study area demonstrated that the clay soils exhibit numerous fissures. This attribute, in conjunction with the shallow overburden and fractured bedrock conditions results in rapid infiltration and interflow similar to that exhibited by a sand or gravel soil. On this basis, it is expected that interflow would also recede quickly, in approximately 6 hours following a storm event.

Based on these observations the following parameter changes were incorporated into the model:

1. Mannings “n” (NSUR) for pervious land segment was reduced. This parameter alteration would increase computed peak flows during runoff events.
2. The interflow recession parameter (IRC) which reflects the slope of the interflow recession limit was reduced to 0.25 from 0.5 (base model), to reflect rapid conveyance of interflow to the receiving watercourse.

3. Lower and upper zone nominal soil storage (LZSN and UZSN) values were increased to 150 mm and 12 mm, respectively, from base model assumptions of 10 mm and 1 mm.
   The use of these values is consistent with the values obtained using the relation (Crawford and Linsley, 1966):

$$\text{LZSN} = 101.6 + 0.125 \text{ (annual precipitation)}$$

   Note: Annual precipitation was assumed to be 900 mm/yr, and the relation UZSN $\approx$ 0.08 LZSN (Crawford and Linsley, 1966) was utilized to estimate these parameters

The foregoing modifications tended to reduce the computed baseflow and peak flows resulting in a good correlation to baseflow/interflow observed on June 1, 1995; (0.015 m^3/s computed, 0.011 m^3/s observed) while maintaining a good correlation to peak flow values under the July and October 1962 storm events as compared to the stormflow calibrated base model (see Table 4.1).

Table 4.1 Peak flow (m^3/s).

Model	March 12, 1962	July 26, 1962	October 4, 1962
Red Hill Creek 1990 Model (Macro)	0.3	12.0	15.0
Rymal 1996 (Micro)	0.1	11.0	13.7

The July 26, and October 4 , 1962 flows generally compare well (within 10%) to the original model results. The March 12, 1962 storm event was observed to exhibit less correlation to the base model. However, peak flow rates for this event are closer to baseflow magnitude and therefore subject to greater variation due to model parameter changes.

4.4.3 Parameterization of Alternatives

The HSP-F datasets developed for existing site conditions were revised to reflect future land use in order to assess the performance of various stormwater management strategies.

4.4.4 Source Controls

The use of source controls primarily relies on providing measures within the context of site development to promote infiltration, thereby reducing runoff volumes as well as reducing or eliminating direct connections between the impervious surfaces and the storm drainage system.

The following source controls were considered for use in the study area:

1. discharge of roof leaders to pervious areas,
2. reduction in minimum lot grades (slopes) from current municipal standards,
3. rear yard ponding of roof discharge (roof discharge only), and
4. rural cross-sections.

The simulation of these techniques was incorporated into the model through the variation of the following input parameters:

1. Increasing the value of the input upper soil zone nominal storage (USZN) from 12 mm to 15 mm, to reflect greater opportunity for infiltration provided by reduced lot grading.
2. Reduced input impervious area contribution to reflect discharge of roofs and road surfaces to pervious areas. Input impervious coverage was reduced to existing land use levels (i.e. 1% directly connected impervious area). However, input pervious infiltration capacity was also reduced from 2.5 mm/hr to 0.75 mm/hr, in accordance with the amount of input impervious coverage, to reflect the lower opportunity for infiltration. The use of a rural road cross-section was also incorporated into the model through this reduction of input impervious coverage.
3. The simulation of rear yard ponding areas/soakaway pits was incorporated using a single reach reservoir in a lumped approach to simulate storage of runoff from the development area. The use of this technique was considered appropriate in this case for the following reasons:
    - baseflow/interflow discharge to the surface water system occurs less than 500 m. downstream of the development area;
    - groundwater flow is primarily horizontal through the upper bedrock layer, parallel to the surface flow system with less vertical conveyance to the deep groundwater system; and
    - interflow response of the subwatershed is rapid, of the order of 6 to 8 hours.

The foregoing characteristics suggest that attenuation and routing characteristics of the reservoir algorithm adequately represent the relatively rapid and shallow groundwater flow regime.

The volume of storage was based on the 20 mm criteria over the roof top area as recommended by the Ontario Ministry of Environment and Energy (MOEE). Outflow from the reservoir was based on a conservative infiltration estimate of 5 mm/hr.

4.4.5 End-of-Pipe Facility

The HSP-F model was revised to incorporate a "wet pond" facility within the vacant quarry located north of Rymal Road (immediately downstream of the development area).

The quarry area was modeled to provide a minimum of approximately 15 mm/ha of storage based on the upstream drainage area, an impervious ratio of 85% and Type 2 habitat as recommended by the Ontario Ministry of Environment and Energy (Marshall Macklin Monaghan, 1994). Given the 46 ha development area, the minimum required storage was calculated to be approximately 6,900 m^3. Of the 6,900 m^3 of total storage, a minimum of approximately 1,840 m^3 would comprise extended detention storage while the remaining 5,060 m^3 would be retained as permanent pool storage. In addition to this base scenario, various levels of extended detention storage ranging from 4 mm/ha to 37.5 mm/ha were also simulated and assessed.

Outflow from the quarry was modeled as a multiple objective pond outlet which included a reverse slope primary discharge outlet with an overflow weir and spillway to accommodate major storm events.

4.4.6 Combination of Source Control and End-of-Pipe facility

A third method of providing water quality and erosion protection was modeled by combining infiltration techniques employed with the source control alternative and an end-of-pipe facility which would collect and treat runoff from roads and parking areas. This management technique was incorporated into the model by routing the impervious land segment within the development area, through an end-of-pipe facility and directing runoff from the pervious land segments to the rear yard ponding areas/soakaway pits.

The impervious fraction which is directed to the water quality facility was approximated as 40% of the total development area.

The stormwater management facility was modeled based on 2,026 m^3 of permanent pool storage, corresponding to 110 m^3/ha as recommended for Type 2 habitat and 85% impervious coverage (Ontario Ministry Environment and Energy, 1994). The facility also includes an extended detention component of 4,534 m^3 or 250 m^3/ha. The drawdown period from the facility was modeled to occur over approximately three days to provide baseflow augmentation benefits.

4.4.7 External Drainage Consideration

The modeling was carried out for two primary methods of external drainage conveyance:

- external flows conveyed separately to their respective outlets (without mixing with urban runoff), and

- external flows mixed with internal drainage and conveyed to an end-of-pipe facility.

The use of an end-of-pipe facility may involve conveyance of external drainage areas through the facility or provisions to convey external runoff separately without mixing to the downstream watercourse.

The provision of a separate conveyance system to convey external flows to their natural outlet was considered to be preferable for the following reasons:

1. mixing of clean external runoff which requires treatment is avoided thus improving the efficiency of treatment;
2. the conveyance of external flow to its natural outlet will tend to preserve the natural function and habitat features of the receiving watercourse reaches; and
3. the overall storage volume of the end-of-pipe facility and conveyance system is reduced.

4.5 Results

4.5.1 Flow Distribution Analysis

The following water quality management strategies were analyzed with respect to potential impacts on base flows within the downstream watercourse:

1. do-nothing - (no development),
2. future uncontrolled development (no mitigation),
3. source controls only,
4. end-of-pipe facility only, and
5. combination of end-of-pipe and source control techniques.

The *do-nothing* alternative, which reflects present conditions within the development area, was used as a benchmark standard by which to compare the various alternatives. The *future uncontrolled* development represents a *worst-case* scenario to assist in the evaluation of the management alternatives.

A number of different *end-of-pipe facility* options were analyzed to determine the relative impacts of extended detention volume and drawdown period on baseflow rates and off-site erosion potential.

Table 4.2a provides a comparison of the various management alternatives in terms of replicating the natural flow class distribution for a typical year (total annual rainfall), while Table 4.2b provides a description of the various stormwater management alternatives. The results are provided for the outlet of the developing area (Node 90.770) which is located downstream of the end-of-pipe facility and includes all drainage from the development area and external drainage which passes through the development area. The results illustrate the impacts associated with uncontrolled development as well as the computed levels of mitigation achieved through the various techniques.

Table 4.2a Flow distribution (percent of annual) for stormwater management alternatives. Node 90.770 (outlet of development area).

Flow Class (l/s)	Existing	Future No Control	Source Controls	Combination Source Control and End-of-Pipe	End of Pipe				
					19 mm /3 day	19 mm /7 day	Mix	4 mm /1.5 day	37.5 mm /1 day
0.0-0.5	5.98	6.53	2.76	2.24	5.84	3.87	3.32	5.12	4.78
0.5-1.0	3.94	1.58	2.98	0.58	1.45	0.84	0.21	1.07	1.17
1.0-5.0	37.12	48.51	52.88	43.40	46.77	39.12	31.87	42.47	42.93
5-10	25.13	21.96	24.80	23.49	18.78	12.79	2.76	18.68	14.23
10-50	25.90	16.40	14.57	27.62	21.52	40.76	59.94	28.94	33.14
50-100	0.94	2.29	1.10	1.61	4.13	1.35	0.86	1.54	2.40
100-300	0.80	1.95	0.55	0.76	1.14	0.88	0.61	1.54	1.02
300-500	0.06	0.27	0.12	0.11	0.12	0.13	0.13	0.25	0.13
500-1000	0.06	0.31	0.12	0.10	0.13	0.13	0.15	0.24	0.13
1000-1500	0.04	0.11	0.05	0.04	0.05	0.05	0.06	0.07	0.05
1500-3000	0.03	0.07	0.05	0.05	0.06	0.07	0.07	0.06	0.02
over 3000	0.01	0.02	0.01	0.01	0.02	0.02	0.02	0.02	0.02

For most flow classes, the use of a combination of *source controls* and an *end-of-pipe* water quality facility best replicated the natural distribution of flow events. However, various end-of-pipe facilities provided varying computed levels of baseflow enhancement through increased computed duration of higher flow class occurrences.

The results of the baseflow analysis indicated that future uncontrolled development would reduce computed baseflow occurrences in the 10 to 50 l/s flow class by approximately 37% and by 15% in the 5 to 10 l/s flow class at this node. A corresponding increase of 31% is noted in the lower flow class range of 1 to 5 l/s.

End-of-pipe water quality facilities provided potential to enhance computed low flow within the 10 to 50 l/s flow class as noted by a 130% increase for an end-of-pipe facility providing conveyance of internal and external drainage to the facility. However, it is noted that computed occurrences of flows within the 5 to 10 l/s and 1 to 5 l/s are reduced by 89% and 14% respectively. This suggests that the effectiveness of end-of-pipe facilities in augmenting computed low flows was largely dependent on the input outflow release rate.

A combination of source control measures such as an end-of-pipe facility to treat road and parking area drainage is noted to be most effective in maintaining the existing baseflow conditions as evidenced by a 17% increase in the computed 1 to 5 l/s flow class, a 6.5% decrease in the 5 to 10 l/s flow class, and a 6.6% increase in the 10 to 50 l/s flow class.

4.5.2 Wet/Dry Year Flow Distribution Analysis

A further flow comparison was undertaken for various land use and management conditions including: existing conditions, future uncontrolled development and the combination alternative for developed land use under representative dry and wet years to verify the effectiveness of the combination of source control and end-of-pipe treatment.

Representative wet and dry years, determined on the basis of high and low annual total rainfall volumes, are represented by the years 1992 and 1963 respectively.

Tables 4.3 and 4.4 provide a summary of flow distribution and total volume of flow at the outlet of development area for representative wet and dry year simulations.

Based on the wet/dry year analysis, the combination alternative appears to be very effective in augmenting computed baseflow during a dry year as evidenced by a 865% increase in the 10 to 50 l/s flow class. During wet/dry years the combination alternative reduces the computed incidence of extreme low flow (0 to 1 l/s) by 23% and 10%, for wet and dry years respectively. Future uncontrolled runoff by contrast, would increase the computed incidence of extreme low flow by 45% and 18% for respective wet and dry years.

Table 4.2b Description of water quality management alternatives.

Alternative	Land Use	Operational Characteristics		
		Storage	Detention Time (Day)	Area Serviced
Existing	Existing conditions, no change in land use	N/A	N/A	46 ha
Future No Control	Full urban development to 80% impervious coverage with no SWM practices	N/A	N/A	46 ha
Source	Development to 70% impervious coverage utilize on-site infiltration and swale drainage system	On-site storage equal to 200 m^3/ha (200 mm from rooftop areas only	1	On-site measures, service 60% of 45 ha development area -Rooftops
Combination	Development to 70% impervious coverage with on-site infiltration of roof drainage and end-of-pipe treatment of parking area and road runoff	On-site storage equal to 200 m^3/ha (20 mm) from rooftop areas only End-of-Pipe facility Total Storage 360 m^3/ha Permanent Pool - 110 m^3/ha Extended detention storage - 250 m^3/ha (19 mm)	1 3	On-site measures, service 60% of 45 ha development area -Rooftops End-of-Pipe facility services 40% Of 46 ha development area - Road and Parking areas
End-of-Pipe 19 mm /3 day	Development to 80% impervious coverage with end-of-pipe facility	Total storage - 300 m^3/ha Permanent pool - 110 m3/ha Extended detention storage - 190 m^3/ha (19 mm)	3	Services development area only (46 ha
End-of-Pipe 19 mm /7 day	Development to 80% impervious coverage with end-of-pipe facility	Total storage - 300 m^3/ha Permanent pool - 110 m3/ha Extended detention storage - 190 m^3/ha (19 mm)	7	Services development area only (46 ha)

Table 4.2b continued Description of water quality management alternatives.

Alternative	Land Use	Operational Characteristics		
		Storage	Detention Time (Day)	Area Serviced
End-of-Pipe Mix	Development to 80% impervious coverage with study area	Total storage - 104 m^3/ha Permanent pool - 110 m3/ha Extended detention storage - 190 m^3/ha (19 mm)	6	Development area and external drainage areas are conveyed to end-of-pipe facility
End-of-Pipe 4 mm/ 1.5 day	Development to 80% impervious coverage with end-of-pipe facility	Total storage - 150 m^3/ha Permanent pool - 110 m3/ha Extended detention storage - 40 m^3/ha (4 mm)	1.5	Services development area only (46 ha)
End-of-Pipe 37.5 mm /1 day	Development to 80% impervious coverage with end-of-pipe facility	Total storage - 485 m^3/ha Permanent pool - 110 m3/ha Extended detention storage - 375 m^3/ha (37.5 mm)	1	Services development area only (46 ha)

During wet year conditions, the combination alternative is also noted to be effective in maintaining the computed natural flow distribution throughout the range of flow classes. Future uncontrolled development would result in increased computed incidence of both extreme high and extreme low flow occurrences.

The volumetric comparison also indicates that the combination alternative is effective in augmenting computed baseflow/interflow at non-erosive flow rates (see Table 4.5), while mitigating erosive flow durations to approximately existing undeveloped land use conditions.

4.5.3 Source Control Infiltration Assessment

The use of on-site source control measures was assessed by relating approximate annual recharge to annual total rainfall. The assessment was based on a duration analysis of the computed outflow rate (infiltration) from the modeled source control measure in order to calculate a total volume of the infiltration attained throughout the year through the on-site source control

Table 4.3 Wet year analysis - Node 90.770 (outlet of development area).

Flow Class (l/s)	Wet Year - 1992 - Percent of Year in Flow Class (%)			Wet Year - 1992 - Total Volume in flow class (m^3)		
	Existing	Future	Combination	Existing	Future	Combination
0 - 1.0	12.04	17.52	9.30	1 898	2 762	1 466
1.0 - 5.0	16.57	19.30	16.97	15 676	18 259	16 054
5 - 10	25.90	26.57	23.43	61 258	62 843	55 416
10 - 50	43.02	32.83	45.69	407 003	310 598	432 263
50 -1 00	1.59	2.91	3.55	37 606	68 827	83 964
100 - 500	0.83	3.24	0.95	78 524	306 529	89 877
500 - 1000	0.02	0.34	0.04	4 730	80 416	9 460
1000 - 1500	0.01	0.07	0.02	3 942	27 594	7 884
1500 - 3000	0.02	0.06	0.02	14 191	42 573	14 191
3000+	0.02	0.05	0.02	18 921	47 304	18 921
	Total Flow Volume below Erosive Threshold (500 l/s)			601 968	769 820	679 043
	Total Flow Volume above Erosive Threshold (500 l/s)			41 785	197 887	50 456

Table 4.4 Dry year analysis - Node 90.770 (outlet of development area).

Flow Class (l/s)	Dry Year - 1963 - Percent of Year in Flow Class (%)			Dry Year - 1963 - Total Volume in Flow Class (m^3)		
	Existing	Future	Combination	Existing	Future	Combination
0 - 1.0	44.17	52.27	39.90	6 965	8 241	6 291
1.0 - 5.0	36.84	41.75	37.61	34 854	39 499	35 582
5 - 10	17.82	2.37	11.40	42 148	5 605	26 963
10 - 50	1.13	1.62	10.91	10 690	15 326	103 217
50 - 100	0.02	0.64	0.07	473	15 137	1 656
100 - 500	0.01	1.09	0.04	946	103 122	3 784
500 - 1000	0.01	0.15	0.01	2 365	35 478	2 365
1000 - 1500	0.00	0.03	0.00	0	11 826	0
1500 - 3000	0.00	0.05	0.00	0	35 478	0
3000+	0.00	0.01	0.00	0	9 460	0
	Total Flow Volume below Erosive Threshold (500 l/s)			96 076	186 930	177 493
	Total Flow Volume above Erosive Threshold (500 l/s)			2 365	56 764	2 365

Table 4.5 Erosive threshold flows.

HSP-F Node	Lower (m^3/s)	Upper (m^3/s)
90.770	0.5	-
30.031	0.3	8.0
80.090	1.5	16.9

facilities.The analysis was based on results of three typical years 1962-average year, 1992-wet and, 1963 dry year according to total annual rainfall volume. The following analysis was based on assessment of the development area only:

Total rainfall depth (average for three years)	= 771 mm
Total rainfall volume (46 ha development area)	= 354 859 m^3
Average natural recharge (150-200 mm)	= 175 mm
Average total recharge volume	= 80500 m^3
Enhanced Volume of recharge through source control measure (HSP-F simulation)	= 43343 m^3
Average Natural Volume of recharge - additional pervious areas (20% of 46 ha)	= 16100 m^3
Average recharge volume (source controls and pervious areas)	= 59 443 m^3
Average recharge (source controls and pervious areas)	= 129 mm

An indication of the effectiveness of source controls in mitigating computed baseflow reductions due to development in the context of the greater watershed area is provided in Table 4.6.

Table 4.6 suggests that source control measures and landscaped areas on development lots would provide 74% of the average naturally occurring infiltration within the developing area. Accordingly, source control measures alone would contribute approximately 57% of the original natural infiltration function within the developing area, providing an effective means of maintaining the recharge function.

The results were considered to be conservative based on the following considerations:

1. Evaporation losses are explicitly included in the reservoir routing function which was been utilized to model the infiltration facilities. Therefore evaporation was subtracted from the infiltrating waters. This was considered to be consistent with the physical characteristics of rear yard ponding areas.
2. The infiltration rate of the soil was conservatively assumed to be approximately 5 mm/hr. It was anticipated that the real soil infiltration rate would be greater than this value providing additional infiltration storage capacity and less frequent occurrence of overflow.

Table 4.6 Infiltration assessment for average annual rainfall (mm) - 1962.

Location	HSP-F Node	Existing Land Use	Future Development No Mitigation	Change in Infiltration over watershed No Mitigation (mm) [%]	Future Land Use with Mitigation	Mitigation of Impacts (% of Objective value)
Developing Area (46 ha)	90.030	175	52	-122 [-70%]	129	74%
Outlet Development Area and External Drainage (133 ha)	90.770	175	133	-42 [-24%]	159	91%
Confluence with West Tributary (403 ha)	30.031	175	161	-14 [-8%]	170	97%

3. Additional contributions to baseflow/interflow regime would occur within local swales and natural watercourse systems. The volume of runoff contribution through this process was not included in the assessment.
4. Seepage into the fractured limestone within the stormwater management facility was also expected to increase baseflow/interflow to the downstream watercourse system.

The foregoing considerations are difficult to quantitatively assess, however it is anticipated that the additional contributions would provide a significant amount of additional infiltration.

4.5.4 Erosion Analysis

The potential impacts of urban development on erosion potential were assessed on a preliminary basis using a duration analysis technique for a typical year. The erosion analysis was undertaken at three locations corresponding to:

1. development area outlet (Node 90.770),
2. confluence with Western Tributary, 500 m downstream of development area (Node 30.031), and
3. Stonechurch Road, 1000 m downstream of development area (Node 80.090).

The analysis results are based on the number of hours during which stream flows are above the erosion threshold flow at each location. The lower threshold reflects the flow rate resulting in a critical velocity, above which channel erosion would be expected to occur. An upper threshold flow corresponding to the bankfull discharge was determined at which point flood flows would spill on to the flood plain and channel velocities would reach a maximum. (An upper threshold flow and velocity was not selected at Node 90.770 due to the steep channel slope, hence all flows above the critical threshold were assessed in the erosion analysis.)

A critical velocity of 1.0 m/s was selected for Nodes 90.770 and 30.031, while a critical velocity of 0.6 m/s was selected for Node 80.090 based on observed channel sloughing and finer grained soils in this vicinity. Table 4.5 provides a summary of the limiting erosive threshold levels for the three locations.

Table 4.7 provides a summary of the erosion analysis results for two cases:

1. Future land use within the development area only, existing land use applied to external areas.
2. Future land use within the development area, as well as future land use applied to external areas (ultimate development conditions).

The results of the duration analysis indicate that future uncontrolled development would significantly increase the erosive flow duration immediately downstream of the development area. Increases in erosive duration were calculated at 263%, 37% and 9% at Node 90.770, 30.031, and 80.090 respectively.

Generally, use of a combination of source controls and an end-of-pipe water quality facility best mitigates increases in erosion impacts at the development area outlet (Node 90.770). However all methods tested were noted to be effective in mitigation of erosion impacts.

It is noted that the various end-of-pipe facilities are also effective in mitigating increases in erosive flows. However their effectiveness is slightly reduced in comparison to source controls or a combination of source controls and an end-of-pipe facility. Of the end-of-pipe facilities evaluated, a 37.5 mm level of extended detention storage in conjunction with a 24-hour drawdown period was computed to be the most effective in mitigating increased erosive flows with a 78% level of effectiveness in mitigating erosion impacts at Node 90.770.

Results for Node 30.031 and 80.090 illustrate similar findings, however the effectiveness of the various stormwater management strategies is influenced by the larger external drainage area outside the development area.

Table 4.8 illustrates the results of an assessment of the combination alternative on the basis of wet, dry and average years combined.

The results indicate that the combination alternative has a 91% effectiveness in mitigating computed increases in erosive flow occurrences.

Table 4.7 Erosion duration analysis. Number of hours above erosive threshold velocity for management alternatives - average rainfall year (1962).

Node	Case	Existing	Future No Control	Source Control	Combination Source Control and End-of-Pipe	End-of-Pipe				
						19 mm /3 day	19mm /7 day	Mix	4 mm /1.5 day	37.5 mm /1 day
90.770		12.3	44.7	20.1	17.5	22.8	23.7	26.3	34.2	19.3
30.031	1	90.2	123.3	77.1	85.0	93.0	90.3	90.3	111.0	89.0
	2	90.2	296.1	238.5	244.0	251.4	247.9	252.3	265.4	248.8
80.090	1	56.1	61.0	50.3	51.0	54.5	54.0	54.8	57.5	53.8
	2	56.1	118.3	112.0	111.0	117.4	115.6	114.8	118.3	116.5

mm = mm of storage over developing area
day = facility drawdown period

Table 4.8 Duration analysis for wet (1992), dry (1963), and average (1962) years (based on total annual rainfall); hours above erosive threshold velocity.

Location	HSP-F Node	Existing	Future Uncon-trolled	Net Change in Duration (hours) [%]	Future Combination Alternative	Mitigation Effectiveness on Erosion Impacts (%)
Outlet of Development Area	90.770	6.4	37.1	30.6 [478%]	9.1	91

4.6 Preferred Stormwater Management Strategy

Based on the technical evaluation process, the stormwater management alternative which incorporates the use of a combination of source controls, as well as an end-of-pipe facility was recommended. This alternative best maintains the computed hydrogeologic, erosion and water quality regime. In addition, opportunities are provided to enhance terrestrial resources in the area and maintain habitat functions through protection and enhancement of on-site watercourses.

Key components of the preferred solution include:

1. Rear yard ponding areas or soakaway pits to promote infiltration. Each ponding area or soakaway pit should provide storage equivalent to 20 mm of runoff from the contributing roof top area.
2 Provisions for an open naturalized watercourse system through the study area. The watercourse system should be designed to incorporate objectives of natural channel design. The watercourse system should be utilized to convey external drainage through the study area, provide a major system outlet for the developing area, and should discharge to its present outlet location, north of Rymal Road. The provision of open natural watercourses will also serve to preserve flood storage throughout the development area.
3. Roads and parking areas within the study area should drain to an end-of-pipe facility via a storm sewer system and should be constructed using an urban curb and gutter cross section. The storm sewer system should be sized according to the drainage requirements for roads and parking areas only, thereby reducing the overall cost of the storm sewer, and concentrating the capture of contaminated runoff, thereby improving water quality treatment.
4. Supplemental use of water quality inlets (such as manhole separators installed within the storm sewer system) within development sites to provide pre-treatment of runoff and spill protection.

5. An end-of-pipe facility (stormwater management pond). This facility should be located in the former quarry location and should include an open water component as well as a shallow wetland area. The pond should be divided into two storage zones consisting of a sediment forebay and a naturalized wetland/open water area. The water storage sizing has been based on results of this study and the Ministry of Environment and Energy Guidelines. The stormwater management facility should provide water quality treatment, erosion control, low flow augmentation, and passive recharge.
   Note: If the natural configuration and elevations within the former quarry are utilized, the permanent pool storage provided would be well in excess of the minimum storage requirements.
6. The end-of-pipe facility should include native wetland plantings such as cattails, and sedges to enhance the natural features of the site. This natural design approach should provide a high aesthetic value in the context of the adjoining environmentally sensitive area (ESA) and watercourse valley owned by the Hamilton Region Conservation Authority.

4.7 Conclusions

The preferred stormwater management strategy includes two primary measures: enhanced infiltration of relatively clean rooftop drainage and treatment of more highly polluted runoff from roads and parking in an end-of-pipe facility.

The preferred stormwater management solution for the 46 hectares of prestige industrial development, was developed using the Municipal Engineers Association (1993) - Class Environmental Assessment process.

The existing hydrogeologic function of a watershed is often difficult and/ or resource-intensive to assess directly. The evaluation methodology, as well as the hydrologic modeling techniques used to simulate the linkage between hydrogeologic and surface flow regimes, provided a simple assessment approach for this unique subwatershed location. The study of the various hydrologic and hydrogeologic aspects of the headwater area determined a significant linkage between surface water interflow and the relatively shallow groundwater flow regime. It has been determined that significant contributions to base flow immediately downstream of the study area occur through this linkage.

The HSP-F modeling approach proved to be an effective tool to evaluate various stormwater management practices with respect to their effectiveness in augmenting base flow/interflow linking the hydrologic and hydrogeologic function of the watershed.

Acknowledgements

We wish to acknowledge the technical and financial support and guidance provided by the Regional Municipality of Hamilton-Wentworth, particularly Mr. E. P. Chajka, P. Eng. In addition, the input provided by Steering Committee members, from the City of Hamilton, Hamilton Region Conservation Authority, and Ministry of Natural Resources, has been most helpful in guiding this project. We also wish to thank the other Project Team members, Mr. C. Portt of C. Portt & Associates and Mr. J. Dougan of Dougan & Associates, for their valuable insights.

References

Bicknell, B.R., Imhoff, J. C., Kittle J. L. and A.S. Donigian, 1993. Hydrological Simulation Program - Fortran (HSP-F) User's Manual for Release 10. U.S. EPA, EPA-6—R93-174.

Bishop, B.E. and R.B. Scheckenberger, 1995. HSP-F Simulation of Constructed Wetland Stormwater Management Practice for Urban Highway Runoff, Chapter 12 in Modern Methods for Modeling the Management of Stormwater Impacts. Edited by William James, Computational Hydraulics International, Guelph, Ontario. ISBN: 0-9697422-X. p.173.

Crawford, N.H., R.K. Linsley, 1966. Digital Simulation in Hydrology: Stanford Watershed Model IV, Technical Report No. 39, Stanford University.

Dames and Moore, Canada, 1993. Environmental Assessment Storm Sewer Outlet, Rymal Road, Hamilton Ontario, Report prepared for the Regional Municipality of Hamilton-Wentworth.

Marshall Macklin Monaghan Limited, 1994. Stormwater Management Practices Planning and Design Manual, Report prepared for the Environmental Sciences and Standards Division - Program Development Branch, Ontario Ministry of Environment and Energy, ISBN 0-7778-2957-6.

Municipal Engineers Association, 1993, Class Enviornmental Assessment for Municipal Water and Wastewater Projects, Ontario Ministry of Environment and Energy File No: MU-0014.

Philips Planning and Engineering Limited, 1993. Dartnall Road Interchange Stormwater Management, Report prepared for the Regional Municipality of Hamilton-Wentworth Special Projects Office.

Philips Planning and Engineering, 1989. Mountain East-West and North-South Transportation Corridor Drainage Study, 1989, Report to Rgeional Municipality of Hamilton-Wentworth Freeway Project Office.

Chapter 5

The Feasibility of Using Continuous SWMM for Water Resources Conservation Planning

Michael F. Schmidt, Brett A. Cunningham, Brian W. Mack

The scope of watershed planning is changing to include not only water quality and flood control, but also management of wetlands and conservation of fresh water in a cost-effective manner. This chapter discusses the feasibility of using an analytical method that uses the United States Environmental Protection Agency (USEPA) Stormwater Management Model (SWMM) for continuous simulation. The potential for fresh water conservation by reducing overdrainage of a sand ridge and wetland system in central Florida is evaluated. The program goals are to conserve fresh water, hydrate wetlands, and increase aquifer recharge by increased infiltration to a sole source aquifer while minimizing flood impacts and maintaining or improving water quality. The analysis includes an average annual mass balance and the evaluation of costs and relative benefits to identify project feasibility. The results of this feasibility study are currently under review by the St. Johns River Water Management District (SJRWMD) and Volusia County, Florida. The technical review process is not complete, and this chapter is offered for conceptual consideration.

5.1 Introduction

The United States (US) Environmental Protection Agency (EPA) Stormwater Management Model (SWMM) was applied to evaluate surface water conservation options for Volusia County, Florida shown in Figure 5.1 and the SJRWMD.

© *Advances in Modeling the Management of Stormwater Impacts - Vol. 5.* W. James, Ed. Pub. by CHI, Guelph, Canada 1997. ISBN 0-9697422-7-4. Fax: +519 767-2770

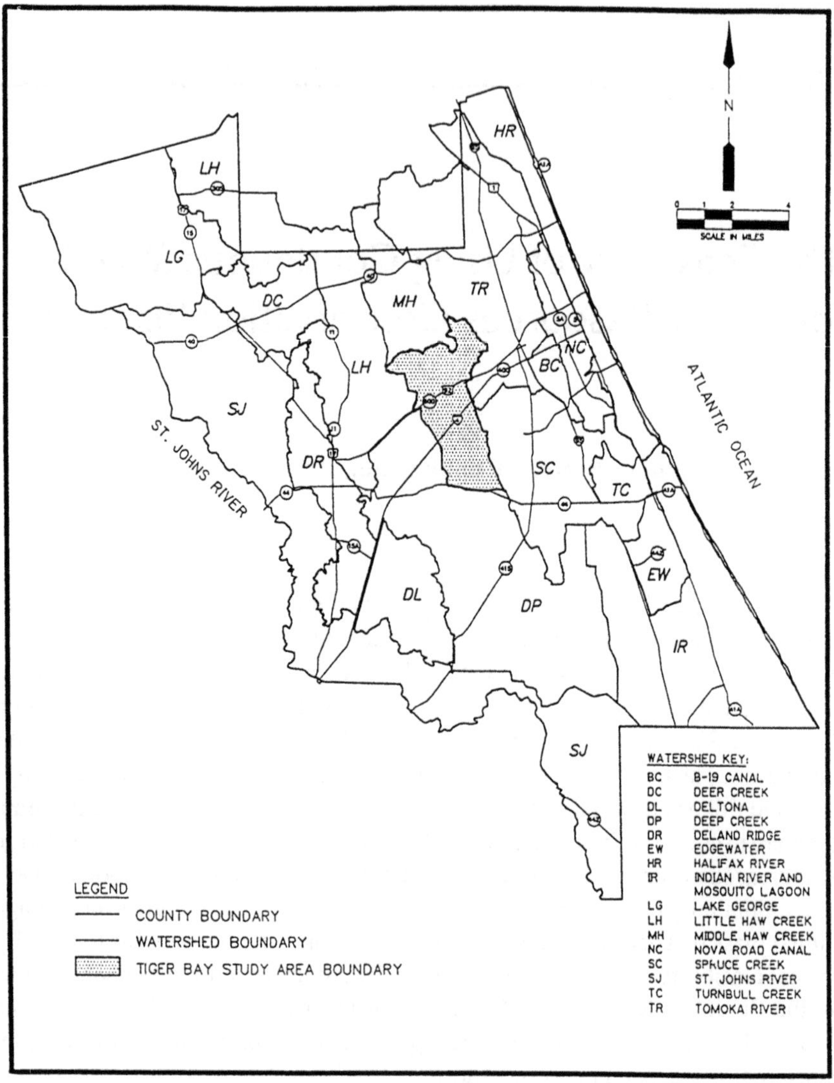

Figure 5.1 The study area, Volusia County, Florida.

Volusia County lies in east central Florida along the coast of the Atlantic Ocean. The study area is in the southwestern portion of the Tomoka River Watershed (28,466 acres or 11,571 ha) called the Tiger Bay area (Figure 5.2), which is named after a series of wetland strands. Volusia County's potable water supplies come from the Floridan Aquifer which is a sole source aquifer in the county due to the Atlantic Ocean, saltwater intrusion, and relic saltwater from the St Johns

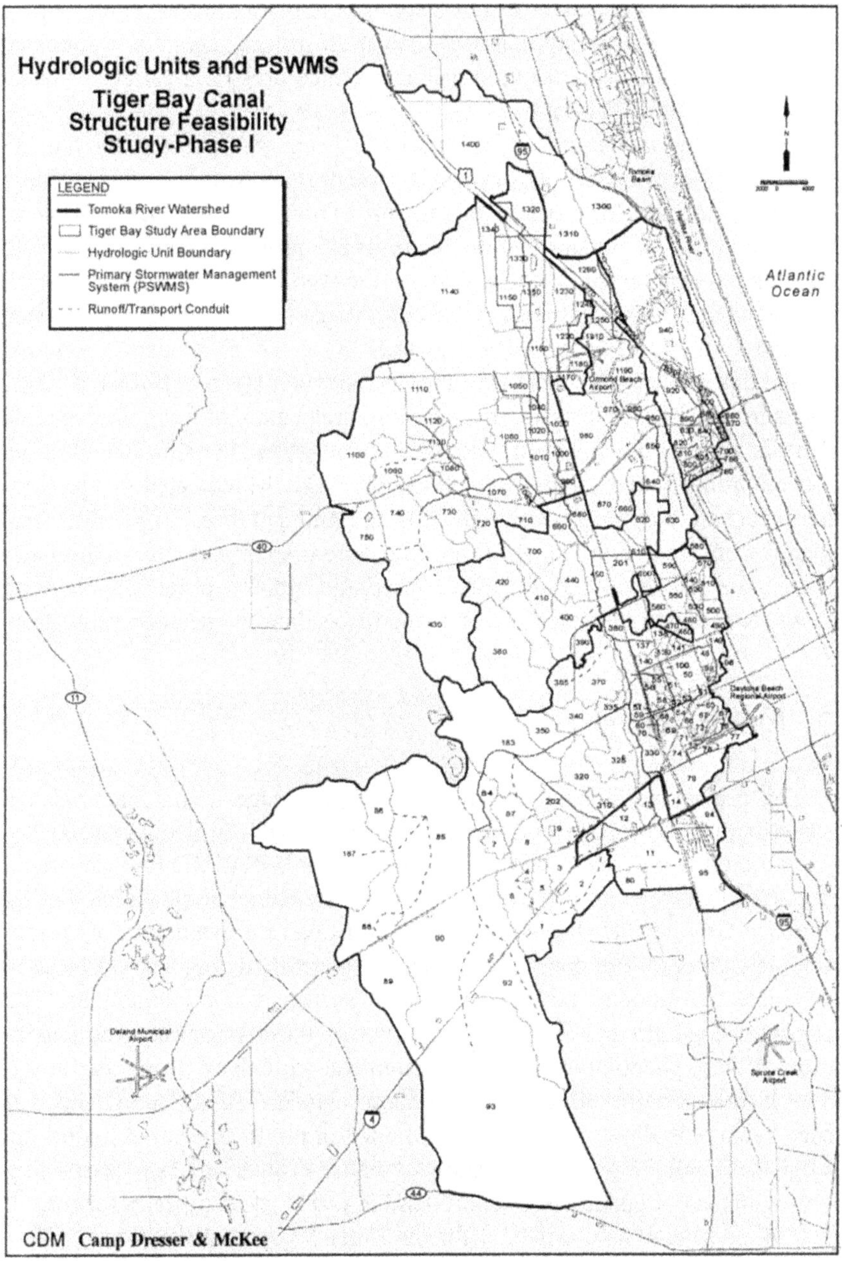

Figure 5.2 The Tiger Bay area.

River (Figure 5.3). Therefore, incident rainfall on the county and its subsequent recharge into the aquifer are essential to the long-term replenishment of potable water. The county is characterized by a series of north-south sand ridges which provide recharge to the aquifer. Between these ridges, there is a series of waterbodies (coastal estuaries) and wetlands which provide a variety of water resource and wildlife habitat benefits.

Over the years, drainage projects have been implemented by the US Department of Interior, US Department of Defense (US Navy), Volusia County mosquito control, and the Florida Department of Transportation (FDOT) to allow development and to protect public health. These projects generally involved ditch cuts through the sand ridges to drain the wetlands. This reduces overall annual recharge to several areas, including Tiger Bay. Tiger Bay has been identified by the US Geological Survey (USGS) as having moderate recharge potential that could be increased with increased water storage. In recent years, impacts from saltwater intrusion have required relocation of municipal potable wellfields westward away from encroaching saltwater intrusion. In addition, recent environmental evaluations by the SJRWMD have indicated that groundwater levels are dropping and that there is an imminent threat to jurisdictional wetlands from the overdrainage. Therefore, there is an opportunity in the Tiger Bay area to establish compatible water conservation goals to potentially increase recharge to mitigate saltwater intrusion and for wetlands vegetation protection.

5.2 Methodology

This project requires the consideration of an average annual mass balance or water budget to properly understand both ground and surface water interactions and their responses to rainfall and runoff. The SJRWMD has prepared a regional groundwater model using the Modular Three-Dimensional Finite-Difference Groundwater Flow model (MODFLOW) for evaluation of current steady state groundwater conditions. A surface water modeling tool was needed to evaluate the random, unsteady nature of surface water interactions. The analysis of long term aquifer recharge and water conservation requires consideration of these phenomena including extreme conditions of droughts, floods, and back-to-back storm effects in a continuous manner. The EPA SWMM was chosen because it allows continuous simulation of physically-based hydrology through the RUNOFF block, and it allows dynamic evaluation of surface routing effects with the EXtended TRANsport (EXTRAN) block, especially for control structure modifications considered for the Tiger Bay Canal study area. This model was set up and calibrated as part of the Tomoka River Watershed Management Plan (WMP) by CDM (Camp Dresser & McKee Inc.,1995), and it was refined further for this study.

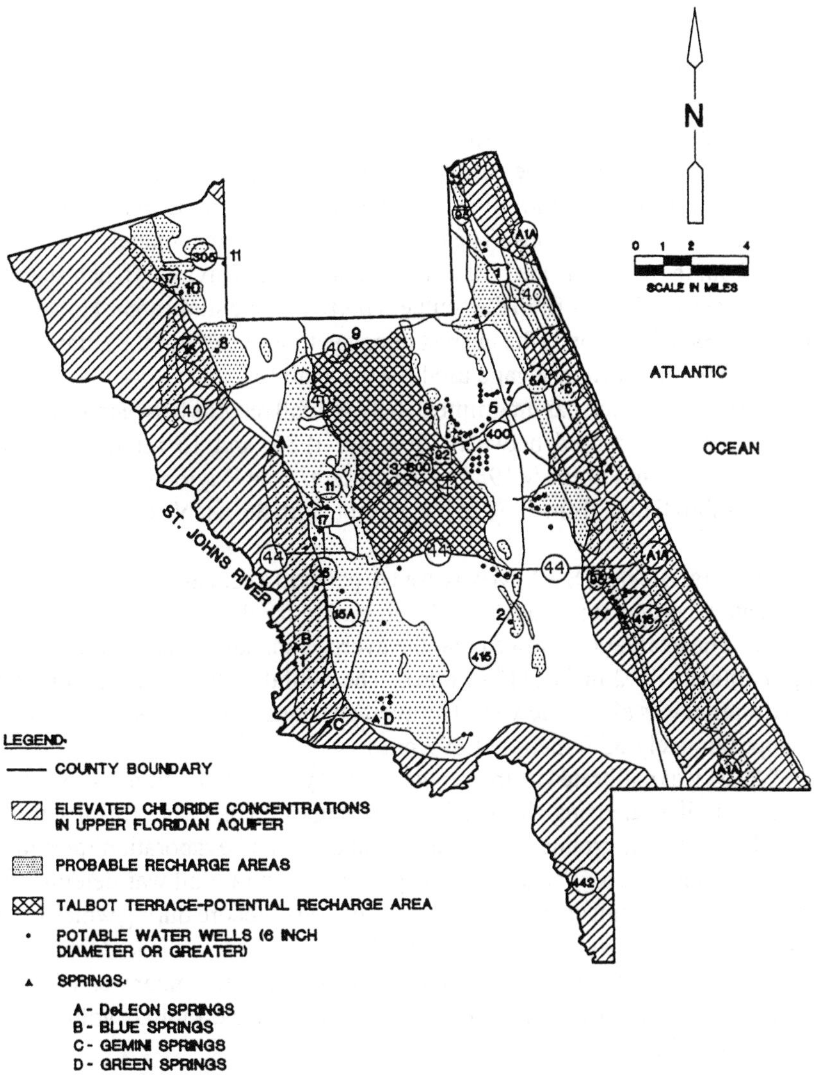

Figure 5.3 Groundwater features of the Tiger Bay study area.

The basic components of the mass balance considered were rainfall, evapotranspiration, runoff, infiltration, groundwater recharge, baseflow, and direct surface discharge. Each of the components in the mass balance is discussed in the following paragraphs.

5.3 Rainfall

Two National Oceanic and Atmosphere Administration (NOAA) weather stations were used to determine the average annual and average monthly rainfall volumes for the Tiger Bay study area. The stations selected included the DeLand and the Daytona Beach stations. The DeLand station is approximately 1.5 miles (2.4 km) from the western side of the study area and the Daytona Beach station is approximately 1.5 miles from the southern portion of the study area. However, the Daytona Beach station is the nearest raingage to over 90% of the study area, so that gage was the primary gage used for the analyses.

The DeLand station has intermittent daily rainfall records between 1900 and 1908, daily rainfall record between 1908 and 1986, and 15 minute interval rainfall records from February 1986 to the present. The Daytona Beach station has hourly rainfall data from 1938 to the present, with the usable data beginning in 1942.

The RAIN Block of SWMM was used to perform a statistical analysis of the rainfall data recorded at the Daytona Beach station for the 1942 to 1994 period of record. The evaluation showed that the average annual rainfall at the station is approximately 48.2 inches (1224 mm). The rainfall data available from the DeLand station was also reviewed. The average annual rainfall was determined to be approximately 55.0 inches (1,397 mm).

SWMM was calibrated to an average rainfall year and to a month within the average rainfall year using the USGS stage gage at Tiger Bay, results from SJRWMD's regional groundwater model, and published evaporation rates in a mass balance approach. The average or typical year of rainfall was determined by reviewing the rainfall data recorded for the period of record during which both rainfall and flow data were recorded. The rainfall data review showed that 1992 most nearly represents a typical year for rainfall volume. September was selected as the calibration month within 1992. A summary of the rainfall volumes is presented in Table 5.1.

The last two weeks in 1991 were also used as a start-up period for the average year since this period was preceded by a dry period. The additional rainfall from this period raised the total for the average annual year simulation period to approximately 47.7 inches (1212 mm) at the Daytona Beach station. A dry antecedent period is desirable since input initial infiltration parameters must be set to dry conditions for a continuous simulation. The reason for this requirement is

Table 5.1 Calibration and simulation rainfall volumes.

Description	Daytona Beach Station (inches)	DeLand Station (inches)
Average Year	48.17	55.04
1992	45.39	54.26
Average September	6.83	6.53
September 1992	7.73	6.57

that the model can only regenerate infiltration parameters back to their original input values. This allows the model to simulate both wet and dry conditions.

5.4 Evapotranspiration

Evapotranspiration (ET) is a combination of evaporation from surfaces, (i.e. evaporation from surface waters), evaporation from subsurface sources (i.e. evaporation from the upper soil column), and transpiration also drawn from soil storage. The total ET term was handled both directly and indirectly. Losses from the soil column (subsurface evaporation and transpiration) were handled indirectly through use of the infiltration regeneration coefficient. The regeneration coefficient was adjusted such that the computed average annual infiltration was approximately equal to 16.5 inches (419 mm) (estimated by the SJRWMD for MODFLOW) in conjunction with the surface evaporation rates, overland flow characteristics, and infiltration parameters previously determined. The final regeneration coefficient used was approximately 0.001. This value is slightly lower than normal to account for reduction from the infiltration term by the recharge term that is sent to the surface. Based on MODFLOW results, computed transpiration and subsurface evaporation in the Tiger Bay study area collectively account for approximately 6.0 to 7.4 inches (152 to 188 mm) of the average annual water budget.

For the Tiger Bay study area, approximately 28.8 inches (732 mm) of rainfall are lost to surface evaporation during an average year, based on rainfall minus infiltration and direct surface runoff. Therefore, surface evaporation is an important term in the overall water budget. Surface evaporation rates were applied in both the RUNOFF and EXTRAN models to account for evaporation lost from overland flow and initial abstraction of surface storage after rainfall events and from ponded areas in the Primary Stormwater Management System (PSWMS), respectively. Surface evaporation in EXTRAN was modeled using

Table 5.2 Monthly surface evaporation.

Month	Monthly Evaporation (inches)
January	1.12
February	1.37
March	1.98
April	2.37
May	2.75
June	2.59
July	2.37
August	2.15
September	1.90
October	1.72
November	1.32
December	1.07
Total	22.71

Note: Evaporation volumes are the combined volumes from the RUNOFF and EXTRAN blocks.

monthly three-point stage-discharge (pump) curves in order to account for fluctuating amounts of inundated surface areas. The evaporation rates were based on pan evaporation values published by NOAA, multiplied by a constant of approximately 0.5, which was based on the calibration. The input monthly evaporation volumes used are listed in Table 5.2.

The total evapotranspiration used is the sum of the components, which are approximately 6.4 inches and 28.8 inches (163 and 732 mm), or 35.2 inches (894 mm). This value is within the expected range of values from this area, which ranges from approximately 25 inches to 40 inches (635 to 1,016 mm) (SJRWMD, personal communication, 1995).

5.5 Runoff

The RUNOFF block of SWMM was used to simulate direct surface runoff in hydrologic units for the Tiger Bay study area. RUNOFF uses a non-linear reservoir solution for overland flow routing of runoff (Manning's equation). The Horton infiltration equation was used to simulate surface infiltration into the soil for shallow pervious area infiltration only. Based on statistics from the USGS

Tiger Bay stream gage from approximately 11 years of data (during which the average annual rainfall was approximately equal to the long-term average of 48 inches (1,219 mm)), total flow from baseflow and direct surface runoff averages approximately 6.3 inches (160 mm) per year. Of the 6.3 inches, 2.9 inches (74 mm) is estimated to be direct surface runoff and 3.4 inches (86 mm) is estimated to be baseflow. Direct surface runoff and baseflow were determined from the measured flow data using a simple hydrograph separation technique.

5.6 Infiltration

The Horton equation uses an initial infiltration rate set to account for water already in the soil, a maximum infiltration rate, and an infiltration decay rate. In addition to the Horton infiltration equation, CDM has added another special feature to SWMM that establishes a maximum infiltration capacity (also called the total soil storage shut-off option). This special feature has been used and successfully calibrated and verified by CDM since 1986, and it will soon be available in the EPA release of SWMM. The maximum infiltration capacity is the same definition as the United States Department of Agriculture (USDA) Soil Conservation Service (SCS) soil storage, S (measured in inches over the pervious area). The total soil storage limits the amount of rainfall that can be infiltrated in the given hydrologic units. This capacity is especially critical during periods of extended rainfall encountered during continuous simulations and larger design storms. Once the maximum infiltration volume or total soil storage is reached, the pervious areas effectively become 100% impervious. The Horton infiltration equation is as follows:

$$f_p = f_c(f_o - f_c)e^{-kt} \tag{5.1}$$

where:

f_p = infiltration capacity into soil, feet/second
f_c = minimum or ultimate value of f_p (WLMIN), feet/second
f_o = maximum or initial value of f_p (WLMAX), feet/second
t = time from beginning of storm, seconds
k = decay coefficient (DECAY), second^{-1}
e = natural log (base e)

Once rainfall has ended (dry hours), the infiltration capacity is recovered according to the following equation:

$$f_p = f_0 - (f_0 - f_c)e^{-k_d(t-t_w)} \tag{5.2}$$

where:

k_d = decay coefficient for the recovery curve, second^{-1}
t_w = hypothetical projected time at which $f_p = f_c$ on the recovery curve, second

The calibrated Horton infiltration parameters used for this evaluation ranged from 4.8 to 6.9 inches (122 to 175 mm) per hour (maximum rate) and from 0.10 to 1.1 inches (2.5 to 27.9 mm) per hour minimum rate. The maximum infiltration capacities (soil storage, S) ranged from 3.3 to 6.0 inches (84 to 152 mm), and the Horton decay coefficient was 2/hr.

Following a rainfall event, the maximum infiltration capacity recovery rate follows the same exponential recovery rate as the Horton infiltration recovery rate. Soil storage becomes an important term in stormwater runoff modeling. Previous calibrations in this watershed for single events have shown that some of the dual class hydrologic soils (e.g. A/D and B/D) tend to become saturated for storm events over 1 to 2 inches (25 to 50 mm). Figure 5.4 shows the USDA SCS soils classification for the study area.

For the Tiger Bay study area, the available soil storage was determined by using the calibrated parameters from the Tomoka River WMP and adjusting the values from AMC II to AMC I. As stated previously, the input soil values had to be adjusted to dry conditions in order to allow the model the possibility of regenerate back to those capacities. A time period of two weeks was simulated prior to the average year to establish equilibrium between model computed and actual soils parameters.

5.7 Groundwater Recharge and Baseflow

For groundwater recharge to the Floridan Aquifer, the EXTRAN block was used to compute equivalent groundwater stage versus discharge curves (simulated as three-point pump curves) and to evaluate the potential change in recharge due to control structure modifications. The stage-discharge curves used to compute groundwater recharge were constructed as follows. The SJRWMD MODFLOW results were used to establish the potentiometric surface for each major wetland area. Using a constant vertical hydraulic conductivity and constant Darcy flow length, the curves were constructed using Darcy's equation, with the driving head being the difference between the potentiometric surface (September, wet season) and the stage in the storage area, and the area equal to the inundated area for the given stage.

The input infiltration parameters (previously discussed) used in the RUNOFF model were adjusted so that the computed annual infiltration volume was approximately the same as the annual infiltration volume used for the SJRWMD

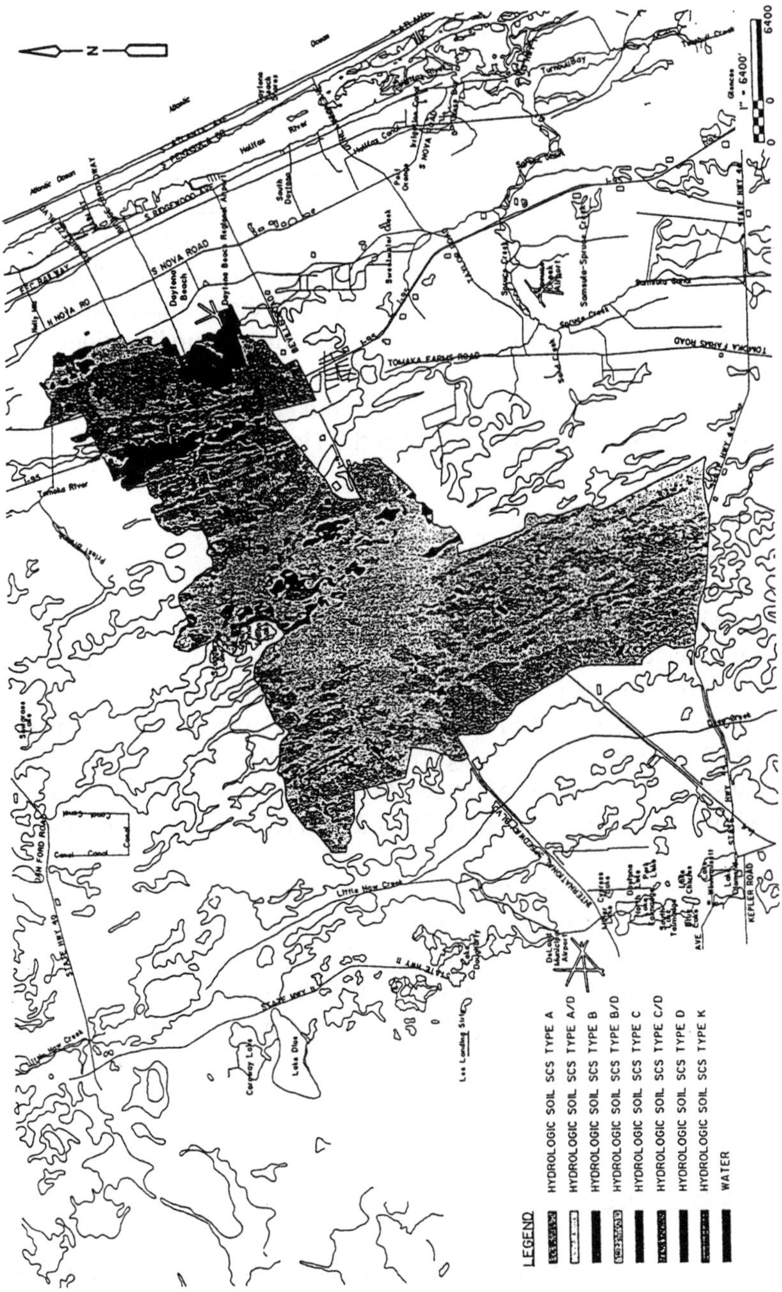

Figure 5.4 The USDA SCS soils classification for the Tiger Bay study area.

MODFLOW evaluation i.e. 16.5 inches (419 mm). The USGS ZONEBUDGET program (part of MODFLOW package) was then used to compute the annual volume of groundwater infiltration that reaches the deep aquifer using the MODFLOW results provided by the SJRWMD. Of the 16.5 inches of annual infiltration volume input to MODFLOW, 6.7 inches (170 mm) were computed to reach the Floridan Aquifer. The remaining 9.8 inches (249 mm) of infiltration are recycled through baseflow and evapotranspiration. As stated above, average annual baseflow was calculated to be approximately 3.4 inches (86 mm), leaving approximately 6.4 inches (163 mm) as evapotranspiration. Using these data, the equivalent groundwater stage-discharge curves were refined in the EXTRAN block to convey 6.7 inches (170 mm) of infiltration into the Floridan Aquifer on an annual basis for existing conditions (no control structure modifications).

Baseflow was modeled as a constant inflow from each hydrologic unit. Evaporation of baseflow from the surface had to be accounted for in order to obtain the net baseflow that was measured at the Tiger Bay stream gage.

5.8 Direct Surface Runoff

Once the infiltration parameters were calibrated and the equivalent groundwater recharge stage-discharge curves were sized, EXTRAN was used to route surface water flows through the Tiger Bay Canal PSWMS which is part of the larger Tomoka River Watershed PSWMS. EXTRAN was used to compute surface water flows, velocities, and stages in the PSWMS as well as surface water volumes transported past points of interest (e.g. out of the system for the mass balance). Figure 5.5 shows the RUNOFF and EXTRAN model schematic to scale.

The month of September 1992 was used to calibrate flows at the USGS Tiger Bay Canal gage. With no adjustments to parameters other than those discussed previously (e.g. adjustment of infiltration parameters to AMC I conditions), the computed flow for the month of September was within 7% of the measured flow. The measured annual flow volume at the Tiger Bay Canal gage was 1,920 ac-ft (2.36×10^6 m^3), and the computed annual flow volume was 2,150 ac-ft (2.64×10^6 m^3).

5.9 Mass Balance

Using data from SJRWMD's MODFLOW model, the Daytona Beach rain station, and the USGS Tiger Bay Canal stream gage, a mass balance or water budget for the average annual year was constructed. A simplified version of the mass balance is shown in Figure 5.6.

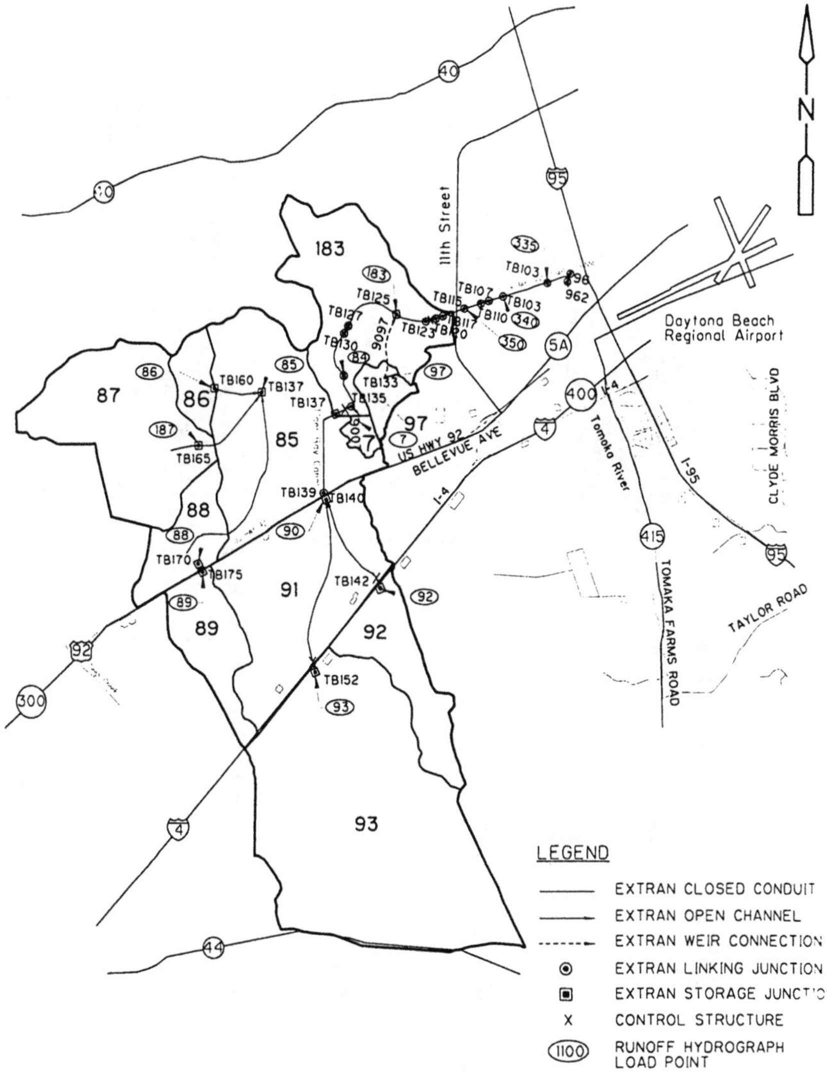

Figure 5.5 RUNOFF and EXTRAN schematic.

5.10 Control Structure Considerations

Hypothetical control structures were strategically sited at four locations within the Tiger Bay Canal PSWMS. The intent of the variable control structures was to estimate the potential impact on aquifer recharge volumes using the calibrated groundwater stage-discharge curves and the potential impact of

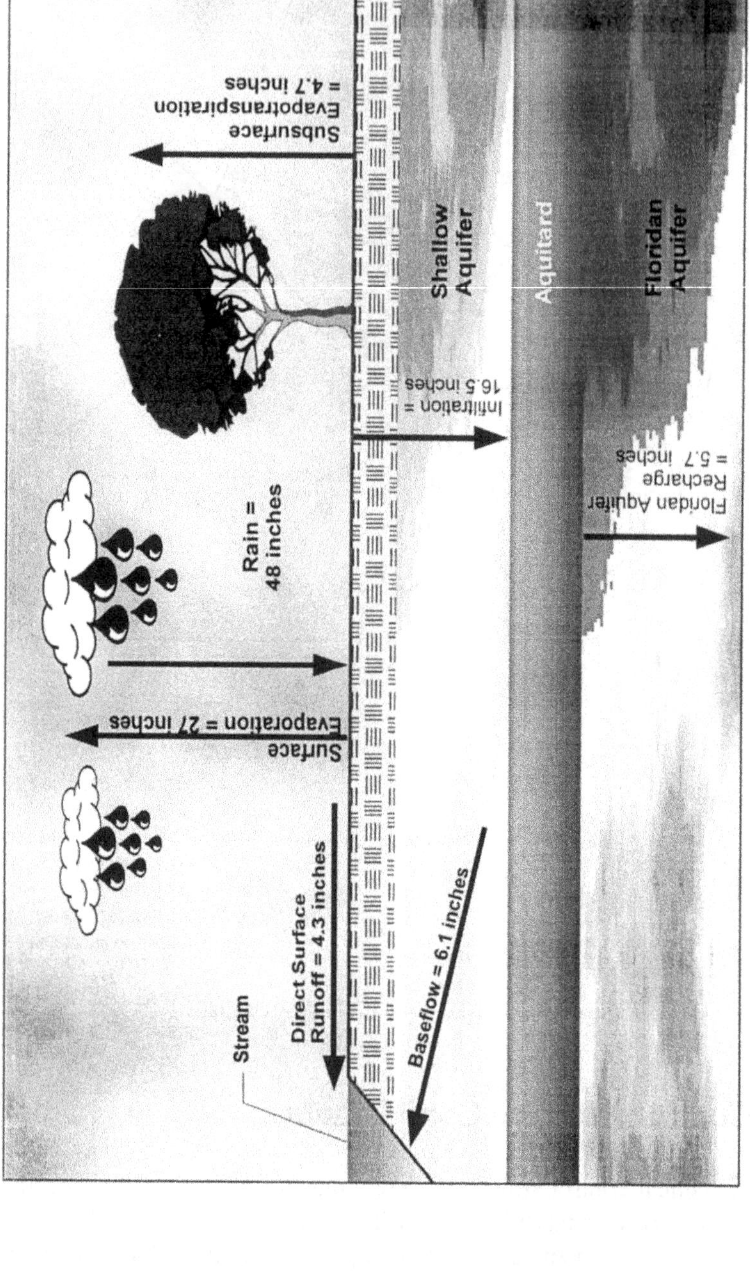

Figure 5.6 Simplified mass balance for the average year.

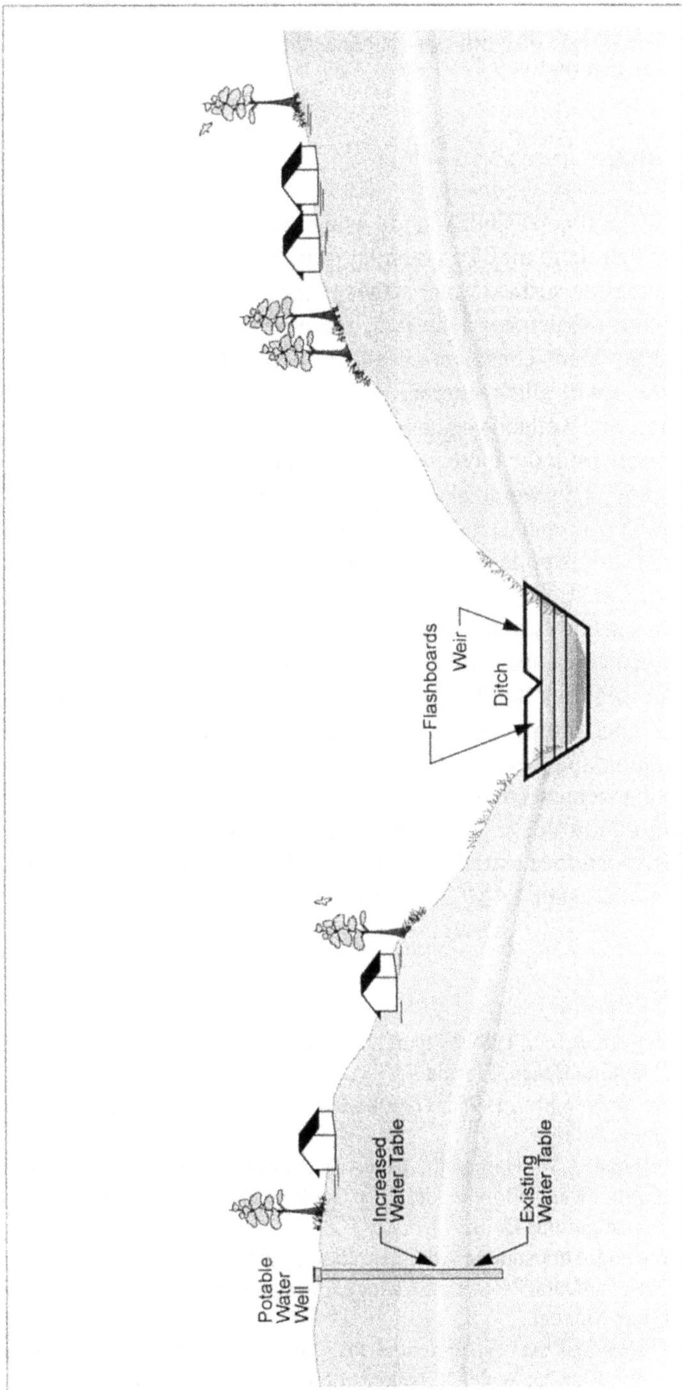

Figure 5.7 Conceptual cross-section view of a weir constructed in a drainage ditch.

increased water surface elevations on adjacent property owners by raising the control structure elevation in the Tiger Bay Canal by 3 feet (0.92 m) and then 5 feet (1.52 m), respectively.

5.11 Results

The results of this feasibility study using the continuous SWMM methodology indicate that there may be potential to increase recharge of the Floridan Aquifer by increasing surface water stages in the Tiger Bay area using variable water level control structures. However, there needs to be a balance between aquifer recharge, water conservation and flood control. The use of variable control structures will allow internal system control of surface water stages, aquifer recharge, and wetland vegetation. Additionally, a variable control system will allow impacts from the increased water levels to be monitored, and system adjustments made. This would also allow for planned maintenance drawdowns. Figure 5.7 shows a conceptual cross-section of a weir constructed in a drainage ditch.

In summary, the application of continuous simulations using the SWMM (as modified by CDM) to consider water conservation appears to work well. It appears that the feasibility of increasing recharge by increasing storage depth and duration can be determined; however, quantifying the benefits from the additional recharge requires additional data and study, as do potential negative impacts from higher groundwater tables and increased durations of inundation of wetland vegetation and private land.

Future enhancements to SWMM are planned to allow more direct interaction between groundwater and surface water. This may include direct evaporation off EXTRAN surface waterbodies and standard interfaces to public domain groundwater models such as MODFLOW.

References

Camp Dresser & McKee Inc., 1989. City of Daytona Beach Stormwater Master Plan, for the City of Daytona Beach, Florida.

Camp Dresser & McKee Inc., 1995. Tomoka River Watershed Management Plan, for Volusia County, Florida.

McDonald, M.G. and A.W. Harbaugh, 1988. A Modular Three-Dimensional Finite-Difference Groundwater-Flow Model. U.S. Geological Survey, Techniques of Water Resources Investigations Book A3.

National Oceanic and Atmospheric Administration., 1982. Technical Report NWS 34.

United States Environmental Protection Agency, 1994. Stormwater Management Model (SWMM) Users Manual.

United States Geological Survey. Potential for Groundwater Development in Central Volusia County, Florida; Water-Resources Investigations Report 90-4010.

St. Johns River Water Management District, 1995. Potentiometric Surface of the Upper Florida Aquifer in the SJRWMD and Vicinity.

Chapter 6

Development and Initial Refinement of a Water Balance Model as a Planning Tool for Stormwater Management Application

Edward I. Graham, H.R. Whiteley and N.R. Thomson

Water balance models can be very useful when establishing the long-term water budget components of an area prior to urban development and the changes in the hydrologic cycle after development. In this case, the water budget components considered include: precipitation; snowmelt and rainfall runoff; evapotranspiration; and surficial and deeper groundwater recharge. Continuous review and refinements of the model component routines and future calibration work allows for expansion of applicability. The model discussed here has been applied to compute the effects of urbanization on the water balance in a high recharge area and to assess specific control measures for mitigating recharge reductions. As part of the first refinement, the evapotranspiration (ET) relation has been enhanced and compared to the original simulation results.

This chapter describes the first phase of the development, refinement, and application of a planning-level water balance model.

6.1 Introduction

The importance of understanding how the surface and groundwater regimes react to land use change or management options it is now well recognized. This is particularly true in head-water areas where groundwater recharge provides

© *Advances in Modeling the Management of Stormwater Impacts - Vol. 5.* W. James, Ed. Pub. by CHI, Guelph, Canada 1997. ISBN 0-9697422-7-4. Fax: +519 767-2770

baseflow to downstream cold water spawning tributaries or water quality dilution zones. As indicated in the Stormwater Management Practices (SWMP) Planning and Design Manual in Ontario (MOEE, 1994), there is no convenient approach to predict the interaction between surface water and groundwater systems before and after urban development, and to determine the relative proportions of groundwater and surface water inputs, other than in a very broad way. Linking surface water models which can simulate surface runoff on a continuous basis with groundwater modeling packages requires substantial effort and can usually be accomplished only if significant groundwater data is available.

This chapter summarizes some of the development features and provides results of an application of the Stormwater Management Practices Water Balance Model (Version 1) produced as an attempt to bridge the gap in the availability of suitable analysis tools. The modeling approach involves the continuous determination of the daily hydrologic budget components of an area as they occur in the hydrologic cycle and as they are modified by urban development and the incorporation of specific stormwater controls. A brief example of its application is given, describing the input and results showing relative changes between different land-uses and control measures. Calibration and validation, or even rigorous comparison of results with other established and detailed models, is left for upcoming development phases. Further review and incorporation of alternative theories and routines is also on-going.

6.2 Methodology

The water budget determination implies quantification of the volumes of storage and rates of water movement from one physical state and location to another. An accounting is made of water entering, leaving, and remaining in storage during a specified time period over a study area which can be defined by topographic, political, or other arbitrary criteria. The water components considered in this modeling approach include precipitation in the form of rainfall or snowfall, rainfall runoff, snowmelt, upper soil infiltration, evapotranspiration, groundwater recharge, and SWMP input to groundwater as a way of controlling reductions in infiltration after development (e.g. infiltration trenches, perforated sewer pipes, reduced lot grading, etc.).

The theories used in the model to compute each of these individual components have been adopted from previously published equations. The model brings the concepts and equations together to simulate their interaction in an attempt to approximate their occurrence. The surface water inputs of rain and snow are applied on a daily basis and calculations made of outputs of runoff, abstraction to surface storage and infiltration to soilwater storage. However, in developing this approach it was recognized that daily time steps are not particularly suited when determining surficial soil infiltration during specific event data due to the tendency to overestimate the infiltration capacity. However, in the interim, for planning purposes, this approach would still compare favourably with other presently-used modeling approaches also applying daily values (Johnstone et al., undated).

The general balance equation used to account for the surface and ground water storage and for the surface abstraction storage for a given time interval is:

$$\Delta S = I - O \tag{6.1}$$

where:

ΔS = change in storage,
I = inflow,
O = outflow.

The water budget over the surface (other than initial abstraction storage) is:

$$\Delta S = R + SM - SI - RO \tag{6.2}$$

where:

ΔS = change in surface water storage (mm),
R = rainfall (mm),
SM = snowmelt (mm),
SI = upper soil infiltration (mm),
RO = surface runoff (mm).

For the purpose of this analysis, no significant evapotranspiration is assumed to occur during rainfall or snowmelt periods (however, this approach may be revised to account for evapotranspiration).

6.2.1 Snowmelt, (SM)

Precipitation occurs as rain or snow. The model algorithm uses the average daily temperature as a base temperature to determine the form of precipitation: if the temperature is above the base (e.g. 0° C), the precipitation occurs as rain. If the temperature is below the base temperature, the precipitation occurs as snow and accumulates in the snowpack.

Snowmelt is calculated using degree-day equations as defined by Environment Canada's Atmospheric Environment Service (AES). The AES defined five different models for several regions in Canada. The Southern Ontario model is given by (Bruce and Clark, 1966):

$$SM = 0.02(Tx - 32) \tag{6.3}$$

where:

SM = snowmelt (inches/day),
Tx = maximum daily air temperature (^{0}F).

A form of Equation 6.3 converted to metric units is used in the model as described by the AES to accumulate snow on the snowpack and to deplete the storage during days with average daily temperature above the base temperature. The algorithm ends when the snowpack is reduced to zero. Evaporation (sublimation) from snow is ignored.

6.2.2 Surface Runoff, (RO)

Similar to other water balance planning approaches (Harbor, 1994) and for this first phase of development, the surface runoff is computed using a modified Soil Conservation Service (SCS) form of the rainfall-runoff equation:

$$RO = \frac{(R + SM + IA)^2}{(R + SM - IA) + SS} \tag{6.4}$$

where:

R = rainfall (mm),
SM = snowmelt (mm/day),
IA = maximum surface abstraction available (mm),
SS = available upper soil storage (mm).

6.2.3 Maximum Surface Abstractions, (IA)

The surface abstractions are accounted separately for pervious and impervious areas. During a precipitation event, the available surface abstractions are filled first, before runoff. The storage is re-established during inter-event dry periods at the rate of evaporation and, for pervious areas, at the added user-defined upper soil infiltration rate.

6.2.4 Available Upper Soil Storage, (SS)

A separate water balance computation is conducted for the amount of storage available in the upper soil layer (upper soil storage). The change in storage is expressed as:

$$\Delta SS = SI - SET - RGI \tag{6.5}$$

where:

ΔSS = change in upper soil storage (mm),
SI = infiltration (from surface to upper soil storage) (mm),

SET = upper soil evapotranspiration (mm),
RGI = recharge (seepage) to groundwater (mm).

The maximum amount of storage available in the upper layer at any one time corresponds to antecedent soilwater condition I (AMC I) in the Soil Conservation Service (SCS) relation. The equivalent depth below the surface which defines the upper soil layer is determined from maximum storage divided by the soil porosity value as listed in Table 6.1.

Table 6.1 Hydrologic capacities of soil texture classes (after Viessmann *et al.*, 1977).

Texture Class	S [%]	G [%]	AWC
Coarse Sand	24.4	17.7	6.7
Coarse Sandy Loam	24.5	15.8	8.7
Sand	32.3	19.0	13.3
Loamy Sand	37.0	26.9	10.1
Loamy Fine Sand	32.6	27.2	5.4
Sandy Loam	30.9	18.6	12.3
Fine Sandy Loam	36.6	23.5	13.1
Very Fine Sandy Loam	32.7	21.0	11.7
Loam	30.0	14.4	15.6
Silt Loam	31.3	11.4	19.9
Sand Clay Loam	25.3	13.4	11.9
Clay Loam	25.7	13.0	12.7
Silty Clay Loam	23.3	8.4	14.9
Sand Clay	19.4	11.6	7.8
Silty Clay	21.4	9.1	12.3
Clay	18.8	7.3	11.5

where: S = total porosity - 15 bar soilwater %,
G = total porosity - 0.3 bar soilwater %
AWC = S - G = porosity drainable only by evapotranspiration.

6.2.5 Upper Soil Infiltration, (SI)

The upper soil infiltration accounts for the water movement between the surface and the upper soil storage. The rate and volume of water moving into ground storage depends on the available storage in the soil. The upper soil infiltration is calculated directly from Equation 6.2 by adding any residual surface water storage (after depletion by evapotranspiration), to the difference (R+SM-RO).

6.2.6 Evapotranspiration, (ET)

Water leaving the watershed through evaporation and transpiration can be very significant over time and continuous modeling requires that estimates of this be incorporated. In most cases, potential evapotranspiration and soilwater conditions are the predominant factors used in calculating evapotranspiration. The potential evapotranspiration (PET) is defined as the amount of "water loss which will occur if at no time there is a deficiency of water in the soil for the use of vegetation" (Viessman et. al. 1977). In this case the PET is taken from AES estimates of daily lake evaporation. The actual ET from the soilwater layer is calculated using a slightly modified version of an equation developed by the U.S. Agricultural Research Service (ARS) which accounts for the vegetation characteristics and soilwater (Viessman, p. 58). The modified ARS form of equation is as follows:

$$ET = GI \times k \times PE\left(\frac{S - SA}{S}\right)^{n} \tag{6.6}$$

where:

ET = actual evapotranspiration (mm/day),
GI = growth index of vegetation as proportion of maturity,
k = the ratio of ET to potential evapotranspiration at full canopy with freely-available water,
PE = lake evaporation taken as the potential (mm/day),
S = as identified in Table 6.1,
SA = available porosity (unfilled by water),
n = an exponent that varies with soil type in the range of 0.1 to 0.25.

In this case after contacting the ARS, the denominator inside the brackets has been modified from 'AWC' (porosity drainable only by evapotranspiration) in the original form of the equation to 'S', and the exponent from 'x' (x=AWC/G ,where AWC is the porosity drainable only by evapotranspiration and G being the moisture freely drained by gravity).

In this application, the growth index (GI) is simplified by a sinusoidal distribution over the summer growing season. Evapotranspiration from surface storage (interception and depression storage) is assumed to occur at the potential rate.

The model calculates actual evapotranspiration ET from the vegetation growth index as a function of the time of year, lake evaporation as PET, and the underlying soil conditions. The resulting ET is a measure of the water loss from surficial soil storage largely through diffusion of water vapour from plant leaves to the atmosphere (transpiration). It should be noted that evapotranspiration below an air temperature of 4.4°C (40°F) was considered to be negligible in these calculations, consistent with other applications (Viessman et. al., 1977).

The algorithm selected first depletes surficial abstraction or interception storage at the potential rate. The PET remaining and 'evapotranspiration opportunity' are used to quantify and extend the water loss from the upper soil storage to the root zone. Evapotranspiration opportunity is defined as the maximum amount of water available for evapotranspiration at a particular location during a prescribed period.

6.3 Example

6.3.1 Meteorologic Data

Depending on the site, meteorologic input data are available through Environment Canada Atmospheric Environment Service (AES) in either hourly or daily format. For this purpose, twenty years of daily precipitation, mean daily dry-bulb temperature, and lake evaporation components have been obtained from the AES from several gauging stations in Ontario.

6.3.2 Site Data

Site data includes all the measured pervious and impervious area parameters such as tributary areas and land uses, percent imperviousness, and measured surface and subsurface hydrologic parameters such as hydraulic conductivities, and porosity. Additional data includes soil cover, vegetation growth indices, base temperatures for snowmelt, and snowmelt factor. For this purpose, these have been selected based on site-specific conditions and professional judgement.

6.3.3 Analysis Approach

A water balance model was applied to assess the impacts of a proposed residential development on the water budget components in a high recharge area.

The study approach first determined water balance components under present, undeveloped conditions. These results provide target values for groundwater recharge after development. The hydrologic and hydrogeologic model parameters affected by the urbanization were then updated to reflect the proposed development conditions without SWMPs. The difference between pre and post-development infiltration volume corresponds to the SWMP infiltration required. Lot-level controls in the form of roof-runoff soakaway pits were introduced and their size sequentially increased up to the maximum recommended size (15 mm of roof runoff). The storage volume is equal to the product of the roof depth sizing criteria (e.g. 5 or 15 mm) and the total roof area.

A perforated pipe system was also incorporated into the model as a reservoir receiving runoff from the roads and its effects evaluated. The perforated pipe storage is implemented by intercepting the excess runoff from the area and discharging only when the storage provided by the pipes is exceeded. The storage volume is set equal to the product of the depth sizing criteria (e.g. 2 mm) and the tributary area. For example, a 2 mm perforated pipe storage (net trench storage below the perforated pipe) corresponds to the product of 2 mm over the 82.5 hectare study area, or 1650 m^3.

6.3.4 Results

The following water balance results are presented in Table 6.2 as average yearly values over the study area. The following assumptions and comments are applicable to the results in Table 6.2:

- Post-development vegetation cover type is assumed to remain as under pre-development conditions. Therefore, changes (reductions) in evapotranspiration between development conditions are due to reduction in vegetated and pervious areas after urbanization.
- Initial abstractions have been decreased to reflect uniform lot grading.
- Although the SWMP sizing criteria recommends a one day drainage time, the model conservatively considers uniform discharge over three days. The two additional days account for decreased infiltration rates resulting from lower overall hydraulic gradient in the confining soil around and beneath infiltration facilities after a wet-weather event (i.e. increase in soilwater content).
- The average runoff coefficient obtained is consistent with typical values expected from similar land-uses and soil conditions.
- The reduction in surficial soil infiltration with soakaway pits and perforated pipes occurs because SWMP infiltration contributes directly to deeper soils. SWMP infiltration has direct benefits for removing surface runoff.

- Computed post-development groundwater recharge decreases about 22% compared to pre-development conditions. This reduction is consistent with the increased imperviousness after development. In comparison, allowing roof areas to drain to pervious surfaces yields a computed 15% reduction in recharge as compared with pre-development levels.
- Soakaway pits sized to infiltrate 5 mm of roof runoff are computed to provide a marginal improvement in total infiltration volume compared to attenuating the roof runoff by dispersing it over the pervious area.
- The perforated pipe system is computed to provide significant improvements to the total groundwater recharge volumes.
- The evapotranspiration values are low in comparison with other sites, particularly those with average infiltration capacity. The values shown in Scenario 8 correspond to an earlier version of the evapotranspiration Equation 6.6 shown previously. However, as in this study area, evapotranspiration values can be affected (reduced) by sands with low soilwater retention capacity (lower irreducible water saturation).
- The refined evapotranspiration relation in Equation 6.6 yields higher average annual ET totals depicted in Table 6.2 (Scenario 9). Further testing with additional model parameters such as the watershed cover factor, resulted in further increases in total evapotranspiration. However, monitoring and calibration will be required to justify and establish more precise values and results.

6.4 Conclusions

The SWMP water balance model produced useful results when establishing relative proportions of pre-development and post-development water budget components. The model provides great flexibility when assessing different development configurations and control measures to achieve specific recharge targets. Additional work is required to refine the water budget routines and extend the applicability of the model.

The refined evapotranspiration relation in Equation 6.6 yields higher average annual ET totals as depicted in Table 6.2 (Scenario 9). Further testing with additional model parameters such as the watershed cover factor, allowed for further increases in total evapotranspiration. However, monitoring and calibration will be required to justify and establish more precise values and results.

Table 6.2 Typical water balance results in a high recharge area (units in mm/year).

Selected Surface and Subsurface Water Budget Components	Pre-Develop	Post-Development Scenarios								
		(1) No SWMPs (*)	**(2)** No SWMPs (**)	**(3)** 5 mm Roof Storage in Soak-Away Pits	**(4)** 15 mm Roof Storage in Soak-Away Pits	**(5)** 2 mm Perforated Pipe Storage (***)	**(6)** 5 mm Perforated Pipe Storage (***)	**(7)** 15 mm Perforated Pipe Storage (***)	**(8)** Combined (4) and (6)	**(9)** (4) & (6) with ET Refined
Runoff	57	228	201	220	198	108	54	8	51	51
Surface Abstractions	191	92	97	90	90	91	91	91	90	90
Surficial Soil Infiltration	659	518	556	544	529	557	557	557	529	529
Groundwater Recharge	557	436	476	478	481	555	610	655	613	575
Evapo-Transpiration	177	119	120	120	120	120	120	120	120	158
Runoff Coefficient	0.07	0.29	0.26	0.28	0.25	0.14	0.07	0.01	0.06	0.06

Notes to Table 6.2

Average yearly precipitation = 785 mm
For checks on water balance (runoff + evapotranspiration + groundwater recharge) = yearly precipitation of 785 mm +/- year to year changes in soilwater storage.
Evapotranspiration amounts in scenarios 1 to 8 have been made with an earlier version of Equation 6.6 and are generally low. Comparative performance of alternative SWMP alternatives is the objective of their presentation.

* Roof runoff discharged directly to sewers (ie. directly connected).

** Roof runoff discharged to pervious areas (ie. not directly connected) - this also applicable to remaining scenarios (3 to 9).

***SWMP sufficient to maintain the ground water recharge rates at pre-development levels.

References

Bruce, J.P. and R.H. Clark, 1966, Introduction to Hydrometeorology, Pergamon Press, Toronto, p 257.

Harbor, J.M., 1994 "A Practical Method for Estimating the Impact of Land-Use Change on Surface Runoff, Groundwater Recharge, and Wetland Hydrology", Journal of the American Planning Association, Vol. 60, No. 1, Winter 1994.

Holtan, H.N., 1970 "USDAHL-74 Revised Model of Watershed Hydrology", U.S. Department of Agriculture, ARS Tech. Bulletin No. 1518, Washington, D.C., 1975. Adapted from England, C.B. "Land Capability: A Hydrologic Response Unit in Agricultural Watersheds", U.S. Department of Agriculture, ARS 41-172. Sept. 1970

Johnstone, K. and Louie, P.Y.T. "Water Balance Tabulations for Canadian Climate Stations". Canadian Climate Centre, Atmospheric Environment Service. Undated.

MOEE, 1970. (Ontario Ministry of Environment and Energy), "Stormwater Management Practices Planning and Design Manual". Jun. 1970

Viessmann, W.; Knapp, J.W. ; Lewis, G.L.; and Harbaugh, T.E., 1977. "Introduction to Hydrology", Second Edition, Harper and Row. 1977.

Chapter 7

SWMM Graphics

Uzair M. Shamsi

This chapter demonstrates applications of personal computer graphics in Storm Water Management Model (SWMM) modeling. Graphics are effective means of bridging the gap between the information and its recipients. It is demonstrated that integration or interfacing of computer graphics in SWMM models, in addition to the traditional tabular output, significantly improves the efficiency of modeling tasks. The graphics provide the modeler with more efficient and cost effective analysis, planning, design, and management tools. The proposed technique requires developing graphical user interfaces (GUI) for SWMM. Twenty five GUI features are recommended from a user's perspective. A software review of the available SWMM GUIs is provided. The advantages and limitations of various GUI techniques are identified and recommendations for future research and development are made from a user's perspective.

7.1 Introduction

Throughout history, tools have shaped almost every physical artifact made by man, and tools usually have had a profound effect on every area of human endeavor. This is no less true with computers and plotters than it was with pencils and vellum, or for that matter, lines in the dirt (Dakan, 1992). The advent of color

© *Advances in Modeling the Management of Stormwater Impacts - Vol.5.* W. James, Ed.
Pub. by CHI, Guelph, Canada 1997. ISBN 0-9697422-7-4. Fax: +519 767-2770

graphics and powerful minicomputers in the 1970s moved mapping from pen-and-ink to keyboard, digitizer and stereographic viewers. The 1980s mark the revolution of the advent of personal computers (PC). In the past decade, powerful workstations and sophisticated software combined to bring mapping capability to any desktop. More recently, PCs have become so powerful that graphics software will now run on off-the-shelf computers. The current technology is already well beyond a simple electronic emulation of pencil lines on paper, yet many see it as merely an electronic drafting table. This technology is, in fact, much more: it is new tools, new design methodologies, new ways of visualization, new thought patterns, and new ways of doing business (Dakan, 1992).

Today, computers are so rapidly changing the way we do business that the day of the mainframe on the desk seems an inevitable reality. In the field of environmental cleanup the software boom has largely come in products that speed up or simplify data processing tasks such as tracking toxic releases and waste site samples; some more novel developments stem from the need to better visualize site information and even provide a peek at what may be lurking underground (Rubin and Powers, 1992). Once the province of accountants and administrators, computers are today infiltrating almost all areas of water and wastewater utility operations and management. From forecasting sewer flows to tracking water quality through the collection system to recording maintenance activities and supplies, computer systems are easing the jobs of supervisory and operational personnel through a broad range of state-of-the-art hardware and software. Since the PC was introduced in the 1980s, there have been many changes in sewer system modeling practice. A PC on every engineer's desk has encouraged even the small utilities to develop sewer system models.

No one said mathematical modeling would be easy, but preparation of input data and interpretation of output data required by the ever changing complexities of a complex computer model like the Storm Water Management Model (SWMM) has been mind boggling. The good news is that the model building and interpretation of the results is now easier than ever before, thanks to advances in computer graphics. An assembly of model input data by traditional manual map measurements is just too time-consuming and difficult to justify now that the cost of computers is coming down and their capabilities and speed are up. It is much easier to graphically view a storm surge progress through a sewer, to pinpoint the areas of flooding and surcharging, than it is to digest reams of computer output, especially for *non-modelers*. The non-modelers are those people who are not the expert modelers but need to know the modeling results, such as the clients, project managers, politicians, and regulatory agencies, etc. The chip makers and the PC are the Henry Ford and Model T of this era and we can no longer afford to use the modern-day equivalent of buggy whips. This chapter demonstrates applications of personal computer graphics to revolutionize the way we perform SWMM modeling.

7.2 SWMM Problems

SWMM is a large and complex model which simulates the movement of precipitation and pollutants from the ground surface through pipe and channel networks, storage treatment units and finally to receiving waters. Both single event and continuous simulation may be performed on sewersheds or natural watersheds for predicting flows and pollution concentrations. SWMM can be used for both planning and design. The planning model is used for studying urban runoff problems and abatement options. The design model performs event simulations using a detailed sewershed schematization and shorter simulation time steps. The SWMM program consists of four computational blocks (RUNOFF, TRANSPORT, EXTRAN, and STORAGE/TREATMENT) and five service blocks (STATISTICS, GRAPH, COMBINE, RAIN, and TEMP).

SWMM is a public-domain computer program, originally developed by the U.S. Environmental Protection Agency (EPA) in 1969-1971 as a mainframe computer program. At that time SWMM was one of the first of such models. Versions 2, 3, and 4 of SWMM were distributed in 1975, 1981, and 1988, respectively. The first batch mode microcomputer version of SWMM (version 3.3) was released by EPA in 1983. The first conversational ("user-friendly") mode PC version of SWMM known as PCSWMM was commercially distributed in 1984 by Computational Hydraulics Inc. of Guelph, Ontario. The first PC version of EPA's SWMM was version 4, which was distributed in 1988 (Huber and Dickinson, 1988; Roesner et al., 1988). An excellent review of SWMM's development history can be found in a book by James (1993).

Most modelers have now become accustomed to the modern computer graphics features, such as pull down menus, spreadsheet data input and editing, color plots, on-line help, etc., which are not currently available in SWMM. SWMM was developed in an era when input files were created on punched cards. After 25 years, SWMM now runs on PCs, but it is still a text-based, non-graphical, DOS program. It reads ASCII input to produce ASCII output which is most suitable for mainframe line printers. SWMM's ASCII format output is long, boring, difficult to interpret, and meaningless for non-modelers.

Creating computer models and reviewing the model results is often slowed by our inability to see the system being modeled. It is up to the modeler to review SWMM's voluminous output and construct a mental image of the physical system being modeled. Often, the limitation in understanding the model output has been the modeler's own comprehension of the output, not the model itself. Quite frequently, it is impossible for the modeler to absorb the large amount of information contained in the model output (TenBroek and Roesner, 1993).

Visualization is the key to understanding the relationships among modeled components. Everyone believes that *a picture is worth a thousand words.* It is about time to realize that *a graph is worth a thousand numbers*. Just as comfortably as an artist can paint a picture on a canvas, a modeler should be able to paint a model on a computer screen.

7.3 Graphical User Interface

A graphical user interface (GUI) is a computer program which acts as an interpreter between the user and a computer. The GUI replaces difficult-to-remember text commands by interactive computer graphics consisting of menus, dialog boxes, input and output windows, and icons. The main goal of GUIs is to develop user-friendly computer applications or to add the user-friendliness to the existing command driven applications. For example, Microsoft Windows is a GUI for DOS, and Netscape is a GUI for the internet.

A GUI can be employed to overcome SWMM deficiencies. A GUI stimulates user interest and facilitates interpretation of model results. There are two types of GUIs. An input interface (also called a front-end interface or pre-processor) usually converts graphics to text. It extracts SWMM input from existing drawings, maps, and databases to create SWMM's traditional ASCII input file. Some GUIs provide graphical tools to draw a network model which is subsequently converted to SWMM's ASCII input file. For example, an input interface may extract the sewer segment lengths and manhole coordinates from existing CAD drawings. An output interface (also called a back-end interface or post-processor) usually converts text to graphics. It transforms SWMM's traditional ASCII output file to graphs, charts, and plots which can be easily understood by the users. GUIs provide the following benefits:

1. Users do not have to memorize the command syntax. The text commands are replaced by interactive graphics.
2. Users do not have to memorize the input format. The user input is facilitated by interactive graphics.
3. GUIs improve understanding and interpretation of model results. Tabular results are converted to meaningful graphs and charts which can be quickly and easily understood both by the modelers and non-modelers. More than just reams of computer paper, models become an automated system evaluation tool. In this way, GUIs bridge the gap between the SWMM output and its recipients.
4. Dynamic model results (e.g. time varying hydraulic gradient line or HGL) can be displayed through video-like animation. Engineers can view a storm surge progress through their system and immediately pinpoint areas of flooding and surcharging.
5. Connectivity data errors are easily detected and corrected. Instabilities in the model output, often the most difficult errors to find, are also easily located.

6. Input preparation, analysis, and output interpretation time is decreased, which reduces the total project cost.
7. Users become more productive. They devote more time to solving the problem and less time to mechanical tasks of data input and checking, program execution, and interpreting the output.
8. GUIs increase the confidence of the project team and reviewers in model configuration.

7.4 Recommended GUI Features

It is recommended that SWMM GUIs should have the following features from a user's perspective:

7.4.1 General

1. *Windows Application:* The GUI should be a Microsoft Windows application. Windows applications provide all the benefits of the Windows computing environment, such as a familiar and user-friendly desktop environment, convenient data transfer to and from other Windows applications via object linking and embedding (OLE), readily available access to a wide variety of display and printer drivers, and efficient utilization of computer resources, such as disk space and extended memory. DOS programs which can be launched from Windows do not qualify as Windows applications.
2. *Integration:* Various GUI modules should be seamlessly integrated in one program. Sequential batch processing of modules should be eliminated. Users should be able to perform various tasks (e.g. input, editing, execution, plotting, etc.) from within one program rather than having to run multiple programs to do various tasks.
3. *User Friendliness*: The GUI should include menus, dialog boxes, icons, and context sensitive on-line help to shorten the learning curve and guide users through the model development, execution, and interpretation without having to consult the users manual frequently.
4. *File Management:* SWMM generates a large number of intermediate files (interface files, hot restart files) which are shared among various SWMM blocks. Manual management of these files is cumbersome. Automatic organization, connectivity and manipulation of these files should be provided.

5. *Input Interface*: Preparation of SWMM's input file should be facilitated by pre-processors which will convert the graphics (e.g. existing CAD drawings) to SWMM's ASCII input file.
6. *Output Interface:* Interpretation of model results should be facilitated by post-processors which will convert SWMM's ASCII output file to graphics.

7.4.2 Model Development

7. *Graphical Input:* On-screen creation of a link-node model using simple drag-and-drop of icons should be provided. Nodes represent point and polygon features, such as sewersheds and manholes. Links connect two nodes of the network and represent features which have length and direction attributes, such as sewers, channel, or diversion. Model input should be facilitated through pop-up dialog boxes and data entry sheets. This approach eliminates network connectivity errors and shortens the model learning curve substantially.
8. *Graphical Editing:* On-screen editing of network elements using a point-and-click interface should be provided. For example, when a network link is clicked on with the mouse, a window resembling a

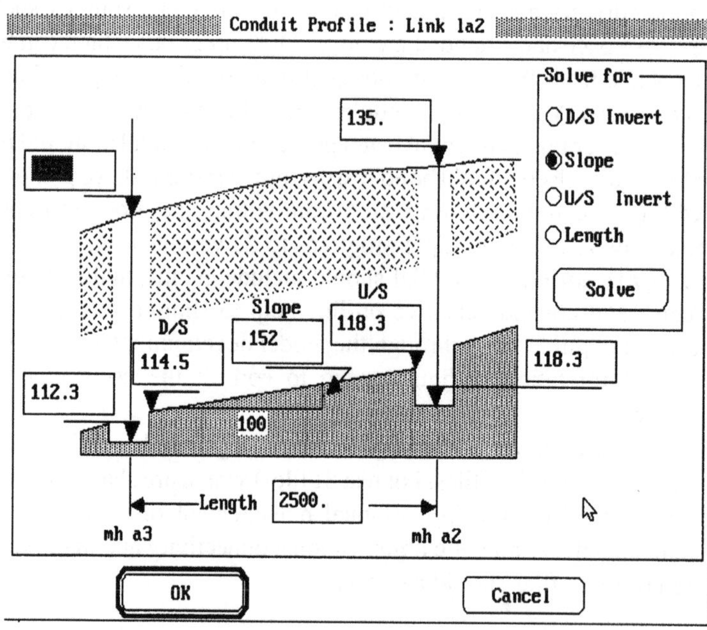

Figure 7.1 Graphical input and editing window in XP-SWMM.

data entry sheet should be displayed. The window should show link input parameters (e.g. length, shape, size, Manning's n, and slope, etc.) and allow on-screen editing. Figure 7.1 shows an example of a graphic editing window.

9. *Spreadsheet Editing:* For experienced users, graphical editing becomes a hurdle rather than an advantage, especially when numerous model parameters must be changed at once. For example, changing Manning's n for all the pipes of a 500 link model by graphical editing may require thousands of point-and-click, pan, zoom, and redrawing actions. To circumvent this problem, users should have an option to edit the input parameters in a spreadsheet with cut, copy, and paste capability.

7.4.3 Display Features

10. *Background Pictures*: Existing CAD drawings, maps, and digital aerial photographs should be imported and displayed as background pictures. Background pictures provide a passive backdrop with familiar surroundings, such as sewers, sewershed boundaries, city streets, buildings, rivers, and contours, etc., on which the network

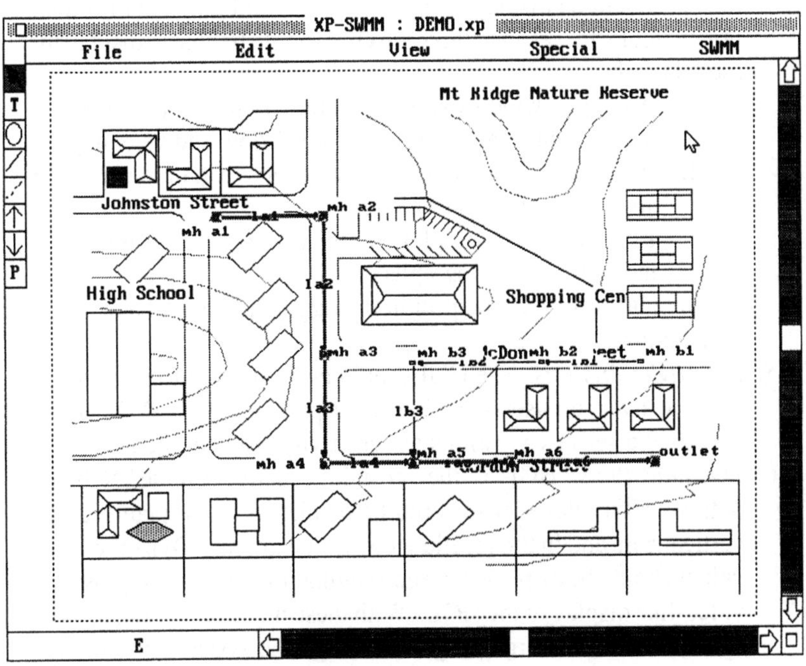

Figure 7.2 Network model overlaid on a background picture in XPSWMM.

model may be overlaid. The background pictures should be organized in CAD-like *layers* for a convenient on-off control. Figure 7.2 shows an example of a network model draped on a background picture.

11. *WYSIWYG (what you see is what you get) Display:* The network model should resemble a scaled drawing or map in which various drawing elements represent the actual physical system components. For example, a node may represent a real manhole and a link may represent a real sewer pipe. This feature eliminates a need to prepare network schematic diagrams to visualize the network connectivity.
12. *Navigation:* CAD quality zoom and pan functions should be provided to navigate the network and background picture. This feature will allow networks of any size to be conveniently modeled and displayed on the computer screen.
13. *Spatial Searching*: This feature prompts the user to enter an ID number for a network element, locates that features, and brings it to the center of the screen.

7.4.4 Model Interpretation

14. *Results Review*: For the user-selected network elements, the model results should be displayed in both tabular and graphical formats. For example, for selected links the GUI should be able to display the model results:
    1. in a popup window showing summary results (e.g. time and amount of maximum computed flow and velocity),
    2. in a table showing intermediate results (e.g. time versus flow hydrograph data), and
    3. in a graph showing plotted results (e.g. plots of inflow and outflow hydrographs).

    The tabular results are helpful in debugging the model and report preparation.
15. *Sensitivity Analysis:* The GUI should perform an automatic sensitivity analysis on user-specified input parameters of SWMM and display the sensitivity results as sensitivity gradient plots.
16. *Dynamic Plots:* HGL is the most informative EXTRAN block output. HGL elevations should be displayed in the plan view as vertical bars and in the profile view as line graphs. Dynamic HGL values should be displayed at each simulation time step to provide a video-like display of the HGL. With these model animations, users can view a storm surge progress through the system and immediately pinpoint areas of flooding and surcharging. Envelopes of peak HGL and flow during the entire simulation should also be provided.

17. *Thematic Plots*: This feature allows graphical encoding (color, type, shape, size, etc.) of network elements according to user specified themes (flow, velocity, surcharge, etc.) and facilitates rapid comprehension of model results. For example, sewers may be color-coded by surcharge for mapping severely surcharged sewers.
18. *Hydrograph Plots*: Time series scaled plots of flow, velocity and depth should be displayed and plotted for all the nodes and links of the model.
19. *Profile Plots:* Model profiles present information in a more natural manner which promotes a thorough understanding of system configuration. Users should be able to define and save a path (a sequence of links and nodes) through the network along which a profile view (section) can be drawn. The saved paths may be recalled later to plot CAD-quality profile plots showing model results (e.g. peak HGL and flows). Figure 7.3 shows a profile plot of HGL.
20. *Calibration Plots*: Observed and modeled hydrographs and pollutographs should be displayed on the same plot for model calibration and verification. Quantitative calibration measures should be provided to compare the observed and modeled

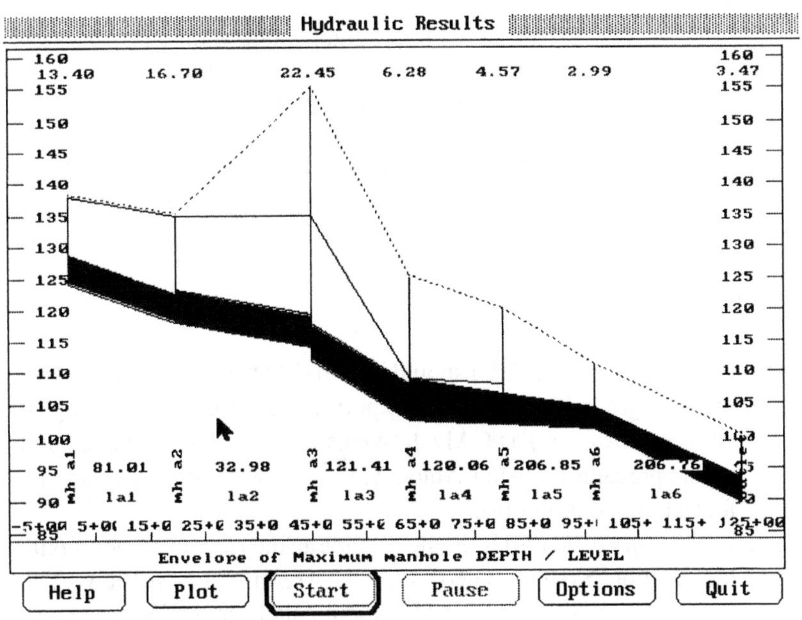

Figure 7.3 HGL profile plot in XP-SWMM.

hydrographs. The degree of calibration can be quantified by comparing three hydrograph parameters: volume, peak flow, and time to peak flow. However, because of the following reasons, this approach is quite subjective:

1. Generally, the three hydrograph parameters do not calibrate equally well. Is calibration acceptable, if the volumes are alike but the peak flows are different? Which parameter should be given a high priority?
2. There is no standard for the adequacy of calibration. For some users a 25% accuracy is adequate while others insist on 10%.

The integral square error (ISE) given by Equation 7.1 is a good measure of goodness-of-fit between observed and modeled hydrographs because it combines all the three hydrograph parameters (Marsalek et al., 1975).

$$ISE = \frac{\left[\sum_{i=1}^{N}(O_i - M_i)^2\right]^{1/2}}{\sum_{i=1}^{N} M_i} \times 100 \qquad (7.1)$$

where:

O_i = observed hydrograph value at time I
M_i = modeled hydrograph value at time i, and
N = number of hydrograph values.

The calibration can be subjectively rated as excellent for $0<\text{ISE}\le 3$, very good for $3<\text{ISE}\le 6$, good for $6<\text{ISE}\le 10$, fair for $10<\text{ISE}\le 25$, and poor for $25<\text{ISE}$.

7.4.5 Interface

21. *CAD Interface*: The GUI should import CAD drawings for displaying background pictures, and export model results (hydrographs, HGL profiles, etc.) as CAD drawings to aid in data exchange and report preparation using AutoCAD™ DXF or other appropriate formats in common use.
22. *GIS Interface*: The GUI should provide an interface to a Geographic Information System (GIS), such as Arc/Info or ArcView. The interface should import SWMM input data from a GIS and export SWMM output data back to a GIS.

23. *AM/FM Interface*: The GUI should provide an interface to an automated mapping/facilities management (AM/FM) system, such as CASS Works or RJN. The interface should be able to import SWMM input data from an AM/FM and export SWMM output data back to an AM/FM.
24. *DBMS Interface:* The GUI should provide an interface to a data base management system (DBMS), such as dBASE. The interface should be able to import SWMM input data from a DBMS and export SWMM output data back to a DBMS.
25. *GIS Integration*: Total GIS integration offers an attractive alternative to developing separate interfaces for CAD, GIS, AM/FM, and DBMS. Most GIS programs have built-in interfaces for CAD, AM/FM, and DBMS packages. Therefore, once a SWMM GUI is an integral part of a GIS, it can share its interfacing functions also. In total integration, SWMM will become an add-on program to a GIS package as a new menu. SWMM networks will be created as a GIS coverage. Existing GIS coverages of streets, collection systems, and watershed boundaries, etc., will serve as excellent background pictures. SWMM execution will be launched from inside the GIS and SWMM output will be displayed as another GIS coverage.

7.5 GUI Review

This section provides a review of the available SWMM GUIs. The review of the first four GUIs (XP-SWMM, MTV, PCSWMM, and WSWMM) is based on the author's personal review. The review of the remaining GUIs is based on the technical and sales literature of the vendors. Products claims and sales hype commonly found in the brochures have been eliminated. Only the straightforward, no-nonsense explanation of each GUI's capabilities has been presented. Table 7.1 provides the address and phone number of the GUI distributors.

7.5.1 XP-SWMM

XP-SWMM was developed in 1988 by WP software of Canberra, Australia as an Apple Macintosh application. A DOS version was released in 1988. This program was developed under the technical guidance provided by Bob Dickinson, one of the co-authors of EPA's SWMM and EXTRAN programs. In the USA, the program is distributed by XP Software of Tampa, Florida (XP, 1995). All the reviewed features of XP-SWMM are listed in Table 7.2 (see section 7.6 Results). Only some special features are discussed below.

Table 7.1 GUI distributors.

No.	GUI	Contact / Address	Phone / Email / WWW
1	XP-SWMM	Robert Dickinson XP Software, 5553 West Waters Ave. #302, Tampa, Florida 33634	800-883-3487 xpsoft@shadow.net http://www.shadow.net/~xpsoft/
2	MTV	Mark TenBroek 10 Brooks Software, 3744 W. Huron River Drive, #200, Ann Arbor, Michigan 48103.	313-761-1511
3	PCSWMM	Robert James Computational Hydraulics Int., 36 Stuart Street, Guelph, Ontario, Canada N1E 4S5.	519-767-0197 info@chi.on.ca http://www.chi.on.ca
4	WSWMM	Ibrahima Goodwin U.S. EPA, Office of Science and Technology 401 M St. S.W. Washington, D.C. 20460	202-260-1308 Goodwin.Ibrahima@ epamail.epa.gov http://earth1.epa.gov/SWMM _WINDOWS/
5	SWMENU	Virgil C. Adderley Portland Bureau of Environmental Services 1120 SW 5th Ave. Portland, OR 97204	503-823-7866
6	CASS WORKS SWMM	Jeff Frauenfelder RJN Group, Inc. 200 West Front Street Wheaton, IL 60187	708-682-4700 (ext. 361)
7	SWMMDUET	Gray Curtis Madrigal Software Corp. P.O. Box 381710 Cambridge, MA 02238	617-876-3379
8	CASCADE	William James School of Engineering University of Guelph Guelph, Ontario Canada N1G 2W1	519-824-4120 (ext. 2433) james@net2eos.uoguelph.ca

XP-SWMM's graphics-based environment is the most user-friendly of all the urban stormwater system design codes (James, 1992). In addition to providing a GUI, XP-SWMM also includes significant modifications and enhancements to EPA's SWMM, such as, entrance/exit energy losses for the pipes. The XP-SWMM GUI has been written in the C programming language. The program offers a built-in decision support and guidance based on an embedded expert system designed to minimize the need for human experts. It incorporates both pre- and post-processors which use the expert knowledge of experienced users. The expert shell acts as an interpreter between the user and the model. XP-SWMM's user interface utilizes the WIMP (Windows, Icons, Menus and Pointing devices) technology as the state-of-the-art intuitive user environment, but is not yet a Windows program. [*written March 1996*]

XP-SWMM utilizes a unique object-oriented graphical expert environment in which the user creates a link-node network interactively on the screen using a mouse and a toolbar. Background pictures may be imported from CAD packages to real world scale and used as a backdrop for laying out the network. Both the background picture and the network may be generated to scale. Once imported to scale, the areas and lengths may be computed directly from the plan view. The plan view may be output to plotters, printers, or a DXF file. Full CAD quality zoom and pan functions are available. Built-in knowledge-based rules continuously filter the user input and issue warning messages if incorrect data are entered. This expert data-checking capability eliminates data errors at the input level rather than the traditional run time diagnostics. Interpretation, identification, and correction of run time errors is much more difficult.

A major advantage of XP-SWMM is the integration of all the SWMM blocks with a common interface that manages all the data for various blocks from a central graphical database. The data in the graphical database is entered and retrieved through graphical dialog boxes or optional text files. XP-SWMM is the only commercial SWMM GUI with graphical model development and editing capability. A Microsoft Windows version is being developed which will allow multitasking and sharing of Windows resources, such as print manager and drivers. As of 1995, the program cost varies from $2,495 for a 100 pipes version to $9,495 for a 5,000 pipes version. The cost of program upgrades is $295.

Figures 7.1 to 7.3 demonstrate examples of various XP-SWMM capabilities. Figure 7.1 is an example of a graphical editing dialog box. Figure 7.2 shows an example of how a network can be overlaid on a background picture. Figure 7.3 shows a profile plot of HGL.

7.5.2 Model Turbo View (MTV)

Model Turbo View (MTV) was developed in 1990 by 10 Brooks Software of Ann Arbor, Michigan (10 Brooks, 1995). MTV is a post-processor and does not provide an input interface for creating SWMM's input file. The code has been written in Pascal programming language. MTV provides two separate GUIs for SWMM's RUNOFF and EXTRAN blocks, called MTVR and MTVE, respectively. This review is based on MTVE. All the reviewed features of MTV are listed in Table 7.2. Only some special features are discussed below.

MTV provides a viewing port to standard SWMM input and output files. The complex model networks are often difficult for the user to visualize. MTV, however, can be used to view the network in a schematic format. MTV allows the user to supply and edit coordinate data for the nodes and the links. This provides the user with the opportunity to view how the system is connected, based on the model input data file. Connectivity errors are often easily detected and solutions to setup problems quickly formulated. MTV can define any path through the network and display the profile for that path. This often proves valuable when seeking errors in the elevation input data (which can produce unusual results). MTV portrays dynamic SWMM results in the form of HGL plan and profile views. Figure 7.4 shows an example of a HGL plan view. One of the major concerns in SWMM EXTRAN is numerical instability which is often caused by very steep pipes or chambers with small storage. In large models it is difficult to isolate and correct the network elements responsible for numerical instabilities. MTV searches the entire network and highlights all unstable links. Any of the MTV screens can be printed to a LaserJet compatible printer. Plan and profile plots can be saved as AutoCAD DXF files. The user can then add additional information to these AutoCAD drawing files for inclusion in the project report. The program cost is $850 and the cost of program upgrades is $180.

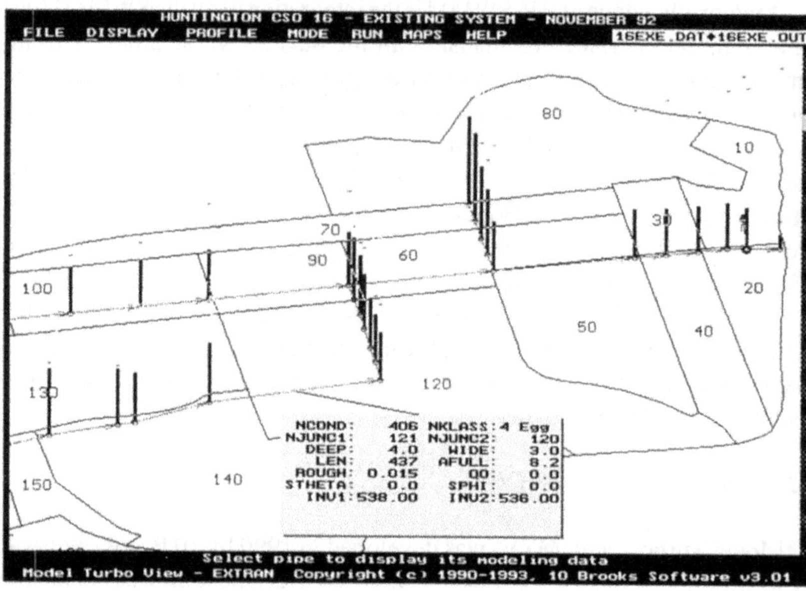

Figure 7.4 Plan view of HGL in MTVE.

7.5.3 PCSWMM

PCSWMM was developed in 1988 by Computational Hydraulics Int. of Guelph, Ontario (Computational 1995). All the reviewed features of PCSWMM are listed in Table 7.2. Only some special features are discussed below.

PCSWMM is a post-processor and does not provide an interface for creating SWMM's input file. PCSWMM is a Microsoft Windows application programmed in Visual Basic which is the language of choice for developing attractive and effective GUIs to meet the specific application needs. Windows Visual Basic applications can communicate with other Windows applications through a mechanism called dynamic data exchange (DDE). While one can easily exchange information by copying and pasting data between applications, DDE automates this process, providing a direct link between the Visual Basic and other DDE applications. This capability of PCSWMM is very valuable for future enhancements and integrations with other CAD, GIS, AM/FM, and DBMS applications.

PCSWMM is developed as a decision support system for SWMM. It is designed to simplify the SWMM environment, speed up execution, and help interpret the SWMM output. It provides fast editing of data files, sophisticated file management, fast and simple graphically oriented execution of SWMM engine, fast and high resolution plotting of hydrographs and pollutographs generated by all blocks of SWMM, automatic sensitivity analysis, calibration and error analysis, field data management, and on-line help. Designed especially for the Microsoft Windows operating system, PCSWMM offers a drag-and-drop icon-oriented environment that greatly simplifies the routine editing, execution, and output interpretation for SWMM, especially long-term simulations. The program cost is $200 and the cost of program upgrades is $100.

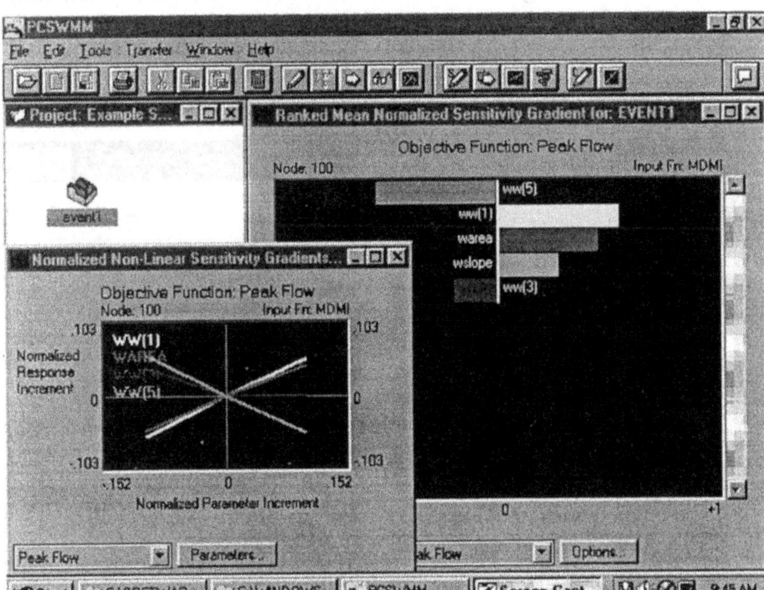

Figure 7.5 PCSWMM sensitivity analysis results.

7.5.4 WSWMM

SWMM for Windows or WSWMM was developed in 1994 for the Office of Science and Technology, Standards and Applied Science Division of EPA to assist them with the Total Maximum Daily Load (TMDL) program (US EPA, 1995). The reviewed features of WSWMM are listed in Table 7.2. Only some special features are discussed here.

The WSWMM GUI was developed to assist the user in data input and model execution, and to make a complex model user-friendly. As the name indicates, this GUI is a Microsoft Windows application. It integrates SWMM and its data handling needs. A pre-processor helps to generate SWMM's input via data sheets and forms. A post-processor is provided which performs the following functions:

1. displays six different types of graphs: hydrograph, pollutograph, loadgraph, flow volume, mass, and landuse;
2. creates calibration plots; and
3. displays summary tables for flow rate (or volume) and pollutant concentrations (or loads) for desired inlets.

WSWMM is public-domain software and can be obtained without cost from EPA.

A key feature of WSWMM is the separation of meteorological data from the RUNOFF Block. A new block called MET has been provided to create and edit meteorological data. The goal of this block is to consolidate user interaction and input of meteorological data into one separate module. Selection of meteorological data for use is a RUNOFF run will occur as part of the RUNOFF block. From

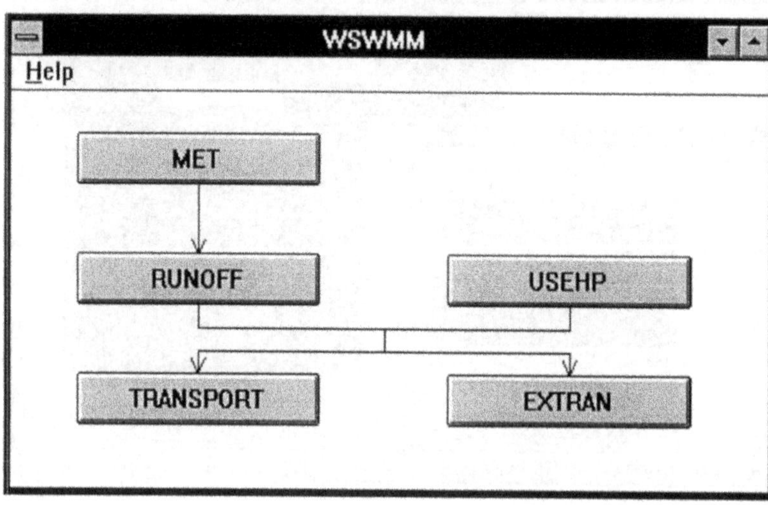

Figure 7.6 Flow chart of WSWMM execution sequence.

a user's perspective all meteorological data will be accessed unambiguously by a single file name. This, therefore, eliminates meteorological data entry in the Runoff input file. Similar consideration made in the TRANSPORT and EXTRAN blocks is the separation of user defined hydrographs and pollutographs from the TRANSPORT and EXTRAN user input (US EPA, 1995). A block called USEHP was developed to handle all user-supplied flows and concentrations. Figure 7.6 shows the execution sequence of WSWMM. Figure 7.7 shows an example of WSWMM input interface using a data sheet.

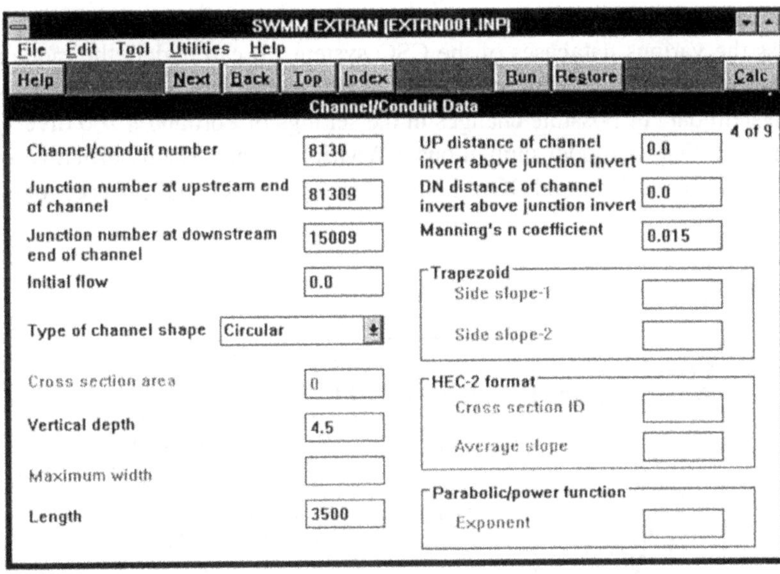

Figure 7.7 WSWMM input interface through data sheets.

7.5.5 SWMENU

SWMENU was developed in 1991 by the Portland Bureau of Environmental Science, Portland, Oregon to model the city's combined sewer overflow (CSO) system using SWMM (Adderley et al., 1994). Due to the intense use and modifications of the CSO models and data, this menu-driven GUI was developed to automate generation and execution of the models, link the model database to other structural databases, and provide for future expansion or linkage to a GIS. Automating the modeling processes required that a relational database system be developed to link the model databases to the independent databases that contain information about the collection system. Using these database links, SWMENU generates SWMM input data files that reflect specific conditions of the collection system.

SWMENU incorporates both pre- and post-processors, and therefore provides interfaces both for creating SWMM's input file and graphing SWMM's output file. SWMENU is a Microsoft Windows application which was developed using Microsoft's Visual Basic for Windows. Database access capability was developed using Q+E/VB by Pioneer Software which supports both dBASE III+ and IV formats. As mentioned previously, Window's Visual Basic programming environment is very valuable for future enhancements and integrations with other CAD, GIS, AM/FM, and DBMS applications.

SWMENU's interface and its relationship to the different components of the Portland CSO modeling system is shown in Figure 7.8. SWMENU is able to access the various databases of the CSO system. It can update the modeling databases and generate new SWMM input files as needed to reflect changes in the system data or possible changes in the settings of Portland's 200 diversion structures. During model simulations, SWMENU launches a modified version of SWMM in a background DOS shell while keeping track of the success of each

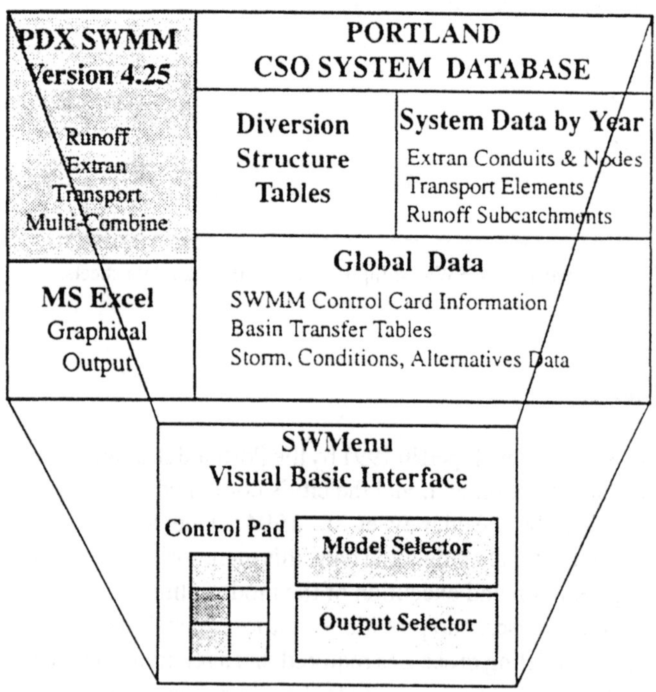

Figure 7.8 SWMENU interface structure.

run. Post-processors and macros are used to import the results into Excel and generate tables, summary statistics, and time series plots of depth and flow. Future enhancements of SWMENU will include integration with Microsoft Access, a relational database management system for Windows. The SWMENU GUI is a public-domain program and is free.

7.5.6 CASS WORKS SWMM

This GUI has been developed by RJN Group, Inc. of Wheaton, Illinois, as a CASS WORKS module. CASS WORKS is an integrated infrastructure management software for water distribution, sanitary sewers, storm drainage, treatment facilities, parks and recreation, GIS, and AM/FM applications. Also known as CASS WORKS' Sewer Hydraulic Modeling Module, the GUI employs SWMM's TRANSPORT block to model network capacity, perform gradient analysis, incorporate time routing of flows, and predict system needs based on future growth or demands. This module can link elements of an SSES type database to hydraulic profile characteristics of a sanitary sewer system. RJN has also developed another module called GeoCAD that can integrate the TRANSPORT module with the leading GIS software programs that use ORACLE. This integration allows the TRANSPORT module to run within GIS programs. Using ANSI SQL RDBMS standards, RJN has integrated CASS WORKS with ARC/INFO and ArcView, the world's leading GIS software programs developed by the Environmental System Research Institute (ESRI). The ability of both systems to access the same database increases the value of the data, eliminates database inconsistencies, allows for both graphic and nongraphic representation of data, and eases implementation issues and costs. The cost of CASS WORKS' TRANSPORT and GeoCAD modules is $4,400 and $2,200, respectively. A $4,400 core module which provides the ORACLE support should also be purchased to run the TRANSPORT and GeoCAD modules.

The CASS WORKS SWMM interface was used by the Massachusetts Water Resources Authority (MWRA) to conduct their Sewerage Analysis and Management System (SAMS) project. The goal of the SAMS project was to provide the MWRA with the capability to efficiently and accurately evaluate both short-term and long-term needs of the MWRA sanitary sewer interceptor system, which serves over two million people in 43 communities in the Boston Metropolitan area. In order to create a geographically correct modeling schematic and simplify the process of geographically representing the modeling results, MWRA plans to link SWMM's EXTRAN block with SAMS inventory database and ARC/INFO GIS. ARC/INFO GIS will be used to create a link between the TRANSPORT data and computerized maps of the MWRA interceptor system and communities serviced.

7.5.7 SWMMDUET

SWMMDUET was developed in 1993 by Madrigal Software Corporation for the Delaware Department of Natural Resources and Environmental Control (DNREC). DNREC's goal was to promote state-wide use of SWMM to obviate the need for Delaware municipalities to make separate efforts to satisfy the requirements of Section 6217 of the Coastal Zone Management Act and the Clean Water Act. SWMMDUET was developed to increase the productive use of SWMM and aid modelers in meeting mandated obligations (Curtis, 1994).

SWMMDUET integrates SWMM with ARC/INFO GIS software, and quite appropriately, has been written in ARC/INFO's native ARC Macro Language (AML). It creates a computing environment that does not require arcane knowledge of SWMM and ARC/INFO. This GUI incorporates both pre- and post-processors, and therefore provides interfaces both for creating SWMM's input file and graphing SWMM's output file. SWMMDUET takes advantage of the graphical paradigm around which ARC/INFO is constructed and utilizes the relational database capabilities to organize the data.

Beyond data management, the program uses expert system logic to assemble data that defines the modeling process, prepares SWMM input, executes the SWMM program, and converts the output into meaningful graphical displays. Data entry sheets and forms eliminate the need for the modeler to know detailed ARC/INFO processing techniques. Similarly, feature selection, spatial joins, and processing commands and options of ARC/INFO are specified and executed for the user. Hyetographs are related to rain gages where rainfall data was recorded. The storms are selected simply by a georeference to the gages. Time series data are stored as sequential records in the database files. SWMMDUET has significantly simplified the management of vast amounts of hydrologic data, allowing hydrologists to concentrate on hydrologic matters. Future development plans of SWMMDUET include ArcView functionality, automatic watershed delineation from USGS Digital Elevation Model (DEM), and a linkage to EPA's WASP (Water Analysis Simulation Program) water body receiving model. The SWMMDUET software is in the public domain and can be obtained for the cost of distribution (Curtis, 1994).

7.5.8 CASCADE and CASCADE2

These programs provide an interface to SWMM utilities. The utility programs aid the development and execution of SWMM models. In the past, the main argument against using continuous SWMM modeling has been the difficulty of managing large amounts of input and output data. CASCADE programs are time series data managers (TSM) for SWMM which facilitate continuous SWMM modeling. They also provide easy data export/import capability to/from other TSMs.

CASCADE was developed by Michael Gregory, a graduate student of the University of Guelph, Guelph, Ontario. CASCADE provided a SWMM interface with the USGS TSM called ANNIE. CASCADE2, which is also a Microsoft Windows application, has been developed by Yiwen Wang, another University of Guelph graduate student. CASCADE2 provides a SWMM interface with the US Army Corps of Engineers' TSM called HECDSS (see Chapter 3).

Both programs also provide a statistical analysis utility to compute storm event statistics by interfacing with SWMM's RAIN and STATS blocks.

7.6 Results

In order to quantify the available GUI features and identify the future improvements, a side-by-side comparison of the available GUIs was performed. Table 7.2 provides a comparison of the first four GUIs discussed in the previous section (XP-SWMM, MTV, PCSWMM, and WSWMM). These GUIs were included in the comparison because detailed information about their features was available from the author's personal experience. The last four GUIs were not included because their review is based on the sales literature rather than the author's personal experience.

The eight reviewed GUIs illustrate the wide variety of functions that can be performed. It is difficult to compare them with one another because of the large number of features and options that each one offers. Therefore, it is very important to assess needs and preferences and compare the various GUI programs to those needs, rather than to one another. The author does not intend to imply that these eight are the only products to be considered when selecting a SWMM GUI. Users are encouraged to supplement this review with the latest information, especially if they are reading this article long after publication.

Table 7.2 is intended to demonstrate a method of comparing the GUIs. The comparison is based on the existing features. Planned future improvements were not included. Due to some limitations of this simple comparison technique, Table 7.2 may not be used for the final GUI selection and purchase. However, as described below, most of these limitations can be easily eliminated by the users. The total score is based on assigning an equal score of one to each feature. In the real world, different features will have a different value to different users. For instance, the graphical model development feature may be the most valuable feature for some users worth 5 or 10 points. Users are encouraged to apply a custom scoring system reflecting their personal modeling needs and preferences. Another limitation of Table 7.2 is its inability to compare the relative quality of a feature in different GUIs. For instance, all the GUIs offering automatic file management were assigned the same score of one regardless of their relative file management capability. Users are encouraged to contact the GUI vendors for

Table 7.2 SWMM GUI comparison.

No.	Feature	XP-SWMM	MTV	PC-SWMM	WSWMM
1	Windows application			●	●
2	Integration	●	●	●	●
3	User friendliness	●	●	●	●
4	File management	●		●	●
5	Input interface	●			●
6	Output interface	●	●	●	●
7	Graphical input	●			
8	Graphical editing	●			
9	Spreadsheet editing				●
10	Background pictures	●	●		
11	WYSIWYG	●	●		
12	Navigation	●	●		
13	Spatial searching	●	●		
14	Results review	●	●	●	●
15	Sensitivity analysis			●	
16	Dynamic Plots	●	●		
17	Thematic plots	●	●		
18	Hydrograph plots	●	●	●	
19	Profile plots	●	●		
20	Calibration plots	●	●	●	
21	CAD interface	●	●		
22	GIS interface				
23	AM/FM interface				
24	DBMS interface				
25	GIS integration				
	Cost	\$2,500 to \$9,500	\$850	\$200	Free

a detailed comparison of the GUI features. Finally, the GUI comparison was based only on the 25 recommended features. Many GUIs provide additional features which may be valuable to some users and should be included in the comparison process.

7.7 Recommendations

1. It is recommended that all the GUIs should move to Microsoft Windows. This will benefit both the developers and the users. Developers will be able to focus on the GUI development rather than creating device drivers and fonts which are already available in Windows to be shared by all the Windows applications. There will be no concerns about the GUI compatibility with a wide variety of monitors, pointers, printers, and plotters. As long as a GUI will be compatible with the Windows standards, it will be compatible with all the Windows device drivers.
2. The programming language should be chosen carefully. It may not be appropriate to continue to program in Fortran 77 to develop the Windows applications. It is recommended that scripting and object orientated programming languages, such as Visual Basic and Visual C++ should be used to provide efficient third party interfacing capability.
3. The GUIs should be flexible and modular to allow third party interfaces. Developers should maximize the use of off-the-shelf software. This would obviate a need to recreate functions which are available in commonly used applications such as Lotus 1-2-3 or Microsoft Excel.
4. Table 7.2 indicates that the GUIs lack GIS, AM/FM, and DBMS interfaces. It is recommended that these features be provided. A batch-oriented approach in which users must write their own programs or use other software (e.g. spreadsheets) to import (export) from (to) SWMM ASCII files is not recommended. The interface feature should be an integral part of the GUI, possibly appearing as *import* and *export* options under the *files* menu. The interface should not be application dependent. The internal structure of the software should allow communication both to and from any standard CAD, GIS, AM/FM, or DBMS. If this is not done, the vendors will have to develop separate interfaces for different packages.
5. As an alternative to developing separate interfaces for CAD, GIS, AM/FM, and DBMS, total GIS integration may be implemented. This feature is not currently available in commercial GUIs. It is important to understand the difference between a GIS interface and a GIS integration.The GIS interface is simply a file menu option in a SWMM GUI to transfer data to/from a GIS. The GIS integration, is a combination of a SWMM GUI and a GIS such that the combined program offers both the GIS and the SWMM functions.

6. Since users will always have more confidence in EPA's official release of SWMM, it is recommended that the GUIs should preserve SWMM's solution algorithms. Users should be able to obtain identical results from SWMM and the GUIs. Using the example problems from the SWMM's distribution diskettes or user's manual, each new version of a GUI should demonstrate that its output is identical to SWMM's output. Any modifications and enhancements made to SWMM's code should be offered as separate user options. This recommendation can be easily implemented if the users are given a choice to use EPA's official SWMM.EXE file in place of the modified SWMM.EXE file provided by the GUI.
7. The GUIs are 100% dependent on the SWMM version. The GUIs read specific data at specific locations in SWMM's ASCII files. For this reason, a GUI must be reprogrammed every time the format of SWMM's input or output file has been changed. For example, even one extra space in just one line of the output file has a potential to upset the normal execution of a GUI. Therefore, each new SWMM version will require a corresponding new GUI version. This limitation does not create a problem because in the last 25 years, EPA has released only four SWMM versions. However, sub-versions (e.g. 4.1, 4.2, etc.) have been released more frequently. It is recommended that, to the maximum extent possible, EPA should not change the input and output formats of the sub-versions. This will avoid the need to purchase a GUI upgrade for each SWMM upgrade.

References

10 Brooks Software. 1995. Model Turbo View - Extran, Users Manual, Version 3.11, 3744 W. Huron River Drive, Suite 200, Ann Arbor, Michigan 48103.

Adderley, V.C., M.A. Liebe and M.M. Vilhauer. 1994. Development of the SWMENU Interface for the Portland CSO Modeling System. in Proceedings, Conference on A Global Perspective for Reducing CSOs: Balancing Technologies, Costs, and Water Quality, Water Environment Federation, Louisville, Kentucky, July 10-13, 1994, pp. 3-47 - 3-48.

Computational Hydraulics Int. 1995. PCSWMM, Users Manual, 36 Stuart Street, Guelph, Ontario, Canada N1E 4S5.

Curtis, T.G. 1994. SWMMDUET: Enabling EPA SWMM with the ARC/INFO Paradigm, ARC News, Environmental Systems Research Institute, Redlands, California, Spring 1994. pp. 20.

Dakan, M.L. 1992. CADalyst Magazine, Aster Publishers, August, p30,

Huber, W.C., and R.E. Dickinson. 1988. Storm Water Management Model. User's Manual, Version 4, Environmental Research Laboratory, U.S. Environmental Protection Agency, Athens, Georgia.

James, W. 1993. Introduction to SWMM. Chapter 1 in New Techniques for Modelling the Management of Stormwater Quality Impacts. Edited by William James, Lewis Publishers, Boca Raton, Florida, pp. 1-28.

Marsalek, J., T.M. Dick, P.E. Winser, and W.G. Clarke. 1975. Comparative Evaluation of Three Urban Runoff Models. Water Resources Bulletin, 11(2), pp. 306-328.

Roesner, L.A., J.A. Aldrich, and R.E. Dickinson. 1988. Storm Water Management Model. EXTRAN Addendum, User's Manual, Version 4, Environmental Research Laboratory, U.S. Environmental Protection Agency, Athens, Georgia.

Rubin, D.K. and Powers, M.B. 1992. How Green Is Our Software, ENR, 229(17), pp. 30-32.

TenBroek, M.J. and L.A. Roesner. 1993. MTV - Analysis Tool for Review of Computer Models. in Proceedings, Conference on Computers in the Water Industry, Water Environment Federation, Santa Clara, California, August 8-11, 1993, pp. 173-181.

U.S. Environmental Protection Agency. 1995. SWMM Windows Interface User's Guide (Draft). Office of Science and Technology, Standards and Applied Science Division, Washington, D.C.

XP Software. 1995. XP-SWMM Storm Water Management Model with XP Graphical Interface, User's Manual, Version 2, Volumes 1 and 2, 5553 West Waters Ave #302, Tampa, Florida 33634.

Chapter 8

Thermal Enrichment of Stormwater by Urban Pavement

William James and Brian Verspagen

Urbanization is known to increase the temperature of surface runoff during storm events and to increase the mean summer monthly temperature of receiving waters downstream (Galli, 1990; Pluhowski, 1970). It affects the temperature of streams as follows: urban construction, comprising roads, parking lots, roofs and sewers, reduces the original forest canopy, and increases impervious areas and, thus, surface runoff. Increased runoff in turn causes wider channels and more surface ponds, both of which lead to more exposure of stormwater to solar radiation, exacerbated by canopy loss. Increased imperviousness also leads to decreased infiltration and baseflow, which reduces the dilution of heated stormwater. Changes in the texture and color of the ground cover are also significant sources of thermal enrichment in an urban watershed. Elevated stream temperature is the inevitable result of these synergistic effects.

Several methods are available to control the thermal enrichment of stormwater - some infiltration approaches include: infiltration basins, infiltration trenches, seepage trenches, filter strips, grassed swales, and permeable pavement. While no single method may be sufficient, combinations of these methods may markedly reduce the impacts of urbanization on receiving waters (Marshall, Macklin, and Monaghan, 1991; Ahmed and James, 1995).

This study is part of our continuing research (Xie and James, 1994; Thompson and James, 1995; Kresin and James, 1996. See also the web: http://www.eos.uoguelph.ca/~james/research.html#porous); this chapter covers the

© *Advances in Modeling the Management of Stormwater Impacts - Vol. 5*. W. James, Ed.
Pub. by CHI, Guelph, Canada 1997. ISBN 0-9697422-7-4. Fax: +519 767-2770

thermal enrichment of surface runoff from impervious asphalt and porous concrete block pavement. Part of the research was conducted in a laboratory setting on pavement samples measuring about 1 x 1 x 0.5 m. Energy for heating the laboratory pavements was provided by either the sun or a 28000 Btu propane heater, and a rainfall simulator was used to generate thermally-enriched surface runoff. Experimental procedures are detailed in a dissertation (Verspagen, 1995) and will be published separately.

For this methodology a spatial resolution of about one hundred metres, approximately the size of a parking lot, is required. At this scale, the temporal resolution is of the order of one or two minutes. Such a resolution is considered to be very fine, even when compared to modern stormwater modeling practice.

We hope that this methodology will encourage designers, engineers, and planners of small urban areas, such as parking lots for shopping centers, to use alternative stormwater management practices, in particular pavement surfaces with environmentally-sensitive thermal characteristics.

8.1 Earlier Research

Related research has focused on long-term thermal enrichment as would result from reduced watershed infiltration, improper implementation of stormwater best management practices (BMPs) and removal of stream canopy (Galli, 1990; Pluhowski, 1970; Weatherbe, 1995). These three writers report significant impacts of thermal enrichment, including an alteration of the general thermal regime of receiving streams and rivers. However, increased temperature of urban stormwater resulting from rainfall on hot pavement has not received the same attention, and very little research has been conducted on the relevant component processes of heat transfer.

Xie and James (1994) evaluated the effectiveness of the Hydrological Simulation Program - FORTRAN (HSPF) for estimating expected thermal enrichment of stormwater runoff. They applied the model to the Speed River in Guelph, conducting additional field work to determine the required field parameters. Several conclusions were drawn:

1. runoff temperature is affected by the rainfall intensity and pavement temperature at the onset of storms;
2. air temperature is indirectly linked to pavement surface temperature;
3. a fine time resolution, of the order of minutes, is necessary when modeling expected stormwater runoff temperature, as storms may occur suddenly and last only a few minutes during hot weather; and
4. expected thermal enrichment in the Speed River receiving waters (early summer) was related to the percentage imperviousness:

$$Turb = 17.0 + 0.01 \times (\% imperviousness) \qquad (8.1)$$

where *Turb* is the expected or mean temperature of urban runoff (°C), and the constants 17.0 and 0.01 were determined by fitting against observed data.

Based on only three experiments using a rainfall simulator on an asphalt parking lot in medium-good condition (lot P10 at the University of Guelph), they tentatively proposed the following relationship between the temperature of the wet paving surface and the expected mean temperature of the surface runoff:

$$T_R = 3.26 + 0.828 \times T_{Pw} \qquad (8.2)$$

where:

T_R = the expected mean temperature of the surface runoff (°C),
T_{Pw} = the temperature of the pavement before wetting (°C).

8.2 Processes of Thermal Enrichment of Stormwater

To clarify our understanding of the underlying processes, a rather crude description is given here. Later, we state why we regard some of these processes as unimportant, in order to develop a simple empirical relation.

Clearly, pavement temperature fluctuates on a daily and seasonal basis. Seasonal cycles have been reported by Oke (1987). In areas such as southern Ontario, the subgrade experiences the warmest temperatures in July and the coldest temperatures in January. Diurnal fluctuations in Figure 8.1 were observed by the present authors in the subgrade of an instrumented asphalt pavement (described by Thompson and James, 1995) at 02:15 and 13:30 on August 23, 1995. This heating and cooling cycle is a commonly-observed, daily process.

Insolation (radiant energy from the sun) has a significant effect on the direction of the thermal gradient in the subgrade. Figure 8.1 is for a clear day where the temperature changes are relatively large. Late at night, heat energy accumulation from the previous day causes an gradient upward, because the pavement surface is cooler than the ground. During late afternoon, the thermal gradient is downward because the pavement surface is now warmer than the subgrade. A similar cycle might be observed in cloudy weather, although perhaps not as extreme.

Late afternoon, evening and early night rainfalls may advance the clear-weather day-night cycle: clouds reduce the amount of radiant energy reaching the pavement surface; relatively cold precipitation contacting the warm pavement surface creates a thermal gradient from the top of the pavement surface to the moving water film on the pavement surface, cooling the pavement surface and, subsequently, the subgrade.

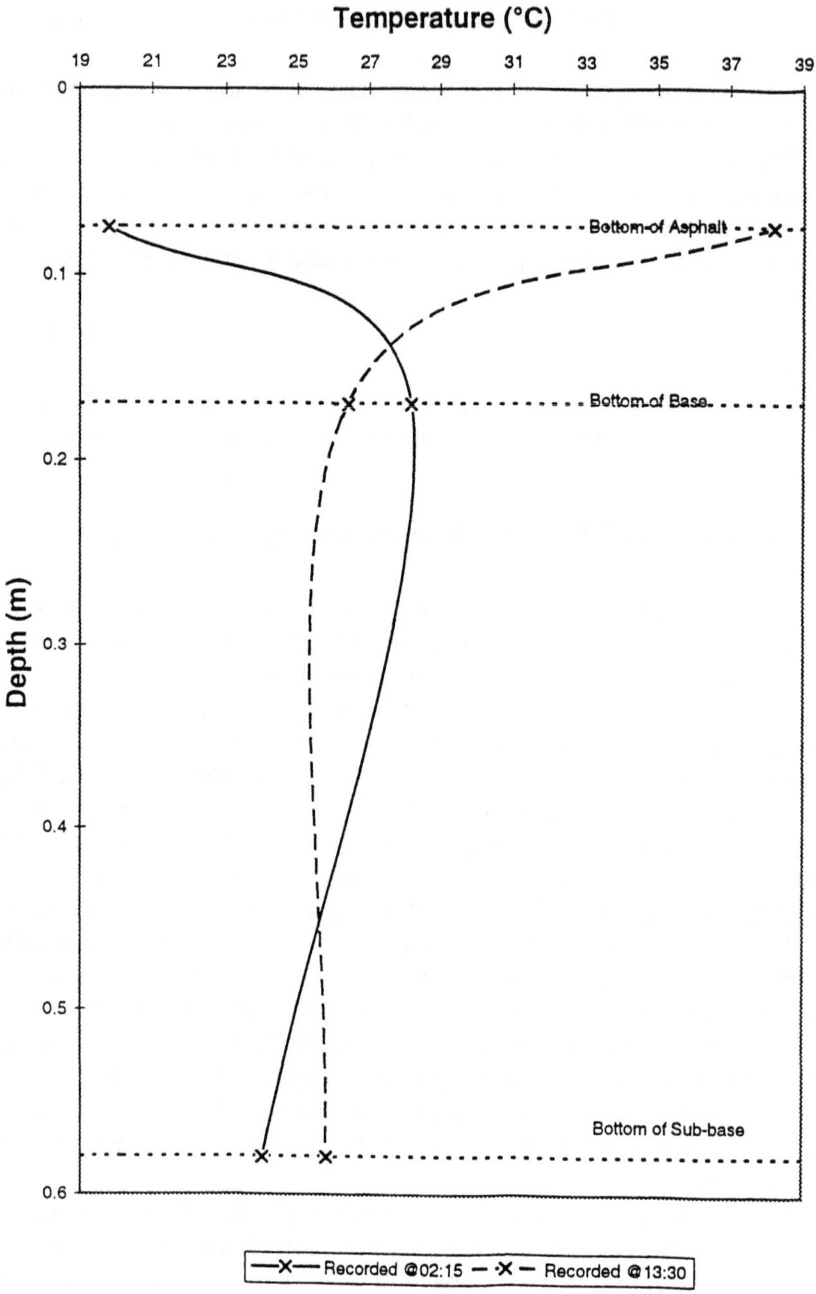

Figure 8.1 Diurnal temperature comparison - instrumented asphalt pavement, parking lot P10, August 23, 1995.

In this chapter we use the term *paving* to describe the hardened upper surface layer, e.g. four inches [10 cm] of asphalt or precast concrete block. *Subgrade* is placed below the paving. *Pavement* describes the entire construction (paving and subgrade) perhaps 3 ft [1 m] thick. *Surface* normally means a face having no thickness, usually the air/paving interface.

Cooling rates are influenced by many physical characteristics of the pavement, including the thermal and infiltration properties of the paving and subgrade. To consider the processes affecting temperature of surface runoff we follow hydrologic principles, expressing what thermodynamicists would consider an imprecise energy budget, for the combined, lumped paving-and-surface-water-film, whose temperature gradients across the paving and water film are not considered, as explained below:

$$\Delta H = (Q_n + Q_r - Q_e - Q_h - Q_g - Q_w) \times \Delta t \tag{8.3}$$

where:

ΔH = change in heat energy of the paving ($J{\cdot}m^{-2}$), averaged over the thickness of the paving, ignoring physical changes to the pavement material;

Q_n = net radiant heat flux at pavement surface ($W{\cdot}m^{-2}$) ignoring effects of the water film;

Q_r = heat transfer between raindrops and the paving surface ($W{\cdot}m^{-2}$), a likely process of pavement cooling;

Q_e = latent heat flux causing evaporation ($W{\cdot}m^{-2}$);

Q_h = convective heat flux, paving surface to air ($W{\cdot}m^{-2}$), conveniently ignoring the film which is considered to be discontinuous and intermittent;

Q_g = heat flux into or out of the ground ($W{\cdot}m^{-2}$);

Q_w = heat transfer between pavement surface and overlying water ($W{\cdot}m^{-2)}$, now admittedly inconsistently considered;

Δt = time interval over which change in stored heat energy is evaluated (s).

In reality the film of water on the pavement is ill-defined and likely to be thin compared to the large rugosity of urban paving, and it will be inherently difficult to identify its transient properties, as it drains across and through the paving. To avoid later difficulty in characterizing the film, our formulation does not separately consider heat flow through the water film to the paving surface. Nor does it allow temperature gradients to be identified within the subgrade, or within the water film. (No doubt this will raise concerns for thermodynamicists - our purpose here is only a consideration of the underlying mechanisms, in order to later derive a simple empirical relation, and not to derive a formal mathematical procedure!) By assuming that the water film is essentially intimately attached to the paving, the conceptual model favours situations where the depth-mean water-film temperature and the paving-mean temperature are similar. The model simply allows net energy in the paving to increase or decrease, depending on the conductivities, specific heats and thermal gradients between the subgrade and paving and between the pavement surface and the air above. It also allows paving to lose heat to the water film, but only paving-mean and water-film-mean (or depth-averaged) temperature is considered. Later we use this description to derive a very simple empirical relation.

As shown in Figure 8.2, where the film is shown to be continuous and uniform, but would in reality forms puddles in depressions in the rough surface, three situations may be described:

1. late on a clear sunny day: the dry pavement surface is warmer than the air above and ground below: Q_n and Q_g are down, Q_h up;
2. night and cloudy days: Q_n is small, ground is warmer than the dry pavement which is warmer than the air above: Q_g is up; and
3. daytime rainfall: cloudy conditions prevail and Q_n may be assumed to be small.

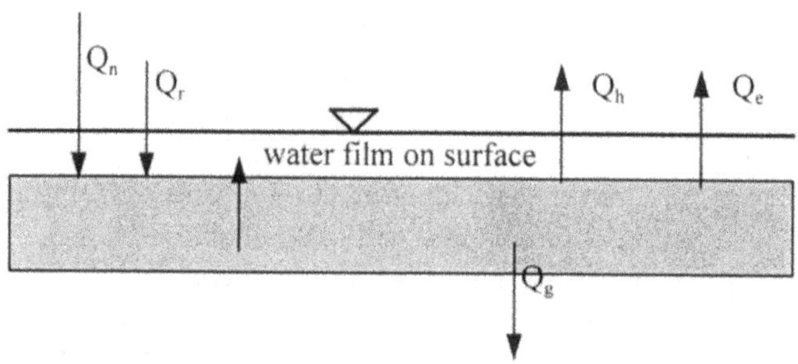

Figure 8.2 Wet pavement energy budget (note energy flux directly past water film, due to thin discontinuous and intermittent film and rough paving surface).

Infiltrating water will also convey heat to or from the paving and the subgrade. Partitioning of energy between overland surface runoff and infiltrating water is given approximately by the ratio of infiltrating water to surface runoff.

Because, during rainfall, the air and surface water temperatures are similar, and the humidity is high, the convective heat loss (Q_h) and the latent heat loss (Q_e) are small. (In reality this may not be true after cessation of rain and during drying of the paving surface; it is common for most of the residual water to gradually evaporate, perhaps within an hour or so in hot weather). If the pavement surface is warmer than the surface water, the heat energy loss to the surface water (Q_w)

will be significant. Energy is supplied from ΔH. If it is assumed that no thermal gradient exists between the bottom of the paving and the top of the subgrade, or if the energy budget is considered over a sufficiently short period of time so that conductive heat transfer from the bottom surface to the top surface of the paving is not significant, $Q_w \times \Delta t$ may be considered to be equal to ΔH.

8.2.1 Paving Energy Components

Net radiation absorbed by a pavement surface, Q_n, may be calculated using the following equation (Oke, 1987):

$$Q_n = K_n + L_n. \tag{8.4}$$

where K_n = net short wave radiation ($W \cdot m^{-2}$).

K_n represents the difference between the short wave radiation energy transmitted by the atmosphere and the short wave radiation energy reflected by the pavement surface:

$$K_n = (1-\alpha)K\downarrow \tag{8.5}$$

where:

α = albedo or reflectivity of the receiving pavement surface,
K = total short wave radiation in direction indicated ($W \cdot m^{-2}$).

L_n represents the net longwave radiation ($W \cdot m^{-2}$), i.e. the difference between the longwave radiation energy emitted by the atmosphere and the longwave radiation emitted by the pavement surface:

$$L_n = \varepsilon_o \varepsilon_a \sigma (T_a + 273)^4 - \varepsilon_o \sigma (T + 273)^4 \tag{8.6}$$

where:

ε_o = emissivity of the pavement surface (dimensionless);
T_o = temperature of the pavement surface (°C);
σ = Stefan-Boltzman constant (5.67×10^{-8} $W \cdot °K^{-1} \cdot m^{-2}$);
T_a = air temperature (°C); and
ε_a = atmospheric emissivity (dimensionless):

$$\varepsilon_a = (0.72 + 0.005T_a)(1 + an_c^2) \tag{8.7}$$

where:
n_c = fraction of cloud cover (n_c=1 is overcast), and
a = cloud constant (under cloudy conditions, a=0.20).

For the purposes of this study, primarily warm-weather rainfall, raindrops are normally cooler than the pavement surface. The temperature of the film of water on the pavement is used to estimate the heat transfer from/to the paving.

Thus, ignoring heat transfer to/from the paving, and assuming complete mixing (overlooking the effect of raindrop splash redistribution), water-film mean temperature may be determined using a heat balance:

$$T'_{sr} = \frac{q_r T_r + q_{sr} T_{sr}}{q_r + q_{sr}} \tag{8.8}$$

where:

T'_{sr} = mean temperature of the film of water on the pavement surface evaluated for the current time step (°C);
q_r = rainfall flow rate ($L \cdot t^{-1}$);
T_r = temperature of the rainfall (°C);
q_{sr} = surface runoff flow rate ($L \cdot t^{-1}$); and
T_{sr} = mean temperature of the surface runoff from the previous time step (°C).

The latent heat flux is:

$$Q_e = L_v \times E \tag{8.9}$$

where:

L_v = latent heat of vaporisation ($J \cdot kg^{-1}{}_{water}$) approx. 2.5×10^6, varies with temperature, and
E = mass flux of water removed through evaporation per unit time ($kg \cdot m^{-2} \cdot s^{-1}$).

Evaporation rates are dependent on the humidity of a relatively thin layer of air above and near the pavement surface. If this layer is very dry (and windy), the evaporation rate will be significant. Wind near the pavement surface promotes mixing of the upper and lower layers of the air above. Traffic also affects local air turbulence. When upper layers of relatively unsaturated air are mixed with lower, saturated, layers, greater evaporation rates result. *E* can be estimated from mass balances of long-term simulations, when it is likely to be significant.

Convective heat flux, Q_h, may be calculated using:

$$Q_h = C_a \frac{(T_o - T_a)}{r_H} \tag{8.10}$$

where:

C_a = volumetric heat capacity of air ($J \cdot m^{-3} \cdot K^{-1}$) (commonly assumed to be 1200 $J \cdot m^{-3} \cdot K^{-1}$, Oke, 1987);
T_0 = temperature of the pavement surface (°C);
T_a = temperature of the air (°C); and
r_H = aerodynamic resistance (s/m) - for more details see Oke (1987).

In calm conditions, the air above forms an insulating barrier due to its low heat capacity. But wind increases convective heat flux: the temperature gradient between the pavement surface and air above is steeper in windy conditions because the insulating barrier is removed and air from upper layers is mixed with air at the surface.

The subsurface heat flux, Q_g, may be first-order estimated using Fourier's law:

$$Q_g = k_s \left.\frac{dT_s}{dz}\right]_{z=0} \approx k_s \frac{T_1 - T_2}{z} \tag{8.11}$$

where:

k_s = thermal conductivity of the paving ($W \cdot m^{-1} \cdot K^{-1}$);
T_1 = temperature at a predetermined depth in the subgrade (°C); and
T_2 = temperature at a small depth z (m) below T_1 (°C) (where Q_g has not changed significantly from its value immediately below the pavement surface).

Change in heat storage ΔH is:

$$\Delta H = C_s \times \frac{\Delta T}{\Delta t} \times \Delta z \tag{8.12}$$

where:

C_s = heat capacity of the paving ($J \cdot m^{-3} \cdot K^{-1}$);
ΔT = mean temperature change (°C) over time;
Δt = change in time (sec); and
Δz = thickness of the layer under consideration (m).

8.2.2 Heat Transfer from Paving to the Overlying, Moving Water Film

Conductive heat transfer is dependent on the medium's mass density, specific heat, heat capacity, thermal conductivity, thickness, and thermal gradient (Arya, 1988). (Convective heat transfer to the fluid moving *around* the paving, e.g. down into the subgrade, is ignored here.) Specific heat is defined as the amount of heat absorbed or released in raising or lowering the temperature of a unit mass of the material by 1°C. The product of mass density and specific heat is called the heat capacity per unit volume. The thermal conductivity of a material is a proportionality constant relating the rate of heat flux or heat transfer in a direction along the temperature gradient. The ratio of thermal conductivity to heat capacity is called the thermal diffusivity ($m^2 \cdot sec^{-1}$). Thermal diffusivity is considered to be the most appropriate measure of how rapidly temperature changes are transmitted to other layers in a medium (Arya, 1988).

It is useful to note that air has the lowest heat capacity and thermal conductivity of all natural materials. On the other hand, the thermal diffusivity of air is very large because of its low density. Conversely, water has the highest heat capacity of known natural materials. Addition of water to an initially dry soil will significantly increase its heat capacity and thermal conductivity, because the water replaces air in the pore spaces of the soil matrix. Heat capacity and thermal conductivity have been found to be monotonically increasing functions of soil moisture content (Arya, 1988).

Somewhat different from heat transfer between soil media, heat transfer from a plane surface to a liquid under turbulent conditions has been given by Holman (1990):

$$Q_w = \bar{h} \times (T_o - T_\infty) \tag{8.13}$$

where:

- Q_w = amount of heat transfer from plate to the liquid ($W{\cdot}m^{-2}$);
- T_0 = temperature of the pavement surface (°C);
- T_∞ = temperature of overlying fluid under free stream conditions (°C) (free stream conditions occur at the point in the fluid where the thermal gradient is no longer influenced by the warm pavement surface over which it flows);
- $\bar{h}$ = average heat transfer coefficient ($W{\cdot}m^{-2}{\cdot}°C^{-1}$) calculated using the Nusselt number:

$$\bar{h} = \overline{Nu_L} \times \frac{k}{L} \tag{8.14}$$

where:

- k = thermal conductivity of the overlying fluid ($W{\cdot}m^{-1}{\cdot}°C^{-1}$);
- L = length of fluid flow over the pavement surface (m); and
- $\overline{Nu_L}$ = average Nusselt number (dimensionless).

The Nusselt number provides a basis for comparing rates of convective heat loss from similar bodies but at different scales (Monteith and Unsworth 1990). Under turbulent conditions, which is what we experience at the paving surface during rainfall, it may be calculated using the following equation (Holman, 1990):

$$\overline{Nu_L} = Pr^{1/3} \times (0.037 \times Re_L^{\;0.8} - 871) \tag{8.15}$$

where Pr is the Prandtl number (dimensionless) whose value for water may be obtained from tables or by:

$$Pr = \frac{c_p \times \mu}{k} \tag{8.16}$$

where:

c_p = specific heat capacity of the paving ($kJ \cdot kg^{-1} \cdot °C^{-1}$);
μ = dynamic viscosity of the fluid ($kg \cdot s^{-1} \cdot m^{-1}$); and
k = thermal conductivity of the fluid ($kW \cdot m^{-1} \cdot °C^{-1}$).

Re_L is the Reynolds number (dimensionless) describing the fluid flow over the pavement surface:

$$Re_L = \frac{\rho \times u_\infty \times L}{\mu} \tag{8.17}$$

where:

ρ = density of the fluid ($kg \cdot m^{-3}$);
u_∞ = velocity of the fluid under free stream conditions ($m \cdot s^{-1}$);
L = length that the fluid flows over the pavement surface (m); and
μ = dynamic viscosity of the fluid ($kg \cdot m^{-1} \cdot s^{-1}$).

Together, these equations permit the net energy flux, its direction, and the periods of warming and cooling of the paving to be estimated. Warming of the surface water can be computed from the energy flux into surface runoff during rainfall if all the variables are known.

The heat capacity of water, i.e. the amount of energy necessary to warm 1 kg of water by 1°C, is approximately 4.2 $kJ \cdot kg^{-1} \cdot °C^{-1}$. Noting that 1 joule is 1 W·s and that 1 L of water may be assumed to have the same mass as 1 kg of water, the change in temperature is:

$$\Delta T_{sr} = \frac{Q_w}{4200 \times q} \tag{8.18}$$

where:

ΔT_{sr} = change in temperature of the surface water (°C);
Q_w = total energy transferred from the pavement surface to the overlying water (kW);
4200 = heat capacity of water ($W \cdot s \cdot kg^{-1} \cdot °C^{-1}$) between 10 and 80°C; and
q = flow rate of water ($kg \cdot s^{-1}$).

8.3 Development of a Design Methodology

Collectively the above expressions form a mathematical model for estimating the temperature of surface runoff, but such a model requires a significant quantity of physical information in the form of input parameters describing the pavement surface, and time series describing rainfall. The following outline suggests how the model could be used.

1. Assume that net radiant heat flux (Q_n) is negligible during rainfall (net longwave back radiation is balanced by weak shortwave influx). Similarly, assume that upward heat flux from the ground does not influence the paving temperature during rainfall (conductive heat process is slow, considering the thickness of the pavement and typical short duration of summer thunderstorms). Thus assume that all energy flux to the water film is the result of thermal energy stored in the paving (ΔH) or thermal energy in the raindrops (Q_r).
2. Measure or otherwise determine the initial pavement surface temperature.
3. Assume that on the pavement surface a film of water exists that does not disappear during rainfall. Further, assume that the temperature of this water is the same as that of the pavement surface.
4. Determine the temperature of the surface water using Equation 8.8 where the rain temperature is known and the surface runoff flow rate equals the rainfall rate from the previous time step. The runoff flow rate here represents both the surface water that has run off from the pavement surface and water that has infiltrated the pavement surface, and may be partitioned to reflect the respective proportions. Thus, a control volume is established for each time step.
5. Using the temperature of the surface water calculated in (4), determine the heat transferred to the surface water from the pavement from Equation 8.19 and Equation 8.13. The temperature of the pavement surface and other parameters such as the Reynolds' number must be known to complete this step.
6. Calculate the runoff temperature by summing T'_{sr} and ΔT_{sr}.
7. Substitute the calculated surface water temperature from (6) into the mass balance in (4).
8. Iterate steps (4) to (7) for the duration of rainfall re-evaluating the rainfall temperature and intensity at each time step.

8.3.1 Sensitivity Analysis

For modeling purposes using varying rainfall intensities as input time series, it would clearly be advantageous to simplify the method, perhaps by eliminating component processes that are less important than others, using environmental parameters to describe pavement materials, and using a reasonable range of paving temperatures as initial or start-up conditions.

Significant parameters may be found by sensitivity analysis:

1. assuming that $\dfrac{\Delta H}{\Delta t} = Q_w$,

2. substituting the energy budget and the heat transfer relationship (Q_w) into the equation converting input heat energy to a change in temperature,
3. examining the sensitivity of the heat storage expression $\left(\frac{\Delta H}{\Delta t}\right)$, and
4. determining the change in pavement surface temperature runoff (ΔT_{sr}) in both situations.

Parameters that would be appropriate when considering an asphalt pavement surface at 35°C were perturbed by adding 10%. The marginal sensitivity coefficient was then determined using the following equation:

$$S_c = \frac{\partial \phi}{\partial \alpha_i} \tag{8.19}$$

where:

S_c = the marginal sensitivity coefficient of the equation;
$\partial \phi$ = the change in the calculated objective function (i.e. the solution obtained using all base parameters minus the solution obtained using base plus the perturbed parameter); and
$\partial \alpha_i$ = the change in the ith parameter.

A spreadsheet used to calculate the sensitivity of the input parameters in the energy budget and heat transfer equations is presented elsewhere (Verspagen, 1995). Table 8.1 presents in rank order the seven most sensitive input parameters.

Table 8.1 Parameter ranking for heat transfer equations.

Rank	Input Parameter
1	T_{sr} surface runoff temperature
2	T_0 pavement surface temperature
3	μ dynamic viscosity of fluid (Reynolds equation)
4	Re_L Reynolds number ρ fluid density u_∞ free stream velocity L flow length (Reynolds number equation)

Thus, from the rather long discussion on processes, we conclude that the dominant processes are associated with the following parameters: *paving temperature*, the *initial water temperature*, the *surface runoff temperature*, the *heat capacity* of the paving, and the *surface runoff flow rate*. Since it is designed to have a high infiltration rate, some difficulty should be anticipated in obtaining the *surface runoff flow rate* for the permeable pavings, and for this reason *rainfall intensity* is used instead. For impervious paving, the surface runoff flow rate is nearly the same as the rainfall rate. Bearing in mind the uncertainty of estimates of flow lengths for the permeable pavers, and of scale-up errors to larger areas with non-planar surfaces, we stress that these parameters merely form the basis of a simplified surface runoff thermal enrichment relationship, one that should be calibrated for local surfaces and conditions. Indeed we recommend that this problem be given further attention.

Moreover, at this point we make a further gross simplification since the *heat capacities* of the pavement materials are not well known parameters, whereas *thermal conductivities* are available (Holman, 1990; Omega, 1992). *Thermal conductivity* provides a measure of the energy transferred per unit thickness per °C gradient, while *heat capacity* provides a measure of the work necessary per kg mass of material per °C. The correlation between *thermal conductivity* and *heat capacity* is expected to be high for paving materials since the materials have similar densities: i.e. as *thermal conductivity* increases, *heat capacity* will also increase. Therefore, for our simplified thermal enrichment model, *thermal conductivity* of the pavement surface is used instead of the little-known *heat capacity*.

In fact, determination of the thermal conductivity of some pavements requires additional research. When dry, the permeable paver consists of very dense concrete interspersed with open drainage cells filled with permeable material. Warming of the pavement surface and heat transfer to the subgrade is affected by the degree of saturation. As the material becomes wetted during rainfall, the thermal properties of the pavement surface change sharply, depending on the rainfall hyetograph and how the infiltration properties of the open drainage cell material have changed with age of the pavement and adjacent land-use. Over months or years, organic matter and dust forms a covering skin in the open drainage cells and significantly changes the infiltration properties from the original design conditions. Qualitative estimates of clogging are difficult to obtain.

8.3.2 Simplified Relations

Heat transfer between paving and the overlying liquid may be described in two ways (joules per second, T_s is a representative temperature of the surface paving material):

either: $$\frac{\Delta H}{\Delta t} = C_s \times \frac{\Delta T_s}{\Delta t} \times \Delta z \qquad (8.20)$$ or: $$Q_w = \bar{h} \times (T_s - T_\infty) \qquad (8.21)$$

Change in temperature of the surface water is based on the known quantity of input energy:

$$\Delta T_{sr} = \frac{Q_w (or \frac{\Delta H}{\Delta t})}{4.2 \times q} \quad (8.22)$$

To describe the temperature trend, these equations should be solved stepwise in time. However, the expression for thermal enrichment during rainfall is expected to be an exponential decay from a high initial starting temperature and slope at the onset of rain. Since cloudy conditions predominate during rain, Q_n may be assumed to be negligible. Because temperature gradients between the paving and the surface water are initially large, the quantity of energy initially transferred from paving to surface runoff will also be large. As the paving temperature decreases, the thermal gradient driving the heat transfer diminishes proportionately. Thus, the paving and the surface runoff temperature drops during the rain. In sufficiently long rains, it would continue to do so until no temperature difference exists between the pavement surface and the runoff, and the paving and surface water runoff temperature will both approach the temperature of the rain. The exponential decay relationship is:

$$T_{sr} = A \ln(t) + B \quad (8.23)$$

where:

T_{sr} = the temperature of the surface runoff;
t = the time following the start of the rainfall (minutes); and
A and B = fitting parameters with units of (°C·time^{-1}) and (°C) respectively; they are functions of the rainfall temperature, pavement surface temperature, heat capacity of the paving, and runoff flow rate.

8.4 The Laboratory and Test Pavements

A central purpose of this study is a comparison of the thermal characteristics of: (a) impervious asphalt and (b) Uni-ecostone (the registered name is given in the acknowledgements at the end) with the 4 inch [10 cm] layer of mixed sand and washed stone. A detailed description of the construction of the test pavements is presented by Thompson and James (1995). Thermocouples were placed at several locations in the subsurface of the lab pavements, which were carefully insulated, as described elsewhere (Verspagen, 1995). Both pavement samples were stored in the laboratory; on sunny days when the samples were to be tested, they were moved outside early in the morning, and remained in the sun until approximately 13:30. Tests rarely continued for longer than 2 hours and the second sample was often brought into the laboratory by 15:30. On average, the samples were exposed to 5 hours of daytime sun.

A propane heater consisting of two 14,000 Btu radiant elements was constructed for use on cloudy or rainy days. The heater was capable of warming the upper layers of the pavement samples to extremely high temperatures in less than two hours. Longer durations were required to heat the full depth of the sample subgrade.

Data was also collected from the external instrumented parking lot pavements, and they showed differences from the lab pavements, as is to be expected. The surface temperature of the lab asphalt pavement was typically 2.3 °C warmer than the surface temperature of the instrumented parking lot asphalt, probably the result of ageing of the parking lot, which had been exposed to weathering since November, 1994. The colour has worn from black to grey by traffic eroding the bituminous coating on the upper surface of the surface aggregate. The laboratory sample is the same age and material, but had been protected and remained dark black. The darker colour of the lab pavement is considered to be the primary factor causing the surface of the test sample to absorb thermal energy more quickly and to reach greater temperatures than the parking lot pavement. Relationships have been developed for the change in emissivity of asphalt as it ages and are available from various asphalt research agencies, such as the asphalt industry and the American Association of State Highway and Transportation Officials (AASHTO).

Comparison of the temperatures immediately below the pavement indicated that the lab pavement was much cooler than the parking lot pavement. This is attributed to the difference in exposure time to direct sunlight. Had this study included another type of subgrade, the thermal conductivity of the subgrade would have had to be included in the analysis, because the thermal conductivity of the subgrade influences the surface temperature.

Surface temperatures for both permeable instrumented pavements were less than that of either asphalt surface.

8.5 Results for Various Rain Intensities

Detailed temperature plots from each experiment for the asphalt and paving stone test samples are presented in Verspagen (1995), and are only briefly summarized here. The plots indicate that the asphalt surface reaches greater surface temperatures than the paving stone surface. The rate of cooling of the asphalt surface runoff is observed to be consistently greater than that observed in the paving stone sample. Both surfaces exhibit similar properties in that the temperature of the surface runoff drops to a relatively constant level before the temperature immediately below the surface equals the temperature at the sub-base/base interface. This indicates that the heat flow to this point remains in the downward direction. A warmer temperature at the sub-base/base interface than immediately below the surface would indicate an upward thermal gradient from the subgrade, contributing to warming of the surface and the surface runoff. Therefore, these results confirm that warming of surface runoff results from release of heat energy solely from the pavement surface. The type of base does not directly influence the cooling portion of the experiment and thus inclusion of the heat capacity of the base is not necessary when developing an expression for thermal enrichment of surface runoff. The *type* of surface medium is less significant under high surface runoff flow rate conditions: the rate of cooling for both samples was observed to be significantly faster for higher than lower intensity tests.

8.5.1 Regression Analysis

The variation of the surface runoff temperature time series was regressed using a logarithmic regression function in Microsoft Excel for each experiment. The regressed equations are of the form:

$$T_{sr} = A\ln(t) + B \tag{8.24}$$

where:

A = decay of the temperature;
B = y intercept (a starting temperature);
t = time (minutes); and
T_{sr} = surface runoff temperature (°C).

Correlation coefficients were calculated to determine the relationship between the fitting parameters A and B and the independent variables: *rainfall intensity*, *thermal conductivity* of the paving, *initial paving surface temperature*, and *initial rainfall temperature*. Initial surface runoff temperature was found to be strongly related to A. Rainfall intensity and initial rainfall temperature were found to have an approximately equal correlation to A and thermal conductivity was found to have the weakest correlation to A. Initial surface temperature was also found to have a strong relationship to the B parameter. Rainfall intensity and initial rainfall temperature parameters were determined to have a smaller correlation to B. Thermal conductivity of the surface was calculated to have a weak relationship to B. Calculated correlation coefficients for the independent parameters and the parameters A and B are presented in Verspagen (1995).

Multiple variable regression analyses were then performed using the dependent parameters A and B and the independent parameters: rainfall intensity, paving thermal conductivity, initial surface temperature, and initial rainfall temperature. The regression was performed using all parameters from 22 of the 31 experiments. Data from the remaining nine experiments were used to verify the accuracy of the regressed expressions. Linear relationships for A and B were then developed:

$$A = 0.0047 \times i - 5.18 \times k_s - 0.13 \times T_{is} + 0.15 \times T_{ir} - 1.55$$

$$B = -0.0294 \times i - 2.26 \times k_s + 0.52 \times T_{is} + 0.07 \times T_{ir} - 14.62$$

where:

A and B = the fitting parameters for the general equation ;
i = the rainfall intensity (mm·hr^{-1});
k_s = the thermal conductivity of the surface (kW·m^{-1}·°C);
T_{is} = the initial surface runoff temperature (°C); and
T_{ir} = the initial rainfall temperature (°C).

The regression analysis indicated that these expressions had R^2 values of 0.88 and 0.79 (F-test: $P \ll 0.001$) respectively.

Values of A and B calculated using the above expressions can be substituted into the general equation (Equation 8.24), to determine the temperature of the surface runoff at any time t (minutes) following the start of the rainfall event.

8.5.2 Accuracy of the Proposed Equations

In total 31 experiments were conducted. Of these experiments, 22 were used to perform a regression analysis to determine linear expressions for the parameters A and B. Data from the remaining 9 experiments were used as an independent data set to validate the accuracy of the proposed equations.

A spreadsheet was written that included the recorded surface runoff temperatures from all experiments. The average absolute error of the estimate was then calculated for each experiment. The average absolute error for all regressed data was determined to be 1.6°C.

The surface runoff temperatures recorded for the data not used in the regression were then compared to the surface temperatures calculated using the regressed expressions for A and B and the general equation. The average absolute error of the estimate was then calculated for each experiment, and the average absolute error for the independent data set was determined to be 1.4 °C.

Average absolute error was plotted against the difference between the initial surface runoff temperature and the initial rainfall temperature, and it indicated that the error for the independent data set is similar to that of the regressed data. The high absolute average error for large differences between the initial surface runoff temperature and the initial rainfall temperature indicates that the regressed expressions should not be used for extreme conditions such as were obtained on days that the propane heater was used and the pavement surface was able to attain very warm temperatures. The rainfall temperature on these days is not considered to be properly representative of an actual day. The error obtained on those days where the pavement samples were warmed naturally, however, generally appears to be within the range of the regressed data.

The error in the estimate of the surface runoff temperature was also plotted for the first hour of those experiments not used in the regression analysis, and it indicated that the surface runoff temperatures in eight of the experiments were initially underestimated. The error in the first 10 minutes is less than 4°C. After ten minutes, the error is generally less than 2°C.

The error analysis indicates that the equations may be used to determine the temperature of the surface runoff to an accuracy of ± 1.5°C. Over the first ten minutes, however, the accuracy is somewhat less and an error margin of ± 4.0°C should be expected.

8.5.3 Sensitivity Analysis of the Proposed Relations

A sensitivity analysis of the regressed equations and the general expression was performed. The results indicate that the initial surface runoff temperature is the most important parameter and must be the most accurately predicted. The initial rainfall temperature and the rainfall intensity are also influential and should be estimated with care. The thermal conductivity is of less importance but indicates that a difference in the temperature of the surface runoff exists between the asphalt and permeable paving.

8.6 Asphalt and Permeable Concrete Pavers Compared

The results of the experiments indicate that the asphalt paving generally reaches greater temperatures than the permeable paving. The regressed expressions were used to compare the asphalt and permeable pavings. Surface temperature data for the instrumented pavements were used. On August 23, 1995 at 13:29 hours, the observed surface temperature of the permeable paving was 36.9 °C and the observed temperature of the asphalt paving was 40.0 °C. These temperatures are considered to be typical of those seen in the summer months of 1995. The expressions developed from the regression analysis were applied using an initial rainfall temperature of 23.0 °C and a rainfall intensity of 115 mm/hr. These initial conditions are very close to the initial conditions used for the experiments run on that day. The initial surface runoff temperature recorded in the laboratory for those experiments were 35.1 and 36.0 °C for the permeable paving and asphalt paving respectively. The applied rainfall temperatures were 22.7 and 23.0 °C. These conditions are considered to be sufficiently similar for an unbiased comparison of the two surfaces.

The calculations were performed using a spreadsheet which is presented in Verspagen (1995). It was apparent that the calculated temperature of asphalt paving is warmer than the calculated temperature of permeable paving throughout the duration of the experiment. These results are consistent with those observed in the experiments conducted on August 23, 1995. The calculated difference is initially 1.9°C and gradually increases through the remainder of the simulation duration. The difference in calculated surface runoff temperatures, however, is within the error margins.

From this example, it may be concluded that surface runoff from the asphalt paving is slightly warmer than surface runoff from the permeable paving. This is consistent with the observed results.

A dominant factor that has not been stressed is infiltration of precipitation into the permeable paving. The temperature of surface runoff from the permeable paving is between 2°C and 4°C cooler than that of the surface runoff from the asphalt paving and itself a noteworthy environmental benefit, but the environmental advantage of the porous pavement is its ability to allow rainfall to infiltrate the surface. Thus, the total thermal loading on receiving waters is reduced.

An example of this can be considered using the data from August 23, 1995. Assuming two parking lots, 70 m on a side (4900 m^2), are equal in all respects except for their pavings. One parking lot has an impervious asphalt paving and the other a permeable paving. Using conservative numbers, a 20% infiltration rate may be assumed for the asphalt and an 80% infiltration rate may be assumed for the porous concrete. Also assume that the parking lots are directly connected to a receiving river flowing at 8.5 m^3/min (5 cfs, approximately that of the Speed River in Guelph) and at a temperature of 20 °C. This parking lot represents a typical parking lot found in many grocery stores or strip malls. Assuming a brief 15 minute storm event occurs in this catchment, the asphalt parking lot would produce 7.5 m^3/min of surface runoff while the porous paving parking lot would produce 1.9 m^3/min of surface runoff.

The impact on the river by the surface runoff from the asphalt, aside from the obvious increase in flow volume, is that the temperature of the water in the river is increased to 26°C for the first two minutes and drops to 23°C after 9 minutes. The impact on the river temperature receiving surface runoff from the permeable paving is that the temperature increases to 22°C and drops to 21°C after 3 minutes. While both surfaces increase the temperature of the surface runoff, the permeable paving is estimated to have a less severe thermal impact on the receiving waters. The stress resulting from a short term, instantaneous, temperature change as estimated for permeable paving is more likely to be within an acceptable tolerance level of aquatic biota. The calculations for this example are presented in Verspagen (1995). Instantaneous changes in temperature such as this are stressors to aquatic life and have an impact on the health of the aquatic ecosystem. The total thermal loading on the receiving water may be considered as the product of the temperature and the surface runoff flow rate: the resulting total thermal loading from the asphalt surface is significantly greater than that of the paving stone surface.

While the thermal impact presented in this example may seem small and to occur over a sufficiently small period of time as to be insignificant, the accumulated impact of many warm, impervious surfaces in a city may be significant.

The aspect of clogging of the porous paving stone surface is often noted. The above example considers the permeable paver to be four times more permeable than the asphalt. Studies regarding the infiltration capacity of aged porous paving stone surfaces are ongoing (Kresin, James and Elrick, 1996).

It is important to note that the relationship in this study was developed for a 1 m by 1 m lab pavement. Full scale testing should be completed to determine the true applicability of this relationship. For example, if a catchment 30 m by 30 m with one central catch basin is considered, this relationship would apply to the outer 1 m perimeter of the catchment. The inner catchment area would receive surface runoff from the outer catchment area. Thus the surface runoff flow rate towards the centre of the catchment could be significantly greater than the outer perimeter. The type of surface and micro-topography of the catchment should be considered. The surface runoff may flow along a narrow strip of the catchment and the relationship would give an accurate estimate of the surface runoff temperature. This is likely to be the case with the permeable paving since the bevelled edges of the individual paving stones encourage flow along the joints rather than the raised surface.

8.7 Conclusions and Recommendations

Further research is required, but the relationship proposed in this study represents a cautious estimate (from an environmental perspective) of the thermal enrichment of surface runoff from the asphalt and paving stone surfaces described in this study. Thermal enrichment of urban stormwater runoff should be considered when new developments are proposed, and thermally-sensitive pavement materials should be used more extensively than is the case now. Specific findings from this study include the following.

- Very little surface water runs off permeable paving in the laboratory.
- Both pavings in this study caused increases in the temperature of the surface runoff, asphalt more than permeable concrete.
- These results confirm that warming of surface runoff results from release of heat energy solely from the pavement surface.
- The rainfall intensity, thermal conductivity of the pavement, initial surface runoff temperature, and initial rainfall temperature are dominant parameters in surface runoff thermal enrichment.

- The expression $\Delta T_{sr} = A\ln(t) + B$ may be used to determine the thermal enrichment of surface runoff from either impervious asphalt or permeable paving stones, where:

$$A = 0.0047 \times i - 5.18 \times k_s - 0.13 \times T_{is} + 0.15 \times T_{ir} - 1.55$$

$$B = -0.0294 \times i - 2.26 \times k_s + 0.52 \times T_{is} + 0.07 \times T_{ir} - 14.62$$

The accuracy of the relationship is ± 4.0 °C in the first 10 minutes after rainfall begins and ± 1.5 °C when averaged over the entire duration of the rainfall event.

Empirical coefficients A and B were regressed for specific conditions, and therefore the results of this study are limited to pavements, climatological and instrumentation conditions encountered in this study. Extrapolation of the results beyond these will require verification. If the initial pavement surface temperatures are extremely warm and the difference in temperature between the rainfall and the pavement surface is large, the evaporative heat component may not be negligible and the relationship developed in this study may not be applicable. Similar limitations apply to continuous modeling methodologies. Research should continue to improve the accuracy of the relationship and further validate the relationship over a range of rainfall intensities.

Application of this or a similar relationship will lead to a better understanding of the thermal impact of new and existing developments. Approval agencies, engineers, and planners may make better informed decisions and choose a more thermally sensitive pavement surface or implement appropriate best management practices to minimize any negative thermal impacts.

Acknowledgements

Research facilities and support were provided by Unilock Canada, Uni-International, and Von Langsdorff Licensing Ltd. Brian. Verspagen did all the hard work and presented the work for an MSc degree at the U of Guelph. Bill Verspagen (his Dad) at the University built the laboratory rigs with considerable ingenuity. Bill James provided the research ideas, supervision, facilities and support funds through an NSERC Grant, and wrote this chapter from Brian's work. UNI-ECOSTONE® is a registered trademark of Von Langsdorff Licencing, Ontario.

References

Ahmed, F. and W. James, 1995. BMP Planner - a Tool for Developing Stormwater Management Plans. *In:* Modern Methods for Modeling the Management of Stormwater Impacts, Computational Hydraulics International. Guelph, ON. pp. 33-50.

Arya, S.P., 1988. Introduction To Micrometeorology. Academic Press, CA. 307 pp.

Galli, J., 1990. Thermal Impacts Associated With Urbanization And Stormwater Management Best Management Practices. Metropolitan Washington Council Of Governments, Washington, D.C. 117 pp.

Holman, J.P., 1990. Heat Transfer. McGraw Hill, Toronto. 714 pp.

Kresin, C., James, W. and Elrick, D. 1996. Observations of infiltration through clogged porous concrete block pavers. In: James, W. (Ed.) Further advances in modelling the management of stormwater impacts. Pub by CHI Guelph. ISBN 0-9697422-7-4. (See chapter x in this book.)

Marshall Macklin and Monaghan Ltd., 1991. Stormwater Quality Best Management Practices. Queen's Printer For Ontario, Toronto, ON. 177 pp.

Monteith, J.L. and M.H. Unsworth, 1990. Principles Of Environmental Physics Second Edition. Chapman And Hall Inc., NY. 283 pp.

Oke, T.R., 1987. Boundary Layer Climates Second Edition. Methuen, NY. 400 pp.

Omega Engineering Inc., 1992. Omega Complete Temperature Measurement Handbook and Encyclopedia, Volume 28. Omega Engineering Inc., Mississauga, ON.

Pluhowski, E.J., 1970. Urbanization And Its Effect On The Temperature Of The Streams On Long Island, NY. Geological Survey Professional Paper 627-D, United States Government Printing Office, Washington, D.C. 58 pp.

Sass, B.H., 1992. A Numerical Model For Prediction Of Road Temperature And Ice. Journal Of Applied Meteorology, American Meteorological Society, Boston, Massachusetts. pp1499-1506.

Shao, J., P.J. Lister., and W.D. Fairmaner, 1994. Numerical Simulations Of Shading Effect And Road Surface State. Meteorological Applications, Cambridge U. Press, NY. 1:209-213.

Shao, J., P.J. Lister, and A. McDonald, 1994. A Surface Temperature Prediction Model For Porous Asphalt Pavement And Its Validation. Meteorological Applications, Cambridge U. Press, NY. 1:129-134.

Thompson, M.K., and W. James, 1995. Provision of Parking-Lot Pavements For Surface Water Pollution Control Studies. *In:* Modern Methods for the Management of Stormwater Impacts. Computational Hydraulics International, Guelph. pp. 335-348.

Verspagen, B. 1995. Experimental investigation of thermal enrichment of stormwater runoff from two paving surfaces. MSc Thesis. U of Guelph. Oct. 1995. Ca 170pp.

Weatherbe, D.G., 1995. A Simplified Stream Temperature Model For Evaluating Urban Drainage Inputs. *In:* Modern Methods For Modelling The Management Of Stormwater Impacts, Computational Hydraulics International, Guelph, ON. pp. 259-274.

Xie, J.D.M. and W. James, 1994. Modelling Solar Thermal Enrichment Of Urban Stormwater. *In:* Current Practices In Modelling The Management Of Stormwater Impacts, CRC Press Inc., Florida. pp. 205-220.

Chapter 9

Contrary to Conventional Wisdom, Street Sweeping Can be an Effective BMP

Roger C. Sutherland and Seth L. Jelen

Recent work suggests that street sweeping programs can be optimized to significantly reduce pollutant washoff from urban streets. The abilities of several different sweeping technologies to pick up accumulated sediment of various sizes were evaluated. In addition, the expected reductions in average annual washoff loads were evaluated using calibrated model simulations of the Simplified Particulate Transport Model (Sutherland and Jelen, 1993) for two stormwater sites in Portland, Oregon.

Results suggest that reductions of up to 80% in annual TSS and associated pollutant washoffs might be achieved using bimonthly to weekly sweepings. Frequencies and associated reductions would vary with patterns of precipitation sediment accumulation and resuspension, but it is clear that sweeping technology can have a profound effect on sweeping results and achieve meaningful runoff quality benefits.

These results stand in sharp contrast to earlier conclusions dating back to December 1983. At that time, street sweeping had been found to be generally ineffective as a technique for improving the quality of urban runoff. This conclusion resulted from the United States Environmental Protection Agency sponsored Nationwide Urban Runoff Program (NURP) in which over 30 million dollars was expended in an intensive three-year investigation of urban runoff quality at 28 locations throughout the United States (USEPA, 1983).

© *Advances in Modeling the Management of Stormwater Impacts - Vol. 5* W. James, Ed.
Pub. by CHI, Guelph, Canada 1997. ISBN 0-9697422-7-4. Fax: +519 767-2770

9.1 Previous Research

The NURP studies of street sweeping effects on stormwater quality (USEPA,1983) concluded that street sweeping was largely ineffective at reducing the event mean concentration (EMC) of pollutants in urban runoff. This conclusion was reached mainly because the street sweepers tested were not able to effectively pick up very fine accumulated sediments that can often be highly contaminated.

In general, street sweeping equipment of the era was unable to effectively pick up the very fine, highly contaminated, sediments that accumulate on impervious areas such as streets, driveways and parking lots. These same sediments, located on paved areas that are directly connected to a city's storm drainage system, have been identified over and over again as the primary source of urban nonpoint pollutants entering the receiving waters of the United States.

Broom sweepers of that era removed litter and large dirt particles well, but contaminants are known to concentrate primarily in the fine particle sizes (e.g. less than 63 microns). However, these finer and much more pollutant-laden particles were largely left behind, and moreover, they were left exposed to be even more readily entrained in washoff since their armoring shelter by larger sediment particles was removed.

However, recent studies by the authors over a period of four years show clearly that the NURP conclusions from the early 80's are no longer valid today. This is largely because of the considerable increase in street sweeping's effectiveness at removing the smallest particles. Examples of this improvement include the following:

1. Even most mechanical sweepers (i.e. broom and conveyor belt) now available are much more effective at picking up fine sediments.
2. Tandem sweeping operations (i.e. mechanical sweeping followed immediately by a vacuum-assisted machine) have been found to be even more effective at fine sediment pickup.
3. Regenerative air sweepers have been refined considerably since their infancy during the NURP era, have also been found to be effective at fine sediment pickup.
4. A revolutionary new vacuum-assisted dry sweeper has greatly advanced the technology of fine sediment pickup and containment.

These considerable advances in sweeping technologies result in a need to re-evaluate the NURP conclusions and incorporate new performance data and benefits that result from more demanding and water-quality-driven sweeping programs.

9.2 Sweeping Technologies

The pickup performance for the NURP era sweepers show typical values based on the authors' previous analysis (Sutherland, 1990) of the Bellevue, Washington NURP data, as summarized by Pitt (1985). Having been a consultant to the City of Bellevue during the NURP study, the author had direct access to the street sweeper pickup performance data collected as part of that study. The sweeper tested at that time was a *Mobil* standard mechanical broom street sweeper, probably manufactured around 1978. It provides the baseline against which several modern street sweeping technologies are compared for immediate pickup rate and expected long-term washoff load reduction.

Against this, the performance of a newer mechanical (i.e. broom and conveyor) sweeper was compared, in order to establish the level of improvement achieved in types of sweepers still in wide use. Data for this comparison was obtained when the authors measured the pickup performance of a newer mechanical sweeper, which was a 1988 *Mobil*, as a result of a Portland study mentioned later.

Research by the authors has identified three promising technologies that may provide significant improvements in performance beyond that observed for NURP era or mechanical sweepers. For each, the sediment pickup from sweepings by each technology was measured in the field by the authors under a variety of conditions. Resulting removals were obtained for each of eight particle size ranges. These show significantly greater removals for each of these new technologies than those typical for sweepers from the early 1980's.

The first technology is the use of a tandem sweeping operation. A tandem operation involves two successive cleaning passes, first by a mechanical (i.e. broom and conveyor belt) sweeper, then immediately followed by a vacuum-assisted sweeper. The pickup performance of a tandem operation using the *Mobil* broom sweeper followed by a *TYMCO* vacuum sweeper was monitored for over a year in a medium-density residential area located in Southeast Portland, Oregon. The detailed description of this study and its results can be found in HDR (1993) and were briefly summarized in Alter (1995).

The second technology is the stand-alone use of a regenerative air sweeper. Regenerative air sweepers blow air onto the pavement and immediately vacuum it back in order to entrain and filter out accumulated sediments. Regenerative air machines were just in their infancy during the NURP era, and to the author's knowledge were not extensively tested at any of the NURP sites. Regenerative air sweepers are generally considered to be good at removing fine sediment, if the accumulated loading is not too great. The authors measured the pickup performance of the Elgin Crosswind regenerative air sweeper in and near Seatac International Airport on April 21, 1995.

The third technology is the stand-alone use of a new, highly effective, vacuum-assisted dry sweeper called the Enviro Whirl I developed and manufactured by Enviro Whirl Technologies Inc., located in Centralia, Illinois. This sweeper applies technology developed and still used to remove spilled coal and coal dust along railroad tracks. The technology has also been applied to clean similar materials from industrial sites where complete removal without leakage of airborne particles is important.

From these demands have evolved a technology that is extremely efficient at removing the finest particles and preventing their escape into the air. In contrast, most other units, especially mechanical types, trail a visible cloud of dust behind in the air and on the street.

The Enviro Whirl I combines the important elements of tandem sweeping into a single unit. It uses rotating sweeper brooms within the powerful vacuum head to provide both mechanical and aerodynamic particulate removal. Data comparing the sweeping performance of this technology to others was measured by the authors on an April 24, 1995 test prepared by the City of Las Vegas, Nevada (during an air quality conference) and in Centralia, Illinois during September 1995.

This data reveals marked improvements in the street sweeping technology that result in much more effective pickup of accumulated sediments. Using the NURP-era broom sweepers as a baseline, performances are compared for improved mechanical sweepers and promising sweeping technologies. As a result, it becomes clear that street sweeping is now capable of removing significant pollutant loads from urban surfaces and effecting significant reductions in urban pollutant washoff.

9.3 Evaluation Procedure

The ability of street sweeping to reduce overall pollutant washoff loads depends on several things. First is the street sweeper's innate ability to remove accumulated sediment. Another is the environmental dynamics of sediment accumulation and resuspension, and of sediment washoff during storm events plus suspended sediment removal by downstream water quality controls.

The Simplified Particulate Transport Model (SIMPTM) can accurately simulate this complicated interaction of accumulation, washoff, and street sweeper pickup that occurs over a period of time (Sutherland, and Jelen, 1993). The remainder of this chapter presents the issues involved in applying the SIMPTM model to successfully evaluate the overall effectiveness of street sweeping technologies and programs as a water quality management practice. The following are addressed:

1. how to model street sweeper pickup performance;
2. how the SIMPTM model compares to real pickup performance data;
3. how various technologies can be compared using their calibrated SIMPTM model parameters; and
4. how technologies can be best compared using their average annual pollutant reductions, as simulated for two example stormwater basin sites in Portland, Oregon.

9.4 Pickup Performance Model

The street sweeping component of the SIMPTM model was based on the results of Pitt's street sweeping study conducted for the USEPA in San Jose, California (Pitt, 1979). This model was confirmed in additional studies conducted in Alameda County, California (Pitt and Shawley, 1982) and in Washoe County, Nevada (Pitt and Sutherland, 1982).

These studies found that sweeping removes little, if any, material below a certain base residual which was found to vary by particle size. Above that base residual, the street sweeper's removal effectiveness was described as a straight line percentage which varied by particle size.

Figure 9.1 illustrates the street cleaning component and equations used by SIMPTM. For each of eight size groups, the amount removed (*Prem*) is related linearly to the initial accumulation (*Po*) using two parameters - a base residual (*SSmin*) and a sweeping efficiency (*SSeff*):

$$Prem = SSeff \times (Po - SSmin) \; for \; Po > SSmin$$

Therefore, to describe a unique street sweeping operation, one simply needs to know the operations *SSmin* and *SSeff* values for each of the eight particle size ranges simulated by SIMPTM. Note that *SSeff* is dimensionless, while that for *SSmin* must match that for accumulation, usually either pounds per curb mile or pounds per paved acre. The initial accumulation *(Po)* is a simulated parameter, or may be measured in the field (from a similar surface near that swept) in order to evaluate the *SSmin* and *SSeff* parameters.

Figure 9.2 shows an example of how this model component actually compares to real pickup performance data for each of the eight particle size groups. The plotted points are the data obtained from monitoring the tandem street sweeping operation on Portland's Sellwood drainage basin (HDR, 1993). Note that the correlation coefficients (R^2) for the fits of the eight particle size

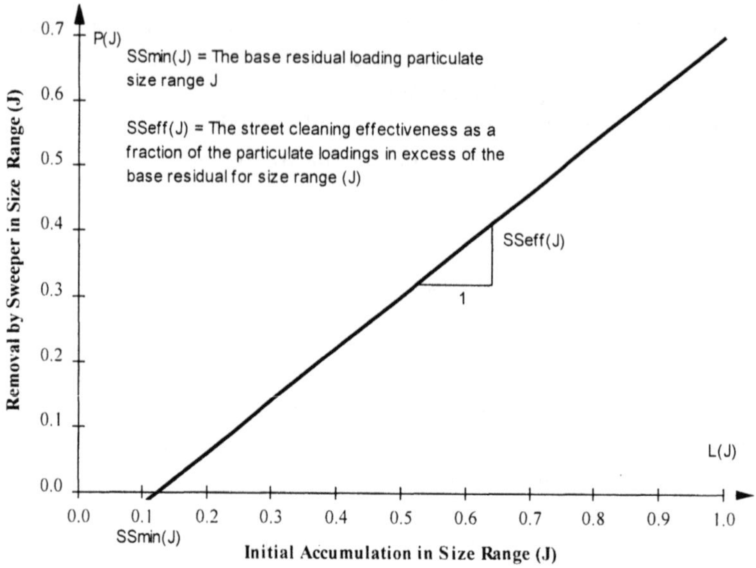

Figure 9.1 Street sweeping model component of SIMPTM.

groups ranged from 94.3% to 99.9%, so the model is doing an excellent job of reproducing the observations. These high R^2 values were typical of all of the model fits to the pickup data from the various sweeping technologies.

Table 9.1 compares the SSmin sweeping parameters calibrated to model each of the five sweeping technologies. It shows dramatic improvements in reducing residual loadings for all the newer technologies when compared to the NURP sweepers. While both tandem sweeping and the Elgin Crosswind regenerative air are very impressive, the across-the-board zero residual loadings for the Enviro Whirl I is the best possible.

Table 9.2 compares the corresponding marginal sweeping rate, SSeff, for sweeping loads that exceed the threshold SSmin. They were also calibrated to model each of the five sweeping technologies. The results mirror those for the SSmin parameter, and show impressive removal efficiencies above the residential loadings. Dramatic improvements are again evident since the NURP era. It must be recognized that this table shows only marginal removal rates. The overall removals must also incorporate the residual loading that always remains after sweeping. Thus although the rates of the Elgin Crosswind (regenerative air) and the Enviro Whirl I for the finer particle size groups may not be impressive, their residual loadings are very low, or even zero, resulting in overall removal efficiencies that are essentially the same as the rate shown. Other technologies with larger SSmin's would be significantly less efficient.

Table 9.1 Calibrated SSmin sweeping residuals for alternative technologies.

Particle Size Group	Size Range *microns*	*Street Sweeping Technology* NURP Mech.	Newer Mech.	Tandem Sweeping	Regenerative Air	Enviro-Whirl
1	<63	9.0	5.8	2.0	0.0	0.0
2	-125	12.0	5.8	2.0	0.0	0.0
3	-250	18.0	5.3	2.3	0.9	0.0
4	-600	18.0	2.5	2.3	1.9	0.0
5	-1000	12.0	0.4	0.8	0.7	0.0
6	-2000	4.2	0.5	0.6	0.7	0.0
7	-6370	3.6	0.3	0.5	0.0	0.0
8	>6370	1.8	0.0	0.0	0.0	0.0

Data from various studies, minimum pounds per paved acre remaining after street sweeping.

Table 9.2 Calibrated SSeff - marginal sweeping efficiencies for alternative technologies.

Particle Size Group	Size Range *microns*	*Street Sweeping Technology* NURP Mech.	Newer Mech.	Tandem Sweeping	Regenerative Air	Enviro-Whirl
1	<63	44%	100%	93%	32%	70%
2	-125	52%	100%	95%	71%	77%
3	-250	47%	92%	93%	94%	84%
4	-600	50%	57%	89%	100%	88%
5	-1000	55%	48%	84%	100%	90%
6	-2000	60%	59%	88%	100%	91%
7	-6370	78%	81%	98%	94%	92%
8	>6370	79%	70%	87%	92%	96%

Data from various studies, marginal removal rate *only* for accumulations *greater* than SSmin.

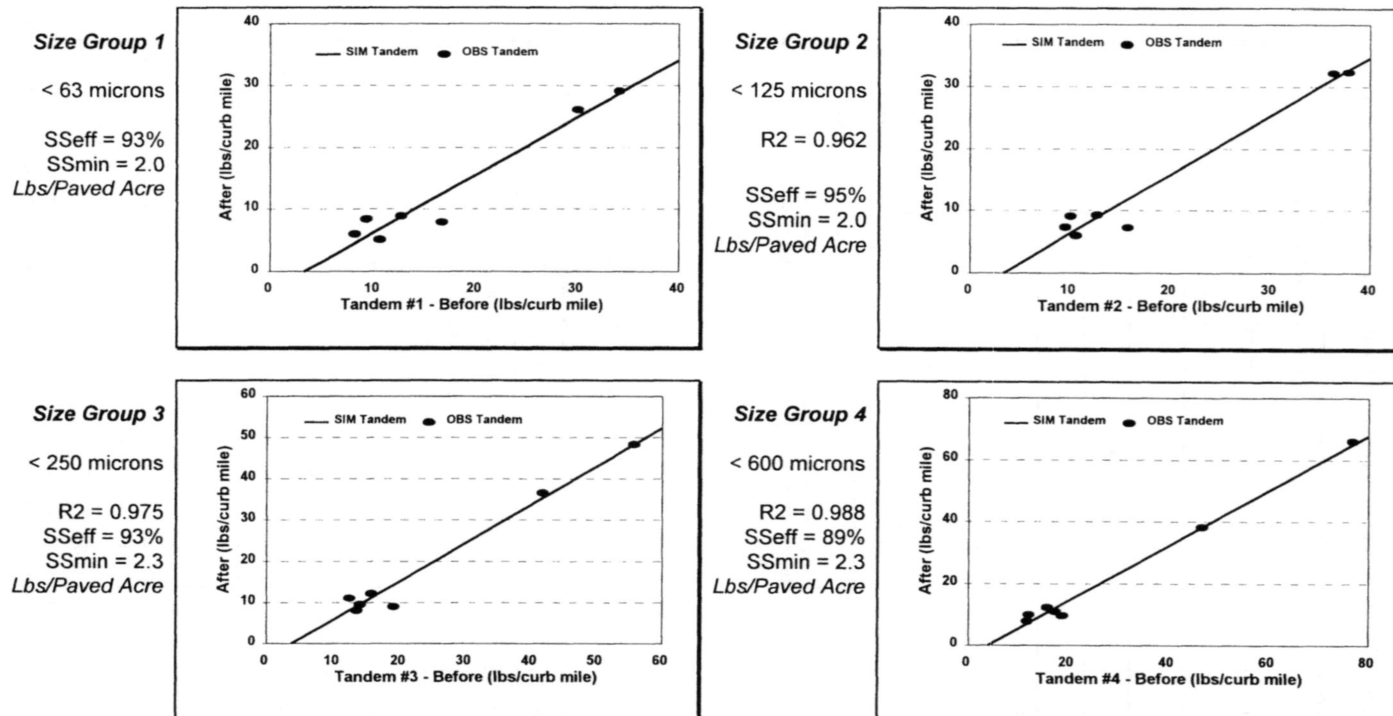

Figure 9.2 Tandem street sweeping model in SIMPTM, size groups 1-4.

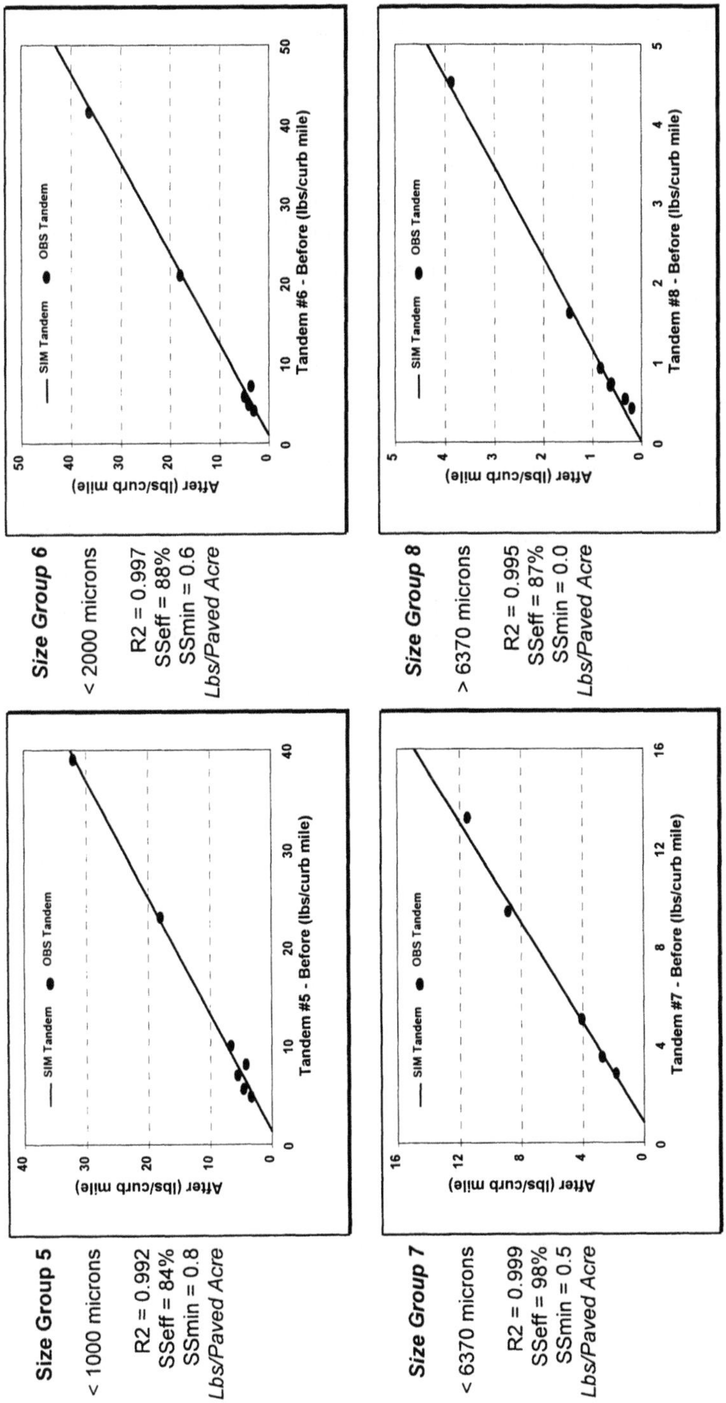

Figure 9.2 (continued) Tandem street sweeping model in SIMPTM, size groups 5-8.

9.5 Pollutant Washoff Reduction Comparison

Working with a calibrated version of the SIMPTM program, the average annual expected reduction in total suspended solids (TSS) washoff from two of Portland's NPDES stormwater sites were projected for varying sweeping frequencies using the NURP era sweepers, the new mechanical sweeper and the three promising sweeping technologies. (For a more detailed description of the SIMPTM program and its calibration to the City of Portland's NPDES monitoring sites, the reader is referred to the program documentation or the study report (Sutherland and Jelen, 1995).

Figure 9.3 shows the resulting curves of expected annual washoff reductions for varied intensity of street sweeping in residential areas by each of the alternative technologies. It clearly shows that all of the newer sweeping technologies would be significantly more effective than the NURP era sweepers in reducing TSS washoff from single family residential areas with curb and gutter drainage in Portland, Oregon. Note that the Enviro Whirl is the best, followed by the Elgin regenerative air and the tandem operation. Even the newer mechanical sweepers will provide reductions in the 20% to 30% range. Also note that weekly or biweekly sweeping appears to be optimum for this type of land use in Portland, Oregon.

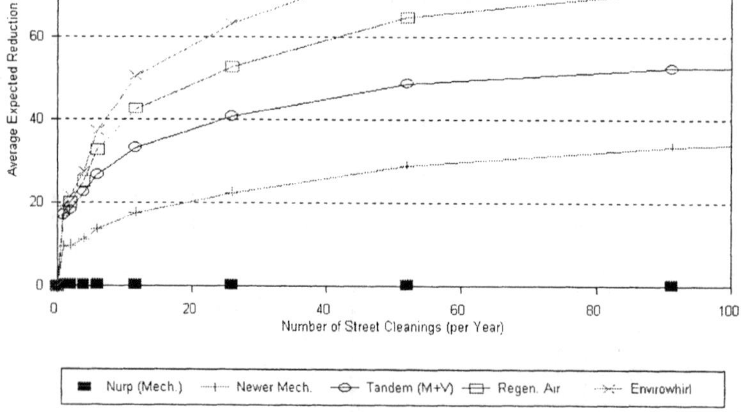

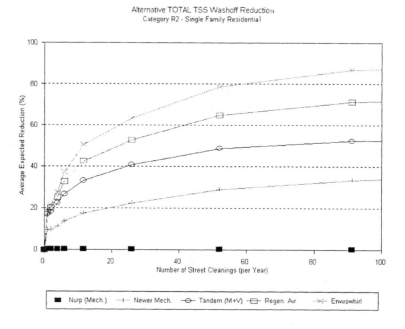

Figure 9.3 Alternative washoff reductions by sweeping residential streets.

Figure 9.4 shows how results change significantly when sweeping is applied to major arterials instead. It even more clearly demonstrates the superiority of the Enviro Whirl I sweeper in reducing TSS washoff from highly impervious major arterials with curb and gutter drainage in Portland, Oregon. The Elgin regenerative air provides some TSS reduction, whereas the other technologies appear to be largely ineffective on this type of land use. This same land use was found to provide the highest pollutant washoffs on a pound per paved acre basis of the six homogenous land uses studied (Sutherland and Jelen, 1995).

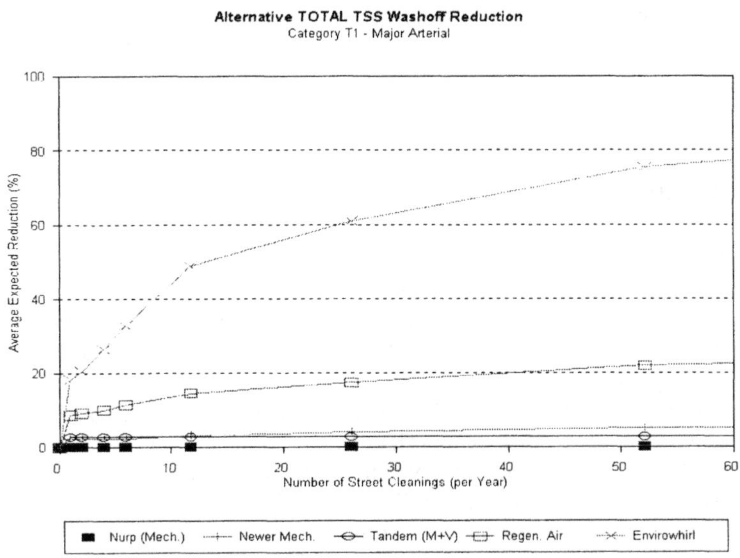

Figure 9.4 Alternative washoff reductions by sweeping major arterials.

Clearly, though, both figures show that the NURP era sweepers were almost totally ineffective in their ability to reduce TSS washoffs from either of the basins simulated. So this confirms the earlier conclusions of the NURP in regard to sweeper performance, while suggesting that significant benefits could now be expected.

9.6 Conclusions

Contrary to conventional wisdom, this chapter clearly demonstrates that street sweeping can be an effective best management practice (BMP). The actual pollutant reduction effectiveness of any given street sweeping operation will depend on characteristics of land use, precipitation, and the accumulation dynamics of contaminated sediments.

The SIMPTM program has been used successfully to account for all of those issues in order to project the potential performance of various street sweeping programs. It was used to evaluate the optimal level of effort to be implemented. Finally, it was used to evaluate the effect of employing updated technologies. In this regard, the Enviro Whirl I sweeper was found to be far superior to the other promising technologies reviewed.

Given the increased concern about the water quality related impacts of urban stormwater pollution throughout the country and the difficulty of identifying and implementing cost-effective BMP's to address them, the pollutant reduction benefits possible from a cost effective street sweeping program must be re-evaluated.

References

Alter, W., 1995. "The Changing Emphasis of Municipal Sweeping . . . May be Tandem," American Sweeper, Volume 4, Number 1, p6. 3pp.

HDR, Inc., 1993. Combined Sewer Overflow SFO Compliance Interim Control Measures Study and Final Report, prepared for the City of Portland, Bureau of Environmental Services, p17-1. 19pp.

Kurahashi and Associates, Inc., 1995. Seatac International Airport Stormwater Quality Characterization, Memorandum to HDR Engineering Inc., 53pp.

Pitt, R.E., 1979. Demonstration of Nonpoint Pollution Abatement Through Improved Street Cleaning Practices, EPA 600/2-79-161, 270pp.

Pitt, R.E.,1985. Characterization, Sources and Control of Urban Runoff by Street and Sewerage Cleaning, Contract Number R-80597012, U.S. Environmental Protection Agency, Offices of Research and Development, 467pp.

Pitt, R.E. and G. Shawley, 1982. A Demonstration of Nonpoint Pollution Management on Castro Valley Creek, Alameda County Flood Control and Water Conservation District, Hayward, California, 173pp.

Pitt, R.E. and R.C. Sutherland,1982. Washoe County Urban Stormwater Management Program - Volume II Street Particulate Data Collection and Analysis, Prepared by CH2M Hill for Washoe Council of Governments, Reno, Nevada, 124pp.

Sutherland, R.C., 1990. Water Quality Related Benefits to the City's Current Street Cleaning Program - Phase 2 Results, letter to Ms. Lori Faha, City of Portland, Bureau of Environmental Services, 12pp.

Sutherland R.C. and S.L. Jelen,1993. Simplified Particulate Transport Model-Users Manual, Version 3.1, 66pp.

Sutherland, R.C. and S.L. Jelen, 1996. Sophisticated Stormwater Quality Modeling Is Worth the Effort. Published in Advances in Modeling the Management of Stormwater Impacts, Edited by Dr. William James, Ann Arbor Press, p1-14.

U.S. Environmental Protection Agency, Water Planning Division, 1983. Results of the Nationwide Urban Runoff Program, Volume 1 - Final report, 186pp.

Chapter 10

Observations of Infiltration Through Clogged Porous Concrete Block Pavers

Christopher Kresin, William James and David Elrick

James and Verspagen (1996), Thompson and James (1995), and Shahin (1994) have observed low runoff volumes from porous concrete paver laboratory test blocks used in their respective research. However, the laboratory test blocks were not subjected to wear or the deposition of pollutants over time on the surface and, therefore, perform under optimum conditions. The purpose of this research is to test the hypothesis that, for a particular permeable paver hereinafter called Uni-ecostone (see acknowledgements at end for trademarks), infiltration capacities decrease with age and certain land uses, and that infiltration capacities may be improved by simply street sweeping and/or vacuuming the surface. The research uses data collected at several Uni-ecostone porous concrete paver installations.

Permeable pavement helps reproduce the pre-development hydrologic regime at urbanized sites (Schueler, 1987). In achieving this, the key is to provide a surface infiltration capacity which allows an adequate volume of stormwater runoff to be captured by the facility. Such an infiltration capacity is dependent upon factors such as surface slope, and surface ponding. There is little difficulty in designing and constructing a system to provide appropriately high infiltration capacities; however, maintaining these infiltration capacities over several years has proven to be challenging.

© *Advances in Modeling the Management of Stormwater Impacts - Vol. 5* W. James, Ed.
Pub. by CHI, Guelph, Canada 1997. ISBN 0-9697422-7-4. Fax: +519 767-2770

10.1 Introductory Background

Permeable pavements, meaning the complete pavement structure from the porous pavement surface to the underlying native soil, provide control of both stormwater quantity and stormwater quality. Stormwater runoff volumes may be reduced by as much as 80% (Schueler, 1987) and peak flow rates lowered. Quality parameters controlled include suspended solids, nutrients, BOD, turbidity, and temperature (Schueler, 1987; James and Verspagen, 1996); however, seasonal effectiveness is low due to higher springtime runoff volumes and constituent concentrations. Schueler (1987) and Pitt (1996) report that the potential for groundwater contamination as a result of stormwater infiltration is slight. Along with providing the advantages associated with stormwater infiltration, permeable pavement installations reduce land consumption and down-size stormwater conveyance systems.

Urbonas and Stahre (1994) describe the following three types of porous pavements: 1. porous asphalt pavement (PAP), 2. porous concrete pavement (PCP), and 3. modular interlocking concrete blocks (MICB) of the internal drainage cell type (MICBIC, e.g. Turfgrass). Not included with the above is the modular interlocking concrete block with external drainage cells (EDCs) (MICBEC, e.g. Uni-ecostone). PAP and PCP may be described as "no fines" asphalt and concrete pavement mixes; they are typically composed of aggregate that has been purged of finer particles; stormwater infiltrates the surface through the resulting voids in the mix (Maryland Department of Natural Resources, 1984; Northern Virginia Planning District Commission, 1992; Marshall Macklin Monaghan, 1994).

The primary disadvantage of a porous pavement is surface clogging. PAP and PCP have been observed (Urbonas and Stahre, 1994) to clog and seal within one to three years of construction. Concrete block paving, however, has overcome the difficulties of surface clogging (Field, 1984, cited in Bedient and Huber, 1992). PAP and PCP are at a further disadvantage in comparison to MICBEC pavements in that once they are sealed they must be replaced in their entirety whereas only the drainage cell material need be replaced in MICBEC installations.

10.2 Previous Research

Little investigation into long-term infiltration capacities provided by the Uni-ecostone has been carried out. Four relevant studies utilized rainfall simulators.

Borgwardt (1994) conducted experiments at two Uni-ecostone installations (2 years and 5 years in age), in much the same manner as this study, in an attempt to determine the infiltration capacities of the sites and investigate the effects of age on these capacities. Borgwardt's findings at a 5-year and a 2-year old site provided infiltration capacities of 1.2 and 2.4 mm/hr. Constant rates were observed to occur at rainfall durations between 10 and 30 minutes.

Muth (1988) conducted laboratory investigations using 2 m x 2 m Uni-ecostone plots under varying slopes and rainfall intensities. No surface runoff was observed for rainfall intensities up to 72 mm/hr at a 0% slope and for rainfall intensities up to 36 mm/hr at a 2.5% slope. After 20 minutes an infiltration capacity of 1.8 mm/hr was observed. Clark (1980) also found, in his investigation of standard porous (e.g. turfstone) pavers, that a change in slope from 1 to 2% has little effect on runoff volumes.

Phalen (1992) investigated permeabilities of different Uni-ecostone drainage cell material types. Phalen's results for sand alone (33.0 - 68.6 mm/hr) compare well with Muth's sand-covered test plot result (71.1 mm/hr); from this it may be concluded that the finer-grained sand controls infiltration capacity (Rollings and Rollings, 1993). Tests of increasingly coarse drainage cell material result in permeabilities ranging from 475.0 to 4368.8 mm/hr (Phalen, 1992).

Shackel (1995) conducted laboratory tests into the suitability of the Uni-ecostone for rainfall intensities higher than those common in Europe. Infiltration under different combinations of five bedding, jointing, and drainage cell materials, ranging from 2 mm sand to 10 mm gravel, was investigated. Results show that uniform and clean 2-5 mm gravel provides the highest infiltration capacities (216 mm/hr) and ensures joint-filling. The addition of fines or sands to the drainage material was also found to substantially reduce infiltration capacities.

10.3 Surface Crusting

The hydraulic conductivity of a soil is often more than that of the surface layer, due to crusting at the surface and its limiting affect on infiltration rates (Bosch and Onstad, 1988; Ferguson, 1994). Mohamoud et al. (1990A) also conclude that surface effects (e.g. residue cover and crusting) are more significant than soil hydraulic properties when considering infiltration capacities. Surface crusts are fine-textured in comparison to the texture of the underlying soil, thus creating larger matric forces. These matric forces may be large enough that flow into the relatively larger pores of the underlying soil is not permitted. Under conditions of large head excepted, the underlying layer does not become saturated due to the limiting effect of the surface crust (Ferguson, 1994).

Crusts are created when soil aggregates break down and enter the soil matrix with infiltrating water, or fine particles are deposited on the surface by runoff and subsequently compacted. Interaggregate macropores become filled as deposition continues; however, once a crust is established, further decline in infiltration capacity is not observed (Behnke, 1969; Norton et al., 1986). Typically, crusts are less than 2 mm thick, appear slick in comparison to the underlying soil (Ferguson, 1994) and consist of a compact skin (0.1 mm thick) and a region of compacted fines (Ahuja and Swartzendruber, 1992). Ferguson (1994) describes two types of soil crusts:

1. the formation of *structural crusts* is affected by soil clay content, mineralogy, organic matter content as well as the chemical constituents in the soil water, and
2. *depositional crusts* form through sediment deposition by surface water flow (Shainberg, 1992).

Of particular importance in the latter process is the deposition of clay particles. Soils are also subject to biological clogging (Allison, 1947); the presence of decomposable organic matter assists this process. Duley (1939), while investigating the effects of surface cover on infiltration capacity, observed an infiltration capacity of 3.05 cm/h decrease to one fifth of this value when the soil surface was exposed to compaction from falling raindrops. A 1 mm thick surface crust had formed which, when removed, returned the soil's infiltration capacity to 4.09 cm/h. In urban settings crust formation is affected by increased mechanical wear and the deposition of rubber, brake dust, and petroleum products due to automobile traffic. To summarize, according to the literature reviewed here, crust formation is influenced by:

- soil composition (e.g. clay content);
- antecedent moisture conditions;
- structural state of the soil;
- rainfall characteristics;
- surface slope;
- surface roughness;
- soil wetting and drying cycles; and
- soil freeze-thaw cycle.

10.4 Spatial Variability and Scale Effects

Spatial variability is inherent in many hydrologic processes; simply stated, a spatially variable parameter is one which differs from point to point. An understanding of spatial variability and scale effects is imperative when investigating hydrologic processes. Thus, it is important to recognize that representing a spatially variable parameter with a value obtained from a point reading or measurement is likely to be incorrect. Several factors jointly affect infiltration capacity making it a highly variable parameter spatially (Smith, 1982). Such parameters must be represented as spatially averaged means (SPAMs). Examples of SPAM parameters used in hydrologic calculations are hydraulic conductivity (K) and rainfall intensity (i). Representing variables as SPAMs should be done carefully. Spatial variations which affect SPAMs at MICBEC installations include compaction (e.g. wheel ruts), vegetative growth, and heavier sediment deposition in low lying areas.

Scale effects are grouped into three sources by Song and James (1992): *Variabilities* include weather, topography, and geology; *discontinuities* amplify the variabilities at boundaries such as those separating soil types; and *processes* (e.g. infiltration) further amplify variabilities. The above sources of scale effects lead to a correlation between scale and heterogeneity; greater heterogeneity requires smaller optimal scales. Scale may be defined as either *laboratory*, *hill-slope*, *catchment*, *basin*, or *continental/global* (Song and James, 1992). When considering infiltration capacities at the test plot scale, the scale must be such that effects of the characteristics of individual plots at each MICBEC installation are not directly reflected in the result; thus, test plots must be randomly assigned. Usual scale effects at the test plot scale relate to processes occurring at the plot boundary and inhomogeneities in the upper soil mantle. Boundary processes include leakage from the test plot and the lateral flow component of infiltration which results during flow from a ponded source. Inhomogeneities typically result from sediment deposition, vegetation, desiccation cracks, as well as structural and thermal deformations.

10.5 Experimental Methodology

Infiltration capacity was measured at a number of plots at installations of different ages. A portable rainfall simulator (shown in Plates 10.1 and 10.2) was used to apply two constant rate rainfalls to 0.7 m^2 test plots (similar to Borgwardt, 1994); 2 cm of head was allowed to pond during both events. Following the second event, observations of decline in head over time were recorded. Applying two rainfalls accounts for initial losses to soil wetting, saturating the drainage cell material. These tests are point measurements; thus, they are highly spatially variable. In an attempt to provide for spatial variabilities and their effect on infiltration capacities, test plots at two plot types were targeted. Travelled (compacted) and untravelled (not compacted) plots were differentiated from each other by qualitatively evaluating the test installations. In order to investigate the potential for regenerating infiltration capacity, the top five mm of EDC material was removed from each EDC within several test plots and the tests repeated. A total of 60 tests were completed at four sites, results from two of the test sites are presented in this chapter.

At Site 1 (Parking lot P10, University of Guelph) adjacent land use consists of asphalt parking areas as well as a small grassed area from which red pine trees overhang the test sites depositing their needles on the MICBEC surface. Site 2 is the Belfountain conservation area parking lot in Belfountain, Ontario. This

Plate 10.1 Portable rainfall simulator, horizontal surface.

Plate 10.2 Portable rainfall simulator, sloping surface.

parking lot is not utilized in the winter and is therefore not subjected to a rigorous winter maintenance program or winter deposition from automobiles. Adjacent land use at this site consists of green space, which includes grass and tree areas, mulch-covered planting beds, and a gravel parking area up-slope from the MICBEC installation. Site 1 was in service for three years and Site 2 for one year.

Infiltration into a porous pavement constructed of Uni-ecostone is limited by the infiltration capacity of the finest layer of material. "*If all sediments are uniform or the deeper sediments are more permeable than those near the surface, and the water table is at considerable depth, the infiltration rate is controlled by the sediments near the surface.*" (Hannon, 1980). Inevitably, EDCs will clog. This is due mainly to surface deposition and the fact that most filtering occurs in the top 50 mm of soil (Ferguson and Debo, 1990). With no maintenance practice in place, a cap which may become impervious, will eventually form. In these cases, it is most likely that the infiltration capacity of the installation will be governed by the infiltration capacity of the EDC cap.

EDCs are delineated by the MICBEC pavers which form them. The MICBEC pavers themselves are impervious; thus, water which infiltrates the EDCs flows vertically through the pavement layer. Therefore, the lateral flow component is considered to be negligible and flow through the EDCs is considered to occur vertically downward. Percolation may be slowed at textural interfaces such as that between the EDC and subgrade materials. However, under saturated and ponded conditions, these effects are accounted for. Once water has percolated through the EDC material and enters the subgrade material it is no longer restricted to vertical flow only, a lateral flow component is introduced. Since the infiltration capacity in this case is limited by material where the flow is vertical, the lateral flow component need not be considered. In order to ensure that the effects of the subgrade are not a factor, the test plot must be either vertically isolated to a depth into the subgrade material or the pavers removed with the drainage cell material kept in-place. Isolating the plot to a depth into the subgrade is not practical as a large area must be excavated and removing a test plot of pavers is inherently risky as the drainage cell material may be disturbed.

Using the data obtained during the field investigations (head versus time) Darcy's infiltration theory was applied to determine infiltration capacities. In order to apply Darcy's theory the following assumptions are made:

- excess rainfall is assumed to act as a storage volume, having uniform depth, over the test plot (Mohamoud et al., 1990);
- matric forces within the drainage cell material are negligible. Prill and Aronson (1978) observed that the moisture content below infiltration basins stabilizes after small ponding periods and Day (1978) employed saturated test plots to ensure baseline soil moisture contents;
- since soil matric forces are negligible, the total hydraulic gradient results from the hydraulic head caused by water ponded on the surface;

- the drainage cell material is assumed to have a homogeneous, stable profile which allows for steady, gravity-induced vertical infiltration approaching a rate equal to Ks. Mein and Larson (1973) and White et al. (1982) assumed similar one dimensional infiltration in the development of their models; and
- flow in the subgrade has a negligible affect on infiltration through the EDCs.

A key factor in this investigation is that the simulated rainfall be applied at a rate greater than the initial infiltration capacity of the test plot, thus ensuring an excess water supply at the surface. Also, since infiltration capacity of the surface crust is assumed to be limiting, the model interprets the crust thickness as the depth of the EDC material. Temperature effects are assumed to be negligible over the ranges encountered throughout the test periods.

Darcy (1856) showed that the specific discharge (flow rate over the cross sectional area of soil through which the flow is occurring) of water through a porous soil media is directly proportional to the difference in hydraulic head over a defined length of soil sample. Darcy also observed that, for different soil types, the specific flow rate differed. This required the introduction of a proportionality constant which essentially provides for matric forces affecting infiltration. Hydraulic conductivity (K) is that constant. Darcy's law is manipulated to reflect falling head infiltration:

$$Ks = [\ln(\frac{(L + Ho)}{(L + H)}) \times \frac{L}{(t - to)}] \quad (10.1)$$

where:

L = crust thickness (length)
H = head at time of concern (length)
Ho = head at previous time (length)
$(t\text{-}to)$ = length of time step (time)
Ks = saturated hydraulic conductivity (length per time)

Ensuring 100% saturation in field studies is difficult as, inevitably, air is trapped within the soil matrix and air flows in an upward direction against the infiltrating water. Thus, Ks is termed the field-saturated hydraulic conductivity (Elrick, 1996), or effective infiltration capacity (fe).

10.6 Results and Discussion

Values of fe determined for the recession period of each individual test plot should, in theory, be equivalent; assumptions of soil homogeneity and saturated flow imply this. Therefore, mean fe values for the recession period are determined for each test plot (these values better represent the effective infiltration capacity of a site during a rainfall event) in order to subsequently determine the overall *f*e values (*f*E) for each plot type. Once *f*E values for each plot type have been determined they are used to produce a value approximating the overall *f*E (*f*Eo) for each site. To do so, the percent travelled and percent untravelled areas at each site must be evaluated so that a weighted average of *f*E-travelled (*f*t) and *f*E-untravelled (*f*u) may be produced. At Site 1, 25 from a total of 81 1 m square grid squares were evaluated during site characterization. Of these grids, thirteen were defined as travelled and twelve as untravelled. Using this information, Site 1 is characterized as 52% travelled and 48% untravelled, similarly Site 2 is characterized as 51% travelled and 49% untravelled. Using the above, *f*Eo for Site 1 is 5.8 mm/h and that for Site 2 is 14.9 mm/h.

Evaluation of EDC material is done so as to provide a comparison of EDC hydraulic conductivities as well as to determine the EDC material installed during construction. This is achieved based on grain-size distribution plots of the EDC material extracted from randomly assigned plots. Bowles (1992) details the methodology used to produce grain-size distribution plots with reference to the American Society for Testing and Materials (ASTM) and the American Association of State Highway and Transportation Officials (AASHTO) standard methods (ASTM D421 and D422 and AASHTO T87 and T88). Prior to grain-size analysis, organic matter (OM) in the EDC material was burned off using a muffle furnace.

Attempts at regenerating infiltration capacities were made by removing the top 5 mm of EDC material (literature suggests that most clogging occurs in the top thin layer of soil), termed the EDC crust, from each EDC within several test plots. The computational methods previously described here also applied to the test data from the regenerated plots. Table 10.1 provides a summary of relevant values for both sites.

Table 10.1 Summary of test results.

Site #	*f*Eo	EDC % passing #200	EDC % OM	Regenerated *f*Eo	Crust % Passing #200	Crust % OM
1	5.8	6.6	0.19	7.7	15.9	0.58
2	14.9	1.9	0.39	40.0	5.5	0.62

Only Borgwardt (1994) has focused on the effects of aging on *f*Eo at Uni-ecostone installations. Borgwardt's findings are in Table 10.2. Both sites targeted by Borgwardt were parking lots,

Comparing these sets of results, it is evident that *f*Eo is more a function of the fines content of the EDC material than of the age of the site. Borgwardt (1994) attributes the lower *f*Eo of the 2 year old site to the higher percent passing the

Table 10.2 Summary of Borgwardt's (1994) findings.

Site Age (yr)	EDC Organic Matter (%)	% Passing #200 Sieve (%)	*f*Eo (mm/h)
2	0.9	7.0	1.2
5	0.5	3.5	2.4

number 200 sieve. Thus, sources of fine materials (e.g. adjacent land uses and winter maintenance practices) are important factors in the decay of *f*Eo. Also of importance is the grinding and crushing action of traffic as this increases fine materials at the surface and the potential for clogging. From Table 10.1 it may also be assumed that, since both the OM content and the percent passing the number 200 sieve for Site 1 are less than those values for Site 2, compaction of the EDC material (increased compaction with increasing age) also affects *f*Eo.

A two-way analysis of variance (ANOVA) of the data from this study may be used to strengthen this assumption. As Table 10.3 indicates, *f*Eo at sites of different ages and different degrees of compaction (travelled and untravelled) differ significantly (6.69 and 6.54 > 4.75). Since the interaction between age and compaction is not significant (2.22 < 4.75), the decrease in *f*Eo is completely explained by the main effects (age and compaction).

Table 10.3 ANOVA table for *f*Eo's at sites 1 and 2.

Source of Variation	Sum of Squares	Degrees of Freedom	Mean Square	Fo	Fcrit*
Age	543.39	1	543.39	6.69	4.75
Compaction	531.19	1	531.19	6.54	4.75
Interaction	180.52	1	180.52	2.22	4.75
Error	974.39	12	81.20		

* Fcrit taken at the 0.05 level (95%)

Results presented by both Muth (1988) and Phalen (1992) may be used to compare *f*Eo's based on the grain-size distribution of the EDC material. Grain-size distribution plots show similar trends, seemingly coinciding with Muth's and Phalen's results where coarser materials are concerned but containing more fines. Thus, values for *f*Eo are lower than those found by the previous researchers. This is also consistent with Shackel (1995) who concluded that the addition of fines or sands to the EDC material reduces *f*Eo. Grain-size analysis in this study suggests that EDC materials used in construction were not to manufacturer specifications.

Knowing where fines collect in the permeable pavement is important. If fines settle in any location other than the top layer of the EDC material (the crust), any degradation in *f*Eo is permanent as efforts to remove fines in other locations is not likely due to expense. Grain-size analyses of the crust material removed from Sites 1B and 2 reveal that this material has both a higher OM and fines content than the EDC material. Plotting these results (Figure 10.1) with those found by Muth (M) (1988) and Phalen (P) (1992), an excellent agreement between P: 33.02 mm/h and Site 2 whereas the higher fines content of Site 1B's crust material obviously affects *f*Eo.

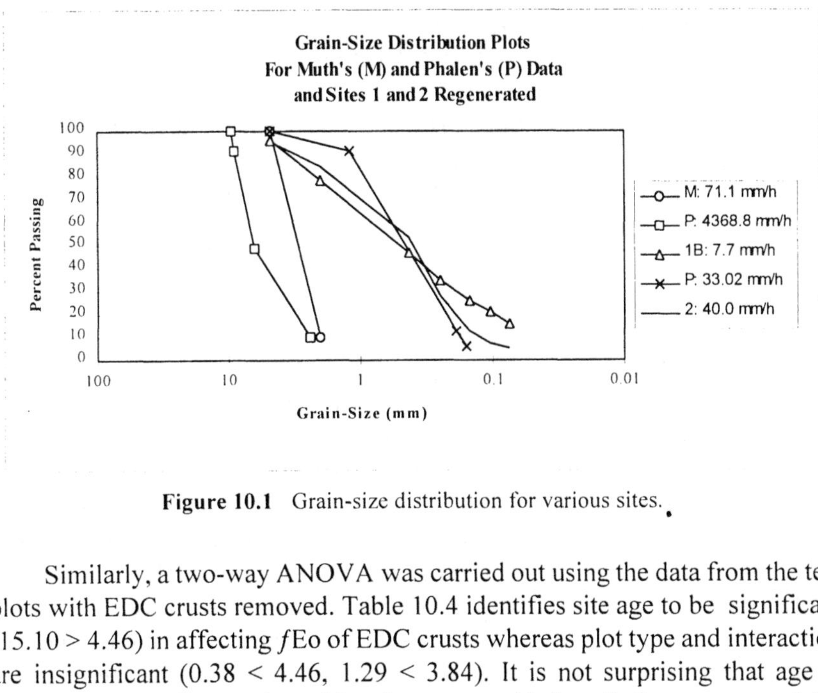

Figure 10.1 Grain-size distribution for various sites.

Similarly, a two-way ANOVA was carried out using the data from the test plots with EDC crusts removed. Table 10.4 identifies site age to be significant ($15.10 > 4.46$) in affecting *f*Eo of EDC crusts whereas plot type and interaction are insignificant ($0.38 < 4.46$, $1.29 < 3.84$). It is not surprising that age is significant as sediment deposition increases with installation age especially when sites (e.g. Site 1) are subjected to winter deposition.

Several researchers conclude that degraded infiltration capacities at PAP/PCP (Schueler, 1987) and MICBEC (Muth, 1988) installations may be regenerated through periodic sweeping and vacuuming of the surface. Schueler (1987) notes that regeneration is not possible if the PAP/PCP surface is clogged with large sediments as sweeping and vacuuming is unable to pull these particles from the pores. Results from this study support reports by Uni-group USA (1993) that sediment accumulation reduces the infiltration capacity provided by Uni-ecostone

Table 10.4 ANOVA table for test plots with EDC crusts removed.

Source of Variation	Sum of Squares	Degrees of Freedom	Mean Square	Fo	Fcrit*
Age	3088.02	1	3088.02	15.10	4.46
Compaction	77.52	1	77.52	0.38	4.46
Interaction	264.14	1	264.14	1.29	3.84
Error	1636.89	8	204.61		

* Fcrit taken at the 0.05 level (95%)

installation (see Table 10.1). Uni-group recommends that the surface be swept and vacuumed on a 4 year cycle; this maintenance procedure has met with success in Germany. Results presented herein show that *f*Eo is able to be regenerated. However, at Site 1 it is evident that the EDC's are clogged to a point where only a fraction of the original *f*Eo could be regenerated, raising questions as to an appropriate maintenance frequency. It should be noted that this site was not constructed according to current (1996) drainage cell material specifications.

10.7 Conclusions

The nationwide urban runoff program (NURP), in the United States, identified the extent of impervious surfaces directly connected to the drainage system as the most important factor affecting runoff volumes after volume of rain (Pitt, 1996). In urban settings runoff volumes are ten-fold those in predevelopment areas (Madison et al., 1979), directly reflecting the NURP findings. Thus, in order to reduce stormwater runoff volumes in urban areas, impervious area must be reduced. The application of permeable pavement is one method. Results from this study show that a significant relationship exists between overall effective infiltration capacity (*f*Eo) and age; as Uni-ecostone installations age, *f*Eo decreases. Also significant is the relationship between *f*Eo and the degree of compaction (defined as travelled or untravelled). *f*Eo can be improved by removal of the top layer of external drainage cell (EDC) material, the EDC crust. Conclusions drawn from the results include:

1. Very little surface water runs off newly laid Uni-ecostone permeable paving. Infiltration capacity of Uni-ecostone modular interlocking concrete blocks with external cells (MICBEC) decreases as the installation ages.
2. Infiltration capacity at Uni-ecostone installations decreases with increased compaction.

3. Infiltration capacity of the EDC crusts, found to be significantly affected by age, limits *f*Eo.
4. *f*Eo may be regenerated, most probably to some fraction of the initial *f*Eo, by street sweeping/vacuuming the Uni-ecostone surface.
5. *f*Eo is affected to a greater extent by EDC fines content than OM content.

Acknowledgements

Unilock Canada and von Langsdorf Licensing supported this project financially. The School of Engineering technical staff was invaluable in helping prepare required equipment and Kristi Rowe offered excellent support during the data collection phase of the study. Brain Hohner at Agriculture Canada provided the portable rainfall simulator.

Uni-ecostone is correctly written UNI ECO-STONE®, a registered trademark of F. Von Langsdorff Licensing Ltd. Canada.

References

Ahuja, L.R. and D. Swartzendruber. 1992. Flow Through Crusted Soils: Analytical and Numerical Approaches. In: *Soil Crusting, Chemical and Physical Processes*. M.E. Sumner and B.A. Stewart, editors. Lewis Publishers, Chelsea, Michigan. pp. 93-122.

Allison, L.E. 1947. Effect of Microorganisms on Permeability of Soil Under Prolonged Submergence. *Soil Science.* Volume 63. pp. 439-450.

Bedient, P.B., and W.C. Huber. 1992. Hydrology and Floodplain Analysis. Addison-Wesley Publishing Company. Don Mills, Ontario, Canada.

Behnke, J.J. 1969. Clogging in Surface Spreading Operations for Artificial Ground-Water Recharge. *Water Resources Research.* Volume 5:4. pp. 870-876.

Borgwardt, S. 1994. Tests on UNI ECO-STONE® Installations of Different Ages: Report for Uni- International. Institute for Planning Green Spaces and for Landscape Architecture, U. of Hanover.

Bosch, D.D. and C.A. Onstad. 1988. Surface Seal Hydraulic Conductivity as Affected by Rainfall. *Transactions of the ASAE.* 31. pp. 1120-1127.

Bowles, J.E. 1992. Engineering Properties of Soils and Their Measurement 4th Edition. McGraw-Hill, Incorporated, Toronto.

Clark, A. J. (1980) Water Penetration Through Newly Laid Concrete Block Paving. Technical Rep. 529, Cement and Concrete Assoc. 7 pp.

Darcy, H. 1856. Determination of the Laws of the Flow of Water Through Sand. In: *Physical Hydrogeology.* R.A. Freeze and W. Black, Editors. Hutchinson Ross Publishing Company. Stroudsburg, Pennsylvania. pp. 14-19.

Day, G.E. 1978. Investigation of Concrete Grid Pavements. Virginia Water Resources Research Center. Virginia Polytechnic Institute and State University. Blacksburg, Virginia.

Duley. F.L. 1939. Surface Factors Affecting the Rate of Intake of Water by Soils. *Soil Science Society of America Proceedings.* Volume 4. pp. 60-64.

Elrick, D.E. 1996. Professor, University of Guelph. Personal Communication.

Ferguson, B.K. 1994. Stormwater Infiltration. Lewis Publishers, CRC Press. Boca Raton, Florida.

Ferguson, B.K. and T.N. Debo. 1990. On-Site Stormwater Management. Applications for Landscape and Engineering. Van Nostrand Reinhold International Company Limited. New York.

James, W. and B. Verspagen. 1996. Thermal Enrichment of Stormwater Runoff by Urban Paving. In: 1996 Advances in Modelling the Management of Stormwater Impacts. Proceedings of the Stormwater and Water Quality Modelling Conference, Toronto, ON, Feb.22-23. Computational Hydraulics International, Guelph. In Press.

Madison, F., J. Arts, S. Berkowitz, E. Salmon, and B. Hagman. 1979. Washington County Project. Environmental Protection Agency Project 905/9-80-003. United States Environmental Protection Agency, Chicago, Illinois.

Marshall Macklin Monaghan. 1994. Stormwater Management Practices Planning and Design Manual. Prepared for: Ontario Ministry of the Environment and Energy. Queen's Printer for Ontario.

Maryland Department of Natural Resources, (MDNR) Water Resources Administration, Stormwater Management Division. 1984. Maryland Standards for Stormwater Management Infiltration Practices. Stormwater Management Division, Tawes State Office Building, Annapolis, MD. 21401.

Mein, R.G. and C.L. Larson. 1973. Modelling Infiltration During Steady Rain. *Water Resources Research.* 9(2). pp 384-394.

Mohamoud, Y.M., L.K. Ewing, and C.W. Boast. 1990. Small Plot Hydrology: I. Rainfall Infiltration and Depression Storage Determination. *Transactions of the ASAE.* 33(4). pp. 1121-1131.

Mohamoud, Y.M., L.K. Ewing, and J.K. Mitchell. 1990A. Small Plot Hydrology: II. Tillage System and Row Direction Effects. *Transactions of the ASAE.* 33(4). pp. 1132-1140.

Muth, W. 1988. Drainage with Interlocking Pavers. Study Commissioned by F. von Langsdorf. Study Completed at the Karlsruhe University of Engineering Research Institute for Water Resources.

Northern Virginia Planning District Commission. 1992. Northern Virginia BMP Handbook. A Guide to Planning and Designing Best Management Practices in Northern Virginia. Published by Northern Virginia Planning District Commission, Annandale, Virginia.

Norton, L.D., S.L. Schroeder and W.C. Moldenhauer. 1986. Differences in Surface Crusting and Soil Loss as Affected by Tillage Methods. In: *Assessment of Soil Surface Sealing and Crusting.* Flanders Research Centre for Soil Erosion and Soil Conservation. Ghert, Belgium. F. Callebaut, D. Gabriels and M. deBoodt, Editors. pp. 64-71.

Phalen, T. 1992. Development of Design Criteria for Flood Control and Ground Water Recharge Utilizing UNI ECO-STONE® and Ecoloc Paving Units. Northeastern University, Boston, Massachusets.

Pitt, R. 1996. Groundwater Contamination from Stormwater Infiltration. Ann Arbor Press Chelsea, MI.

Prill R.C. and D.A. Aronson. 1978. Ponding-Test Procedure for Assessing the Infiltration Capacity of Storm-Water Basins, Nassau County, New York. Geological Survey Water-Supply Paper 2049. United States Government Printing Office, Washington D.C.

Schueler, T.R. 1987. Controlling Urban Runoff: A Practical Manual for Planning and Designing Urban BMPs. Department of Environmental Programs, Metropolitan Washington Council of Governments. Water Resources Planning Board. Publication Number, 877703.

Shackel, B. 1995. Infiltration and Structural Tests of UNI Eco-Loc and UNI ECO-STONE® Paving. School of Civil Engineering, University of New South Wales. Australia.

Shahin, R. 1994. The Leaching of Pollutants from Four Pavements Using Laboratory Apparatus. Master of Science Thesis, School of Engineering, University of Guelph. Guelph, Ontario, Canada.

Shainberg, I. 1992. Chemical and Mineralogical Components of Crusting. In: *Soil Crusting, Chemical and Physical Processes*. M.E. Sumner and B.A. Stewart, Eds. Lewis Publishers, Chelsea MI. pp. 33-53.

Smith, R.E. 1982. Rational Models of Infiltration Hydraulics. In: Modelling Components of Hydrologic Cycle. *Proceedings of the International Symposium on Rainfall Runoff Modelling*. Held in Mississippi, USA. Edited by V. P. Singh. Water Resources Publications, Littleton, Colorado.

Song, Z. and L.D. James. 1992. An Objective Test for Hydrologic Scale. *Water Resources Bulletin*. Published by ASCE. 28(5). pp. 833-844. Paper No. 91063.

Thompson, M. K. and W. James. 1995. Provision of Parking Lot Pavements for Surface Water Pollution Control Studies. Chapter 24 in: Modern Methods for Modelling the Management of Stormwater Impacts. Proceedings of the Stormwater and Water Quality Modelling Conference. Toronto, ON, Mar 3-4. Computational Hydraulics International, Guelph.

UNI-GROUP USA. 1993. UNI ECO-STONE®: The Environmentally Beneficial Paving System. UNI-GROUP USA, Palm Beach Gardens, Florida.

Urbonas, B.R. and P Stahre. 1994. Stormwater: Best Management Practices and Detention for Water Quality, Drainage and CSO Management. Prentice-Hall, New Jersey.

White, I., B.E. Clothier, and D.E. Smiles. 1982. Pre-Ponding Constant-Rate Rainfall Infiltration. In: Modelling Components of Hydrologic Cycle. *Proceedings of the International Symposium on Rainfall Runoff Modelling*. Held in Mississippi, USA. Edited by V. P. Singh. Water Resources Publications, Littleton, Colorado.

Chapter 11

Contaminants from Four New Pervious and Impervious Pavements in a Parking-lot

William James and Michael K. Thompson

A previous account (Thompson and James, 1994) described the design, construction and instrumentation of four different pavements - asphalt (AS), concrete brick (CP), and three -inch and four-inch thick concrete paver stones with infiltration cells (E3 and E4) - in both a typical parking-lot and in a laboratory. Sampling of runoff from both sets of four pavements was subsequently carried out, and the analytical results used to estimate the flux of 23 contaminants including heat (Thompson, 1995). This chapter reports the interim conclusions obtained from the parking-lot pavements for the first year after installation.

11.1 Introductory Background

Impervious paving such as asphalt, when it replace a previously pervious surface, increases contaminant loads to receiving waters, especially pollutants arising from highway pavement, vehicular traffic, air deposition, surrounding land-uses and heated surfaces. Asphalt paving, concrete curbs and impervious gutters are efficient collectors and conveyors of contaminants and stressors to stormwater systems - they may appeal to some people but are generally bad for aquatic ecosystems. Appropriately-designed permeable concrete block pavers, on the other hand, and other infiltration practices, reduce the quantity of these pollutants that reach receiving waters, by filtering stormwater through the upper soil zones.

© *Advances in Modeling the Management of Stormwater Impacts - Vol. 5* W. James, Ed. Pub. by CHI, Guelph, Canada 1997. ISBN 0-9697422-7-4. Fax: +519 767-2770

Contaminants of interest include oils and greases, heavy metals, nutrients, bacteria, suspended material and elevated temperatures. But the altered flow regime itself is an important stressor: increased impervious surfaces lead to increased flood flows and depleted low flows, because of reduced infiltration and reduced groundwater replenishment. Increased frequencies of these flow rates at both extremes degrade natural riparian habitats and aquatic ecosystems.

The purpose of the present study is a comparison of the performance of four different pavements, denoted AS, CP, EC3 and EC4, from the perspective of surface water pollution control. It forms part of a continuing study on permeable pavements, current details of which may be found on the Web at http://www.eos.uoguelph.ca /~james, under "Research". Abstracts of dissertations, eleven of which form the backbone of this work, may be found at the web site.

Furthermore, a long list of literature on this topic is also available at the web site, by searching under "Biblio96". Using Netscape's "Find" facility, for example, keywords in context such as title words, authors, and sources can be located. Abstracts are not included in that list.

11.2 Previous Work

Short reviews are given by the authors in previous volumes in this series (Xie and James, 1993; Thompson and James, 1994) and elsewhere (James and Verspagen, 1996; Kresin and James, 1996). These focus on roads and parking lots, which make up the largest percentage of man-made impervious surfaces (Hade, 1987). Compared to asphalt, permeable paving reduces runoff (see e.g. Thalen et al., 1972) and contaminants reaching receiving waters (Nawang and Saad, 1993; Pratt et al., 1989). When downstream impacts and aesthetic benefits are included, permeable pavement has been found to be more economical than conventional roads (Thalen et al., 1972).

Runoff from asphalt parking lots has been studied at several places, including Sweden by Spangberg and Niemcynowicz (1993), who found increased turbidity, pH, conductivity and concentrations of adverse chemicals, compared to natural areas. Being a petroleum product, the types of substances found in runoff and adjacent areas, from the asphalt, range from polyaromatic hydrocarbons to benzopyrene, and heavy metals (Munch, 1992). There is the potential, but incomplete evidence, that asphalt is carcinogenic to humans (International Agency for Research on Cancer, 1985).

Xie and James (1994) discuss the importance of solar thermal enrichment of stormwater by heated pavement, and further relationships have since been developed by James and Verspagen (1996). Evidently the thermal enrichment of small rivers by modest parking lots can be some 5°C, sufficient to degrade cold water fisheries.

Collectively the literature provides a case for careful environmental review of our current pavement practices, and for these reasons the test pavements were installed and instrumented at the University of Guelph. Readers are referred to the earlier publication (Thompson and James, 1994) for details of the installations. Kresin and James (1996) discuss the long-term performance of permeable concrete pavers in another chapter in this book.

11.3 Pollutants in Pavement Runoff and their Pathways

Snodgrass et al. (1994) found that particulates in the form of heavy metals comprise a large component of the particulate matter found in highway runoff. Friction and automobile deterioration are significant contributors of heavy metals, and deicing salts may contribute to the deterioration of automobiles and highway structures. Fossil fuels contribute petroleum hydrocarbons (PHCs) and incomplete combustion can contribute to the formation of polycyclic aromatic hydrocarbons (PAHs). In fact the most significant pollution loads are due to the use of petroleum-based fuels and lubricants in vehicles. Even though they are weakly volatile, they remain on paving until washed into the drainage system to the receiving waters (Barnes et al., 1979).

The buildup and removal of pollutants is a continuous process occurring over both dry and wet periods, with rainfall washing off the pollutants that have built up. Also, rainfall deposits its own pollutants including low pH (high acid levels). Rainfall also transfers particulates from adjacent pervious areas, and this runoff contains high levels of bacteria, phenols (from insecticides), total and suspended solids, etc. During dry periods, pollutants accumulate until they are blown away.

Contributing factors to a mass balance include the following:

atmospheric deposition, AD (includes rain)
vehicle input, VI
pavement degradation, PD
pavement leachates, PL
population input, PI
vegetation input, VeI
biological removal, BR
vehicle removal, VR
wind removal, WR
intentional removal, IR.

Net accumulation *NAC* is:

$$NAC = AD + VI + PD + PL + PI + VeI - BR - VR - WR - IR \quad (11.1)$$

Atmospheric Deposition, AD: The analysis of atmospheric deposition or atmospheric fallout is beyond the scope of this study. It is, however, an important consideration. Atmospheric deposition deals with the dust fall over the area and depends primarily on climatic and physiographic conditions (James and Boregowda, 1986). Pollutants in rain are also important.

Vehicle Input, VI: Highway runoff pollutants include metals, nutrients and hydrocarbons that are contributed by motor vehicle traffic. Common sources of origin and pathways include parking lots, streets and highways, industrial areas, construction sites, and unpaved roads. Traffic by-products include: petroleum products, vehicle exhaust, tire and brake wear and metallic corrosion (James and Boregowda, 1986). The quantities of vehicle input can be quantified:

$$L_v = NPLM \tag{11.2}$$

where:

L_v = mass of dust and dirt produced by vehicles,
N = number of vehicle axles active each day in a subcatchment,
P = population to vehicle ratio,
L = mass of dust and dirt produced per vehicle per km travelled,
M = total length of road in subcatchment.

Pavement Degradation, PD: Pavement degradation is the deterioration of the pavement surface over time. This is dependent on the vehicles, pavement, weather, environment and pollution input conditions.

Pavement Leachates, PL: Pavement leachate is the removal of material through percolation of water as it infiltrates through the pavement surface and percolates through the bedding material. The rate and amount of pavement leachate depends on: (i) bedding material, (ii) slope, (iii) nature of surface material, (iv) impact energy of raindrop as it hits the surface, (v) rain chemistry, and (vi) air temperature.

Population Input, PI: Population input includes litter such as solid wastes deposited on surfaces, animal and bird faecal droppings and other deposits. Deposits depend upon number of personal or individual littering or amount of animals (and birds) found within the area. For humans:

$$L_p = PL \tag{11.3}$$

where:

L_p = total mass of dust and dirt due to population,
P = total population in a subcatchment,
L = per capita mass of pollutants (James and Boregowda, 1986).

Vegetation Input, VeI: Adjacent to the study area used in this study is a vegetated area. For most of the study period, the area was vegetated with weeds and shrubs. During the month of August, the weeds and shrubs were removed and part of the area was seeded with grass and fertilized. Most of the area was left with fertilizer and mulch. Vegetation input rate depends on the time of year and vegetation density. Vegetation contributes solids, nutrients, and chemicals (James and Boregowda, 1986).

Biological Removal, BR: With the exception of nitrates and phosphates, organic pollutants decompose into simple substances with or without oxygen. This reduces the quantity of pollutants at the time of washoff. Nitrates and phosphates normally increase with the addition of decomposed end-products (James and Boregowda, 1986). Removal of solids by biological decomposition may be estimated:

$$R_b = PF(1 - e^{-kt}) \tag{11.4}$$

where:

R_b = total mass of dust and dirt removed by biological decomposition,
P = average mass of dust and dirt available on the surface,
F = fraction of decomposable dust and dirt,
k = decay coefficient,
t = number of days.

Vehicle Removal, VR: Eddies generated by vehicles transport pollutants to ineffective areas from where they may not be washed off during storm events (James and Boregowda, 1986). This results in a reduction of solids being washed off.

Wind Removal, WR: Natural wind transport process reduces the amount of solids being washed off. Contributing factors to this include wind speed, curbs, and adjacent pervious areas which act as a pollutant sink (James and Boregowda, 1986).

Intentional Removal, IR: Main sources of intentional removal include various types of street cleaning. For this study, street cleaning was conducted only during the month of March 1994.

Net Accumulation, NAC: Calculation of net accumulation is the combination of all the above input and removal processes. At the scale of this study, the equation is considered at only an empirical level.

11.4 Contaminants Examined

Runoff was sampled and samples analyzed using methods documented in Thompson (1995) for the chemical, biological, and physical parameters shown in Table 11.1.

Table 11.1 Contaminants investigated.

Physical Properties:			
		Conductivity	
		Solids	Residue, Total (Total Solids)
			Residue, Particulate (Suspended Solids)
		Temperature	
Chemical Constituents:			
	Inorganic	Heavy Metals	Copper
			Nickel
			Lead
			Zinc
			Iron
			Cadmium
			Chromium
		Chlorides	Chloride
		Nutrients	Phosphorus
			Phosphates
			Nitrogen, Total Kjeldhal
			Ammonium, Total
			Nitrates
			Nitrite
	Organic:	Refractory	Phenolics
		Biodegradable:	BOD
			COD
			E.Coli
			Solvent Extractable (Oils & Grease)

11.5 Results

Results are presented under two headings: *temperature* and *contaminants*. Temperatures were measured continuously from June 22 to October 2, 1994. As summarized in Table 11.2, contaminants were measured for a total of nine rainfall events.

11.5.1 Temperature

Xie (1993) stated that the level of watershed development had the single greatest anthropogenic influence on the temperature regime of urban, headwater streams. As the urban landscape heats up on warm summer days, it tends to impart a great deal of heat to any runoff passing over it. Because base and sub-base temperatures were lower than surface temperatures, infiltrated water would be affected by this lower temperature as it passed through a porous pavement. Infiltration is the foremost urban BMP for temperature reduction (Galli, 1990).

Table 11.2 Collected event summary.

Location	Event 1 Jul. 15	Event 2 Jul. 21	Event 3 Jul. 29	Event 4 Aug. 9	Event 5 Sep. 15	Event 6 Sep. 25	Event 7 Sep. 27	Event 8 Sep. 28	Event 9 Sep. 30
Rain	T,F,CE	T,F,C	F,CE	F,C	T,F,C	T,F,C	T,F,CE	T,F,C	T,F,C
AS Surface	T,F,C	T,FE,C	F,C	FE,C	T,FE,C	T,FE,C	T,F,C	T,FE,C	T,F,C
AS Base	T,FO	T,FO	N/A	N/A	T,FO	T,FO	T,FO	T,FO	T,FO
AS Sub-base	T,FO	T,FO	N/A	N/A	T,FO	T,FO	T,FO	T,FO	T,FO
CP Surface	T,F,C	T,F,C	C,F	FE,C	T,F,C	T,F,C	T	T,FE,C	T,F,C
CP Base	T,FO	T,FO	N/A	N/A	T,FO	T,FO	T	T,FO	T,FO
CP Sub-base	T,FO	T,FO	N/A	N/A	T,FO	T,FO	T	T,FO	T,FO
E4 Surface	T,FO	T,F,C	C,FE	FE,C	T,FO	T,F,C	T	T,FE,C	T,F,C
E4 Base	T,FO	T,FO	C,FE	FE,C	T,FO	T,FO	T	T,FE,C	T,F,C
E4 Sub-base	T,FO	T,FO	N/A	N/A	T,FO	T,FO	T	T,FO	T,F,C
E3 Surface	T,FO	T,FO	C,FE	FE,C	T,FO	T,FO	T	T,FE,C	T,F,C
E3 Base	T,FO	T,FO	C,FE	N/A	T,FO	T,FO	T	T,FE,C	T,F,C
E3 Sub-base	T,FO	T,FO	N/A	N/A	T,FO	T,FO	T	T,FO	T,FO

Note:

Locations

AS - Asphalt
CP - Hollandstone paver
E4 - Uni eco-stone 4”
E3 - Uni eco-stone 3”

Data

T - Temperature data collected
F - Flow data collected
FE - Flow data not collected but estimated
FO - No water entered the collection system
C - Contaminant data available
CE - Rainfall contaminant data based on average concentration of available samples collected
N/A - No temperature, flow or contaminant data collected

Temperature data was collected from twelve thermocouples, one for each pavement and layer. Data for temperature was collected from the pavements between May and October 1994 and may be compared to short wave radiation, which is the amount of direct sunlight. At night the incoming short wave radiation is zero. Short wave radiation also decreases during periods with cloud cover. Our short wave radiation sensors measured photosynthetically active radiation (PAR) in the 400 to 700 nm waveband where PAR is in units of micromoles per second per square metre (mmol $s^{-1}m^{-2}$) (Pettit, 1994).

A comparison of the three layers for the asphalt (AS) pavements is shown in Figure 11.1 for the period of September 21 to October 2, 1994. Air temperature was measured by a shielded sensor designed to attempt to reduce the influence from the adjacent greenhouse. The surface layer recorded the highest and lowest temperatures as well as the greatest variations in maximum and minimum temperature for that pavement. The black top of the AS pavement allows the absorption of short wave radiation from the sun on clear sunny days, so the AS surface temperature is the greatest of all the pavements. During clear, dry nights, the AS surface pavement temperatures drop below the observed air temperature, due to back short wave radiation from the pavement. Figure 11.1 also shows that sub-base temperatures had very little diurnal variation for that period. Sub-base temperatures for AS also had the lowest average temperatures compared to the other pavements for that period. Temperatures from the base level had greater ranges (higher maximum and lower minimum) than the sub-base, but had less range than the surface for that period. Average temperatures from the base layer were greater than the sub-base and less than the surface. This statement is true till September 23, when the asphalt sub-base average daily temperature becomes the greatest temperature of the three levels. This is due to two factors: the average air

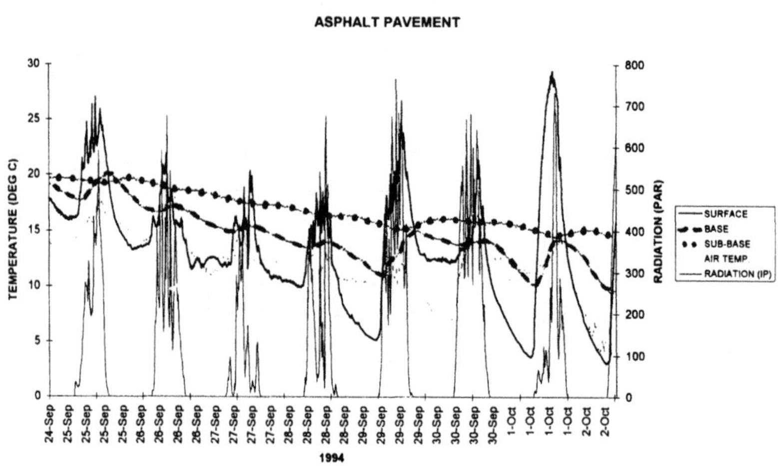

Figure 11.1 Temperatures of layers in the asphalt pavement.

temperature has been decreasing and rainfall occurred almost daily which affected the surface temperatures. Visually, PAR values are evidently related to surface temperatures. This relation decreased at the base and decreased even further at the sub-base level.

Surface temperatures from the asphalt were higher than all the other pavements for the period.

Maximum, minimum and average daily temperatures for the duration of the study are summarized by Thompson (1995). During days with direct sunlight, the AS reached the highest surface temperature. During the night, AS tended to have the lowest temperature, compared to the other pavement surfaces. Average surface temperatures were generally similar. The three porous surfaces tended to have the same temperatures at any time of the day. Results from the comparison of the four surface temperatures show that asphalt generated higher temperatures than concrete pavements. This is probably due to the low albedo of the AS pavement compared to the other paving materials.

Rainfall occurring during daylight hours affected the asphalt surface temperatures more than the other surfaces. This was observed on September 27, where PAR values decreased rapidly during rainfall. AS temperatures dropped to a low of 10.6°C from a high of 20.3°C as rainfall occurred during the daylight hours.

For the period, temperatures at the base of the pavements displayed less temperature difference between each pavement and at almost any time of the day. Asphalt consistently had the lowest minimum base temperatures. Base temperatures also had lower average temperatures than other layers. Temperatures from the four sub-base levels showed minor variations in temperature statistics.

11.5.2 Contaminant Load Results

Loads are based on both flow rates and concentrations. This section discusses the results from the individual flow and concentrations, then summarizes the computed loads. Table 11.3 summarizes the rain events sampled showing the storm duration, volume, and intensities.

11.5.3 Flow Results

Surface runoff flow rates were collected for the nine events from June 22 to October 2, 1994. The product of flow and contaminant concentrations produced load results. Information was collected from the tipping bucket runoff gages (TBRGs) and covered containers in the instrumentation chamber. As flow was not measured for all events due to instrumentation errors, missing flow data was estimated. Estimates of flows were based on results from laboratory experiments on infiltration rates and capacities for each pavement.

Table 11.3 Rainfall volume summary.

Event	Date	Duration (min)	Vol (mm)	Peak Intensity (mm/hr)	Antecedent Period (hours)
1	Jul 14	20.0	1.6	12.0	118.0
2	Jul 21	95.0	5.3	18.0	122.0
3	Jul 29	220.0	19.0	51.6	65.0
4	Aug 9	45.0	6.6	21.6	105.0
5	Sept 15	125.0	10.9	18.0	38.3
6	Sept 25	600.0	4.0	12.0	257.0
7	Sept 27	90.0	8.7	18.0	4.0
8	Sept 28	100.0	14.5	45.6	12.0
9	Sept 30	L/A	787.0	L/A	L/A

Note: L/A Data Not Available

Table 11.4 is a comparison of surface runoff as a percentage of rainfall from Event 9. It shows that AS had the greatest percentage of surface runoff relative to the rainfall. E4 and E3 produced less surface runoff than the other pavements.

Table 11.4 Runoff as percentage of rainfall.

Location	Pavement			
	AS	CP	E4	E3
Surface	100	80	61	38

AS = asphalt; CP = concrete brick paver; E = Ecostone;
3 and 4 denotes thickness in inches of base course.
CP, E3 and E4 are concrete products.

11.5.4 Contaminant Results

Samples for the various contaminants were collected for nine events over the four month period from June 22 to October 2, 1994. Storm event data were collected from the automatic samplers, containers and the TBRGs. Events were of various sizes, volumes and peak intensity.

11.5.5 Contaminant Load Analysis

Tables 11.5 and 11.6 summarize the results, based on total loads from the nine events. Table 11.5 is a summary of the total surface runoff loads generated for the nine events as well as loads from rainfall. For most of the contaminates, AS generated the greatest surface total load. Pavements E4 and E3 generated the least total load for most of the contaminants. This is due to the asphalt surface being almost 100% impervious. Reductions in surface discharge due to porous pavement show a reduced pollutant load leaving the site as noted by Pratt et al. (1989).

Table 11.5 Surface load summary.

Contaminant	Unit	Pavement				
		Rain	AS	CP	E4	E3
Chloride	mg	0.00e+00	2.65e+04	1.27e+04	2.00e+03	8.82e+02
Copper	mg	8.95e+01	1.17e+02	6.03e+01	3.97e+01	2.08e+01
Nickel	mg	0.00e+00	1.81e+00	2.12e-01	0.00e+00	0.00e+00
Zinc	mg	3.22e+02	7.88e+02	3.31e+02	1.06e+02	1.35e+02
Cadmium	mg	1.02e+00	1.21e+01	1.89e-02	0.00e+00	0.00e+00
Chromium	mg	0.00e+00	4.94e+01	6.42e+00	2.66e+00	2.68e-01
Lead	mg	2.98e+01	2.39e+02	5.11e+01	2.28e+01	2.56e+01
Iron	mg	0.00e+00	4.95e+03	1.51e+03	5.29e+02	3.87e+02
COD	mg	3.09e+04	5.25e+05	1.39e+05	4.58e+04	3.69e+04
Phenolics	μg	9.06e+03	1.41e+04	8.05e+03	2.34e+03	1.57e+03
BOD	mg	0.00e+00	3.57e+04	1.58e+04	3.74e+03	3.69e+02
Solvent Extractable	mg	6.60e+02	1.36e+04	8.92e+03	1.51e+03	1.35e+03
Ammonium	mg	2.25e+03	2.58e+03	6.91e+02	4.84e+02	3.01e+02
Nitrite	mg	5.97e+01	2.55e+02	1.64e+02	3.92e+01	4.81e+01
Nitrate	mg	4.88e+03	3.65e+03	2.47e+03	9.72e+02	6.77e+02
Phosphate	mg	9.30e+00	7.91e+02	7.22e+01	1.50e+01	2.04e+01
Phosphorus	mg	5.58e+01	3.31e+03	1.13e+03	1.59e+02	2.72e+02
TKN	mg	3.11e+03	8.23e+04	5.02e+03	1.19e+03	7.91e+02
E.Coli	CFU	3.64e+04	2.88e+08	7.49e+07	3.18e+07	1.89e+06
Residue Total	mg	6.29e+04	3.04e+06	5.06e+05	1.55e+05	1.10e+05
Residue Particulate	mg	2.25e+04	2.69e+06	3.18e+05	8.72e+04	6.82e+04

Infiltration and percolation loads from the base and sub-base levels were generated only for E3 and E4 due to their perviousness and the results are shown in Table 11.6. No runoff was generated from the base layer of the CP during the course of this study. At the start of this study, before the catchment system was installed, it was observed that runoff was generated from the base layer but the runoff volume decreased till no runoff was generated from the base layer. Properties of the CP are such that it is expected to go to 100% impervious as the voids between the stones lock together with sand and dust (King and Smith, 1991). Because the base and sub-base runoff sampled was a very small (even insignificant) part of the infiltration, estimates of their pollutant loads are likely to be very inaccurate.

Table 11.6 is a summary of the calculations of the estimated loads for the sub-surface. The results are not considered reliable enough to form a major finding.

Table 11.6 Total sub-surface load summary.

Contaminant	Unit	Pavement		
		E4 Base	E4 Sub-base	E3 Base
Chloride	mg	2.54e+04	4.70e+03	1.99e+04
Copper	mg	6.09e+01	9.76e+00	4.92e+01
Nickel	mg	3.29e+00	0.00e+00	8.15e+00
Zinc	mg	7.77e+02	1.40e+02	8.62e+02
Cadmium	mg	5.87e+00	0.00e+00	0.00e+00
Chromium	mg	2.20e+01	3.11e+00	1.08e+01
Lead	mg	1.96e+02	2.81e+01	1.79e+02
Iron	mg	2.03e+03	2.20e+02	1.63e+03
COD	mg	1.16e+05	1.49e+04	1.20e+05
Phenolics	µg	7.45e+03	0.00e+00	8.16e+03
BOD	mg	8.84e+03	0.00e+00	1.72e+04
Solvent Extractable	mg	7.59e+03	5.49e+02	9.28e+03
Ammonium	mg	2.08e+03	0.00e+00	1.94e+03
Nitrite	mg	3.55e+02	2.20e+01	3.69e+02
Nitrate	mg	4.42e+03	9.64e+02	3.77e+03
Phosphate	mg	7.08e+02	4.88e+01	2.61e+02
Phosphorus	mg	1.28e+03	1.34e+02	1.73e+03
TKN	mg	8.55e+03	3.66e+02	4.33e+03
E.Coli	CFU	2.28e+08	6.10e+05	6.54e+07
Residue Total	mg	5.15e+05	2.57e+05	1.71e+06
Residue Particulate	mg	5.60e+05	7.81e+04	1.44e+06

Loads generated from porous pavements discharged partly through the subsurface which reduced the amount of direct runoff from the surface to receiving waters. More contaminant loads infiltrated below the surface, therefore, the surface runoff had less contaminants, showing that porous pavement exhibits excellent performance with respect to runoff reduction and pollution abatement (Sztruhar and Wheater, 1993). Of course there remains the potential contribution to groundwater contamination, especially by soluble metals and pesticides.

Rainfall contaminant data "washoff" was collected from the wet portion of the wet/dry collector. In the case when contaminant data was not available (events 1, 3 and 7), the average of all the available contaminant data was taken and this value was used with the rainfall volume collected for that event. For events which did not have available contaminant data, Table 11.7 shows the ratio of the individual rainfall volume to all the rainfall volume. This table indicates the relative size of the individual estimated loads.

Table 11.7 Ratio of event rain to total rainfall.

Source	Volume (L)	Event/Total Rainfall
Total Rainfall (9 Events)	7340.0	1.00
Event 1	130.0	0.0177
Event 3	1540.0	0.210
Event 7	705.0	0.0960

Atmospheric dryfall was collected from the dry portion of the wet/dry collector. Analysis of the atmospheric dryfall was not conducted; however, the mass of the accumulation was examined. Between July 21 and September 16, 1994, 5.01 g of atmospheric dryfall was collected. Between September 16 and September 27, 1994 0.70 g was collected.

Between September 27 and September 29 1994, only 0.01 g of atmospheric dryfall was collected. Total atmospheric dryfall for the length of the study was 5.71 g or 0.147 g/day. Results from the wet/dry collector show that additional work is necessary on the atmospheric dryfall collected.

11.6 Conclusions

Very little previous work has been done to compare pollutant-generation of impervious asphalt to various permeable concrete block pavers in a parking lot. Four instrumented pavements were constructed at the University of Guelph in

1993, and after the first year of runoff sampling, we tentatively draw the following conclusions:

1. pavement surface temperatures are directly related to weather conditions, asphalt showing the highest maximum and lowest minimum daily temperatures;
2. average daily summer temperatures were similar for all four pavings;
3. subgrade temperatures measured about 150 mm below the surface showed a lower diurnal range and lower daily maxima than the surface temperatures;
4. subgrade temperatures measured about 600 mm below the surface showed little diurnal temperature fluctuation;
5. for the 23 contaminants examined, loads from the asphalt surface were always greater than from the other pavings, mostly because the asphalt was 100% impervious; and
6. permeable pavement, particularly Uni-Ecostone (a commercial concrete paver), significantly reduces surface runoff contaminant loads.

Postscript: We add the following comments in response to an anonymous and helpful reviewer.

Heat transmission through pavement is clearly related to the thermal characteristics of the sub-base, as well as the paving surface. In particular the moisture content of the sub-base is expected to play an important role. This of course begs the larger question of the changed thermal behaviour of the earth's urbanized areas, whereby the seasonal heat sink arising from infiltrating water has been compromised. Our work did not determine these effects (see the chapter by James and Verspagen).

We have not considered an important variable for pollutant interactions and mobilities, the pH/redox gradient. We understand that even 0.5 unit changes can be significant within 150 mm depth profiles, particularly for metals. Our work should be seen as merely a start on the problem of estimating the temperature effects of various processes.

Our focus has been on the reduction of surface runoff, and the trapping of pollutants associated with particulates in the upper permeable paving surface, and not the contentious question of the effective retention of soluble metals, herbicides and PAH fractions. We do not expect permeable pavement to retain dissolved pollutants unless the sub-base and top layers have been especially designed and treated for this purpose, which is clearly necessary where important groundwater resources are of concern. The provision of permeable parking lots as sand filtration facilities for removal of stormwater pollutants is an interesting idea, especially for areas of water scarcity, but clearly requires scheduled maintenance operations. Our further research will move in this direction.

Acknowledgements

Financial support from Unilock Ltd Canada and Von Langsdorff Licensing is gratefully acknowledged. They also provided lunch many times. Help in installing the test pavements and associated facilities was invaluable: staff of Operations and Maintenance and at the School of Engineering at the U of Guelph were always cooperative. Our colleagues Reem Shahin, Brian Verspagen and Chris Kresin are included. Most of the chemical and bio-chemical analysis was carried out by the Ontario Ministry of the Environment.

References

Barnes, D., Lacey, D.T., Goronzy, M.C. and Brown, J.D. 1979. Nature of pollutants in pavement drainage. Highway Engineering in Australia. June 1979. Pp 15-22.

Galli, J. 1990. Thermal impacts associated with urbanization and stormwater best management practices. Metropolitan Washington Council of Governments, DC. Report 15799. 157 pp.

Hade, J.D. 1987. Determining the runoff coefficient for compressed concrete unit pavements in situ. MSc dissertation. Ball State University, Munice. IN. May 1987. 47 pp.

International Agency for Research on Cancer. 1985. Polynuclear aromatic compounds, Part 4, Bitumens, Coal-tars and derived products, shale-oils and soots. IARC Monographs on the Evaluation of Carcinogenic Risks to Humans, World Health Organization. Vol 35. 81pp.

James, W. and Boregowda, S. 1986. Continuous mass balance of pollutant buildup processes. Urban Runoff Pollution. NATO ASI Series. Vol G10 pp 244-271.

James, W. and Verspagen, B. 1995. Thermal enrichment of stormwater by urban pavement. In press.

King, P.M. and Smith, D. R. 1991. Interlocking concrete airfield pavements. Concrete International. Dec. 1991. Pp 24-27.

Kresin, C. and James, W. 1996. Long-term stormwater infiltration through concrete pavers. In press.

Munch, D. 1992. Soil contamination beneath asphalt roads by polynuclear aromatic hydrocarbons, zinc, lead and cadmium. The Science of the Total environment, 126:49-60.

Nawang, W.M. and Saad, S.M. 1993. Stormwater infiltration investigation using porous pavement. Proc of the 6th Int. Conf. On Urban Storm Drainage, Niagara Falls, ON. Pp 1092-1097.

Pratt, C.J., Mantle, D.G., and Schofield, P.A. 1989. Urban stormwater reduction and quality improvement through the use of permeable pavements. Water Science and Technology, Water Pollution Research and Control. UK. 21:769-778.

Snodgrass, W.J., Thomson, N.R., McBean, E.A., and Monstreko, I.B. 1994. Characterization of stormwater runoff from highways. In: Current Practices in Modelling the Management of Stormwater Impacts. CRC Press. ISBN 1-56670-052-3. Pp 141-157.

Spangberg, A. and Niemcynowicz, J. 1993. Measurement of pollution washoff from an asphalt surface. Proc of 6th Int. Conf. on Urban Storm Drainage, Niagara Falls, Ontario. pp 1098-1103.

Sztruhar, D. and Wheater, H. S. 1993. Experimental and numerical study of stormwater infiltration through pervious parking lots. Proceedings of the 6th International Conf. on Urban Storm Drainage, Niagara Falls, ON. Pp. 1098-1103.

Thalen, E., Grover, W.C., Hoiberg, A.J., and Haigh, T.I. 1972. Investigation of porous pavements for urban runoff control. Office of Research and Monitoring. US Environmental Protection Agency. Project #11034DUY. March 1972. 142 pp.

Thompson, M.K. 1995. Design and installation of test sections of porous pavements for improved quality of parking-lot runoff. MSc thesis, Univ of Guelph. Jan. 1995. 160 pp + 2 diskettes.

Thompson, M.K. and James, W. 1994. Provision of instrumented parking-lot pavements for surface water pollution control studies. Chap. 24 in: Modern methods for modeling the management of stormwater impacts. Pub by CHI, Guelph. ISBN 0-9697422-X. pp 381-398.

Xie, J.D.M., and James, W. 1993. Modelling solar thermal enrichment of urban stormwater. In: Current practices in modelling the management of stormwater impacts. CRC Press. ISBN 1-56670-052-3. Pp 205-220.

Chapter 12

Energy Losses in a Tangential Helicoidal-Ramp Inlet for Dropstructures

Matahel Ansar and Subhash C. Jain

The results of a series of experiments that relate the piezometric head in the inflow pipe to the discharge are used to derive an empirical relationship for the energy loss coefficient for a tangential helicoidal-ramp inlet for vortex-flow dropstructures. The energy loss coefficient in the closed-conduit flow regime was found to decrease exponentially with the ratio of the hydraulic diameter of the ramp to the diameter of the inflow pipe.

12.1 Introduction

Due to rapid growth and urbanization of most of the major cities in the world, the capacities of their wastewater systems often are no longer adequate to handle the large volumes of wastewater discharged from houses and factories and from surface runoff during large storms. The resulting overflows of stormwater and raw sewage create a host of problems. Among them one can cite the blowout and loss of manhole or dropshaft covers in existing sewer systems (Yen, 1986) causing traffic delays. Overflows of sewer systems also cause flooding and pollution (Guo, 1991). A number of metropolitan cities use dropstructures to collect sewage and stormwater from surface to underground tunnel interceptors before conveying them to the wastewater treatment plants. These dropstructures also dissipate the energy associated with the fall.

© *Advances in Modeling the Management of Stormwater Impacts - Vol. 5* W. James, Ed.
Pub. by CHI, Guelph, Canada 1997. ISBN 0-9697422-7-4. Fax: +519 767-2770

A review by Jain and Kennedy (1983) shows that dropstructures with a vortex-flow inlet are generally considered to be superior to those with a plunge-flow inlet. Existing vortex-flow inlets used in these types of dropstructures are constructed using an open cut construction. Site constraints governing the layout of the dropstructures may make an open cut construction very expensive, or even impossible. Consequently, there is a need for inlets that can be constructed using tunneling alone. Such inlets, termed radial and tangential helicoidal-ramp inlets, were recently developed by Jain et al. (1993). The hydraulic characteristics of the radial helicoidal-ramp inlet are described elsewhere (Ansar and Jain, 1996; Ansar, 1993). In this chapter, an empirical relationship for the coefficient of energy loss for tangential helicoidal-ramp inlets is developed from head-discharge measurements when the inflow pipe is flowing full.

12.2 Experiments

The experimental set-up of the dropstructure model is depicted in Figure 12.1. The main components of the model were the tangential helicoidal-ramp inlet, the dropshaft, the outflow pipe, the conveyance tunnel and the air measuring chamber. All model components, except the air-measuring chamber, were constructed of lucite to permit observation and recording of flow patterns. Most of the components were constructed in sections joined by flanges or metallic straps at their ends, so that modifications to the model could easily be made.

A schematic view of the tangential helicoidal-ramp inlet is shown in Figure 12.2. The inlet consists of an inlet conduit connected tangentially to a vertical dropshaft and a helicoidal ramp installed in the dropshaft. The helicoidal ramp was located at a predetermined distance below the invert of the inflow pipe. This ramp had a center circular hollow column through which some of the entrained air is vented. A three-revolution ramp was used in all runs. The ramp was made of steel and was painted to prevent rusting. Its characteristics (width, pitch, slope, etc.) varied from run to run. Its width and pitch varied respectively between 70 to 114 mm and 76 to 127 mm. A pair of screws, located such that they did not inhibit the incoming flow, were used to hold the ramp at the desired position. The ramp had a guide wall at its upper end to prevent spill of water over the ramp end. The diameter D_0 of the inlet conduit at its junction with the dropshaft can either be equal to, or larger than, the diameter D_I of the inflow pipe. In the latter case, the inlet conduit has a sloping bottom which can be constructed easily in the field by filling with concrete the bottom region of the inlet tunnel. The diameter of the inflow pipe varied between 102 mm and 152 mm depending on the particular run. Its length was larger than ten times its diameter so that the flow at the entrance of the dropshaft was fully developed and well-established.

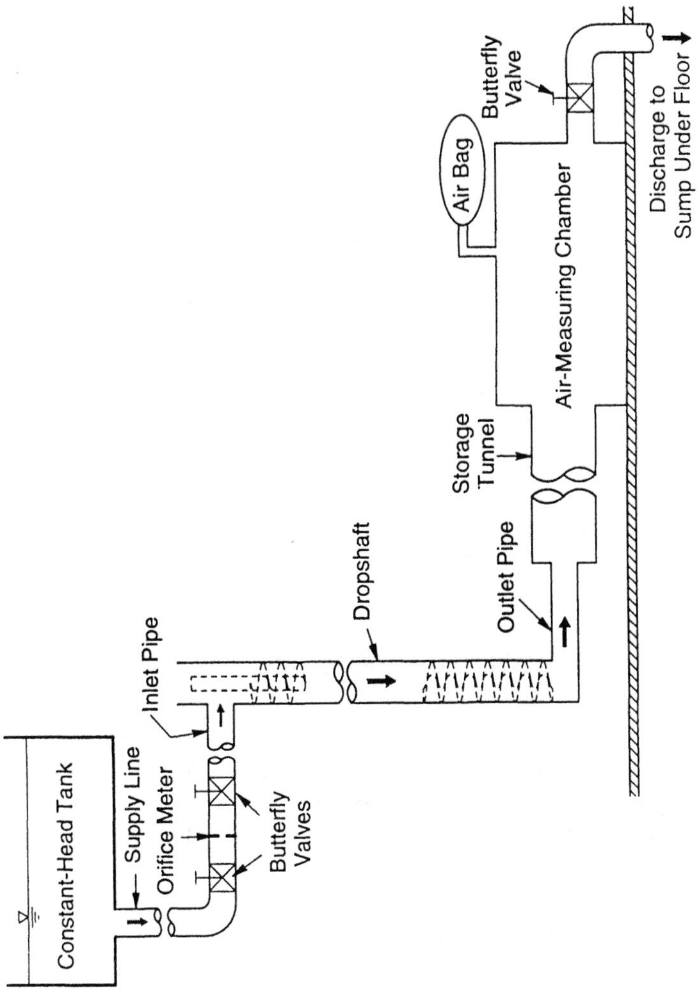

Figure 12.1 Schematic of the experimental set-up.

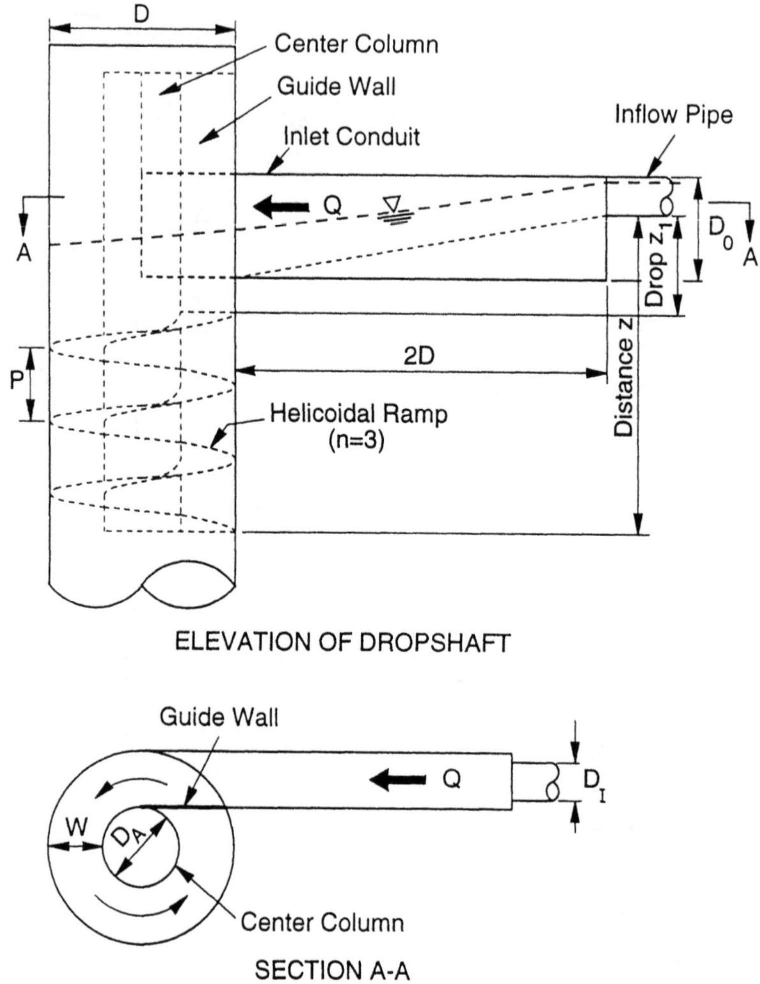

Figure 12.2 Schematic of the tangential helicoidal-ramp inlet.

A flow-straightener section was installed at the upstream end of the inflow pipe to avoid secondary velocities and insure a smooth flow in the inflow pipe. A valve and a calibrated orifice meter were installed in the supply line upstream of the inflow pipe and were used to control and measure the discharge in the inlet section. The dropshaft was 292 mm in diameter in all runs and consisted of two or more sections of different lengths connected by bolted flanges at the ends of the sections. Its length was approximately 3.3 m. This vortex-type inlet induced an angular motion to the water as it entered the dropshaft causing a swirling flow before the water reached the ramp.

Flow to the model was supplied from a constant-head tank, and was measured by means of a calibrated orifice meter in the supply line, and a precision two-tube water manometer. The piezometric head in the inflow pipe was measured by means of a piezometer connected to a pressure tap at the bottom of the inflow pipe, about 1.4 m upstream of the junction of the inflow pipe and the dropshaft.

Each experiment consisted of establishing the desired discharge in the model, and then, after the flow became steady, measuring the piezometric head in the inflow pipe and observing the flow patterns.

12.3 Energy Loss Coefficient

The inlet configurations tested in the study are summarized in Table 12.1. Tests were conducted for two values of ramp pitch P, three values of ramp width W and six values of drop Z_1. Z_1 is the distance between the invert of the inflow pipe and the top of the ramp.

Table 12.1 Tangential helicoidal-ramp inlet configurations.

Run	Diameter		Helical Channel						Remarks*
	Inflow pipe D_I (mm)	Inlet Conduit D_O (mm)	Pitch P (mm)	Width W (mm)	Slope ϕ (deg.)	Area (cm^2)	Distance Z (mm)	Drop Z_1 (mm)	
T1	152	152	127	70	10	87.6	508	127	
T2	152	152	76	114	8	85.8	228	0	NGW
T3	152	152	76	114	8	85.8	380	152	NGW
T4	152	152	127	95	12	118.0	381	0	
T5	152	152	127	95	12	118.0	483	102	
T6	102	152	127	95	12	118.0	534	102	
T7	102	152	127	95	12	118.0	432	0	
T8	152	152	127	95	12	118.0	483	102	NGW

*NGW: No guide wall was used

The test results, depicted in Figure 12.3, present head-discharge curves for runs T1 through T8. A typical head-discharge curve comprises three flow regions (a critical-depth flow region, a transition flow region, and a closed-conduit flow region) and is depicted in Figure 12.4. In the critical-depth flow region the flow in the inflow pipe is controlled by the downstream section of the inlet conduit. The flow depth in the inflow pipe is close to critical in this region. In the transition flow region the flow is between the critical-depth flow regime driven by gravity (open channel flow) and the closed-conduit flow regime driven by pressure gradients. In this region the flow conditions at the downstream section of the inlet conduit are affected by the flow conditions in the ramp, which in turn affect the flow depth in the inflow pipe. In this range of discharges the helical channel was flowing with an air-water mixture. In the closed-conduit flow region both the inflow pipe and the helical channel flowed full, without air, like a closed conduit. The analysis for the results in the critical-depth flow region are reported elsewhere (Ansar 1993). The analysis for the results in the closed-conduit flow region is described below, with particular attention to the energy loss coefficient.

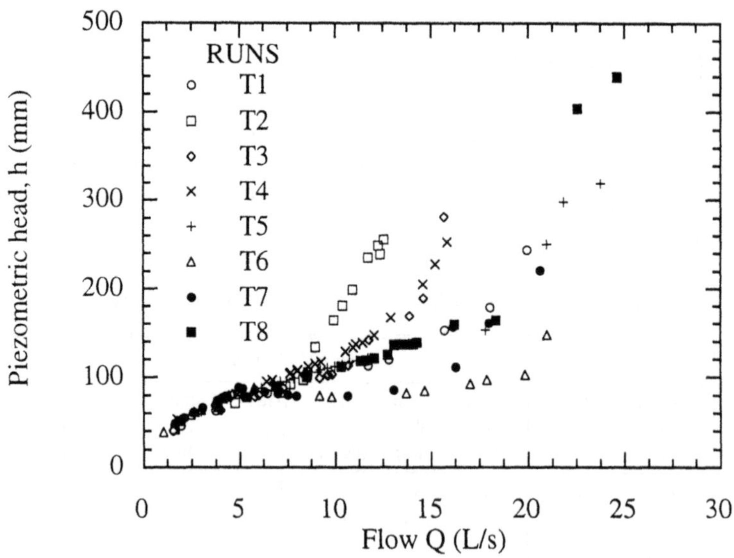

Figure 12.3 Head-discharge curves for the tangential helicoidal-ramp inlet.

In the closed-conduit flow region, the coefficient accounting for energy losses (essentially form losses) can be estimated from the energy equation. This equation, written between the section in the inlet pipe where the piezometric head measurements were taken and the exit of the ramp (Ansar and Jain, 1996; Ansar, 1993), is:

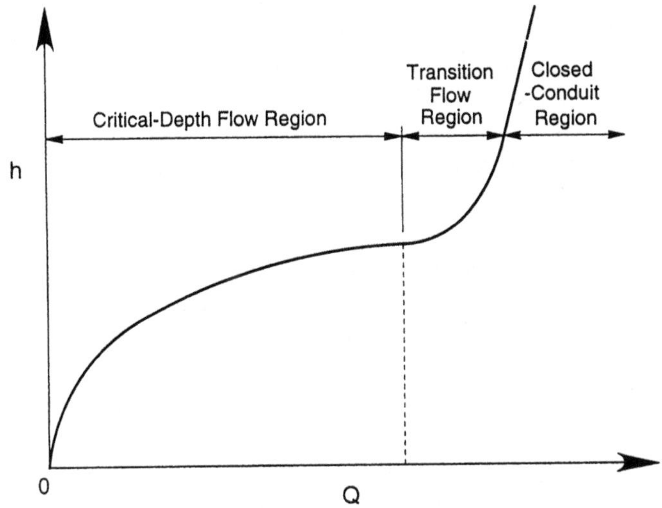

Figure 12.4 Typical head-discharge curve.

$$h + \frac{V_I^2}{2g} + Z = \frac{3}{2}\left(\frac{Q^2}{gW^2}\right)^{\frac{1}{3}} \cos\phi + f_p \frac{L_p}{D_I}\frac{V_I^2}{2g} + K\frac{V_I^2}{2g} \qquad (12.1)$$

where:

V_I = velocity of the flow at the inlet section;
Z = distance between the invert of the inflow pipe and the bottom of the ramp;
W = width of the flow at the exit of the ramp;
ϕ = slope of the ramp;
f_P = Darcy-Weisbach friction factor for the inflow pipe;
L_P = length of the inflow pipe;
K = energy loss coefficient;
g = gravitational acceleration;
h = piezometric head in the inflow pipe;
D_I = inflow pipe diameter; and
Q = discharge.

The three terms on the left hand side of Equation 12.1 together represent the total head at the inlet section. The first term on the right hand side is the total head at the critical depth in a rectangular channel of width and slope the same as those of the ramp. The flow at the exit of the ramp, which previously has been restrained by the ramp, is suddenly released in the form of an annular jet. Such a flow in an open channel is always critical at or near the point of release (Henderson, 1966).

It was therefore assumed that the total head at the exit of the ramp is equal to the total head at the critical depth. It should be noted that initially the total head at the exit was taken to be equal to the sum of the pressure and velocity heads at the exit of the ramp, and the pressure at the exit was assumed to be atmospheric. This initial attempt failed to yield a satisfactory correlation between the loss coefficient and the geometrical parameters of the inlet. The second term on the right hand side of Equation 12.1 accounts for frictional loss in the inlet pipe. The frictional loss in the ramp was neglected because the number of revolutions of the ramp did not affect the head-discharge relation (Ansar, 1993). The last term containing K accounts for the energy losses due to expansion as the flow enters the dropshaft and due to contraction as the flow enters the ramp. The friction factor f_p was obtained from the Moody diagram and the length of the inflow pipe L_p was 1.40 m. In each run the only unknown is K which is obtained from Equation 12.1. The values of the loss coefficient K, which were determined from the experimental data and Equation 12.1, are used to develop an empirical correlation between K and the geometrical parameters of the inlet.

The energy loss coefficient obtained from Equation 12.1 is plotted as a function of the ratio of the hydraulic diameter of the ramp to the inflow pipe diameter in Figure 12.5. This figure shows that the energy loss coefficient decreases as the hydraulic diameter of the ramp increases as expected. In this figure the curve representing the energy loss coefficient can be described by the following relation:

$$K = 96.6 \exp\left(-2.59 \frac{D_h}{D_I}\right) \tag{12.2}$$

where D_h is the hydraulic diameter of the ramp, given by:

$$D_h = \frac{4WP\cos\phi}{P\cos\phi + 2W} \tag{12.3}$$

and, P is the pitch of the ramp, as shown in Figure 12.2.

Equation 12.2 fitted the experimental data with an accuracy of -12.1% to +7.4%.

The inlet configurations in all tests were such that the flow goes through an expansion as water enters the dropshaft, and then a contraction as it flows through the helical channel. The energy losses in the inlet can therefore be expected to behave similarly to those in a circular pipe that suddenly expands and then suddenly contracts. For both sudden expansion and sudden contraction in circular pipes, the energy loss coefficient decreases with increasing diameter ratios (diameter of the smaller pipe over the diameter of the bigger pipe) (White, 1986). A similar pattern is observed with the energy loss coefficient for the tangential helicoidal-ramp inlet as shown in Figure 12.5.

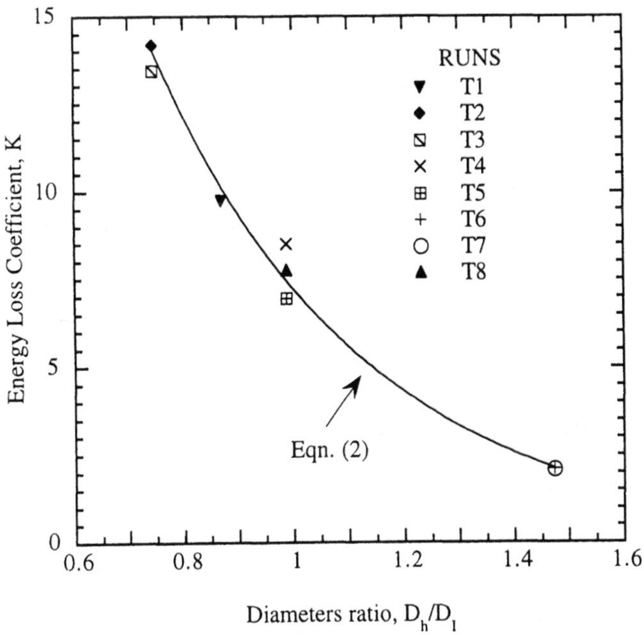

Figure 12.5 Energy loss coefficient for the tangential helicoidal-ramp inlet.

12.4 Conclusion

A vortex-flow inlet to dropstructures, called the tangential helicoidal-ramp inlet, was developed and tested in a laboratory study. This inlet can be constructed entirely by tunneling. An empirical relation for the energy loss coefficient for this inlet is derived from head-discharge measurements in the inflow pipe. The energy loss coefficient decreases exponentially with the ratio of the hydraulic diameter of the ramp to the diameter of the inflow pipe.

Acknowledgments

This work was sponsored by Nihon Suiko Sekkei Company, Ltd., Tokyo, Japan. The guidance of Dr. Tatsuaki Nakato while conducting the experiments and the help of Mike Kundert in making the figures are gratefully acknowledged.

Notation

D	dropshaft diameter (mm)
D_A	center column diameter (mm)
D_h	hydraulic diameter of the ramp (mm)
D_I	inflow pipe diameter (mm)
D_o	inlet conduit diameter (mm)
f_p	Darcy-Weisbach friction factor in the inflow pipe
g	gravitational acceleration (m/s^2)
h	flow depth in the inlet section, piezometric head (mm)
K	energy losses (form losses) coefficient
L_p	length of the inflow pipe (m)
n	number of revolutions of the helicoidal ramp
P	helicoidal ramp pitch (mm)
Q	discharge (m^3/s)
V_I	velocity of the flow at the inlet section (m/s)
W	helicoidal ramp width (mm)
Z	distance between the invert of the inflow pipe and the bottom of the ramp (mm)
Z_1	distance between the invert of the inflow pipe and the top of the ramp (mm)
ϕ	helicoidal ramp slope (degrees)

References

Ansar, M. (1993). Hydraulic Characteristics of Helicoidal-Ramp Inlets for Vortex Flow Dropstructures . Master 's thesis submitted to the University of Iowa, Iowa Institute of Hydraulic Research, pp. 42-44.

Ansar, M. and Jain, S. C. (1995). Helicoidal-Ramp Vortex-Flow Inlet for Dropstructures. *Manuscript submitted for possible publication to the IAHR Journal of Hydraulic Research.*

Henderson, F.M. (1966). Open Channel Flow. The MacMillan Company, New York.

Jain, S. C., Hayden, W. S. and Ansar, M. (1993). Novel Truncated-Ramp Dropstructures. *Report No 206,* Iowa Inst. of Hydr. Res., Univ. of Iowa, Iowa City, Iowa.

Jain, S. C. and Kennedy, J. F. (1983). Vortex-Flow Dropstructures for The Milwaukee Metropolitan Sewerage District Inline Storage System. *Report No. 264,* Iowa Inst. of Hydr. Res., Univ. of Iowa, Iowa City, Iowa.

Guo, Q. and Song, C.S. (1991). Dropshaft Hydrodynamics Under Transient Conditions. *J. Hydr. Engrg.*, ASCE, 117 (8), pp. 1042-1055.

White, F. M. (1986). Fluid Mechanics. McGraw-Hill, Inc. New York. pp. 336.

Yen, B.C. (1986). Hydraulics of sewers. *Advances in Hydroscience,* B.C. Yen, ed. , 14, Academic Press, Orlando, Florida., pp. 1-122.

Chapter 13

Uncertainties in Metering Stormwater Flows

Steven J. Wright

Modern stormwater management often involves the development of numerical hydrologic/hydraulic models in order to provide decision making tools regarding placement of retention facilities, in-system storage, etc. Since many existing combined sewer systems are inadequately characterized and metered, flow meters are often installed in order to provide data sets for calibrating such models. Primary metering devices such as Parshall flumes or venturi meters are often unacceptable because of their high losses and the need to avoid surcharging during high flow conditions. Low loss meters typically utilize velocity measurements to estimate discharge and the accuracy of these types of meters may be less than that associated with primary meters which is reported to be 0.5 to 1% for orifice and venturi meters (Pomroy, 1996) and 5% for Parshall flumes (Parshall, 1926). In a recent hydraulic model development for the city of Detroit, Michigan, the stated goals for the model calibration were to predict the peak discharge to within 20% and the total flow volume to within 10% of the values recorded at select metering stations (Camp, Dresser & McKee, 1993a). In this particular application, two semi-independent methods were available for estimating the metered discharge. An analysis of the discharges predicted by these two methods indicated significant discrepancies under certain flow conditions, raising questions regarding the ability of the meters to resolve flow rates and volumes to within the stated calibration goals. The subsequent investigation of

© *Advances in Modeling the Management of Stormwater Impacts - Vol. 5* W. James, Ed.
Pub. by CHI, Guelph, Canada 1997. ISBN 0-9697422-7-4. Fax: +519 767-2770

meter uncertainty is discussed in this chapter. A review of the metering systems indicates the possibilities of considerable differences between the independent discharge measurements; this is verified by a comparison for selected discharge events.

13.1 Sources of Metering Uncertainty

Formal methodology is generally available for making estimates of measurement uncertainty (ISO 5168), however the types of data available in many applications is insufficient to fully implement the specified procedures. Nevertheless, the basic concepts may still be applied in order to derive an estimate of the uncertainty associated with a measurement or set of measurements.

The purpose of a general uncertainty analysis is to obtain an estimate of the largest error expected to remain with the measurement of a particular quantity such as flow rate. Measurement errors may be subdivided into precision and bias errors. Precision errors may be considered to be random deviations from the true value of a measured quantity that are associated with limitations in the ability of the measurement apparatus or procedure to resolve the measurement. Bias, on the other hand, refers to a constant or systematic deviation between the true and measured values of a particular quantity. Bias error is particularly difficult to quantify in many applications because feasible methodology is not available for determination of the "true" value of the quantity of interest; a master meter or some other independent (and less uncertain) means of measuring the flow would be required to estimate the bias in a metering application. Precision uncertainty, however, may be estimated by repeated measurements at the same condition with the sample standard deviation S used to represent the precision or repeatability of the measurement. Even precision uncertainty is difficult to estimate in the metering of stormwater flows as the discharge is typically unsteady and it cannot be guaranteed that the meter is recording the same flow rate during the course of repeated measurements.

Measurement errors can be categorized into three elemental sources namely (i) calibration errors; (ii) data acquisition errors; and (iii) data reduction errors, each of which has a bias and a precision component. Data reduction errors would most likely be associated with limitations in the ability of a calibration equation to "fit" observed data over a range of conditions due to systematic deviations or bias between the calibration equations and the data. Calibration errors, on the other hand, may be associated with both bias and precision errors in components of both the master meter used to define the "true" discharge and the meter being calibrated. Again, it is important to recognize that it may be impossible to define the true discharge in a specific application and therefore the bias error cannot be accurately estimated. Estimates of precision errors during calibration may be determined by repeating each calibration flow several times and computing the sample standard deviation although this approach is generally not followed in a typical calibration. In the case where the calibration is obtained under the assumption of a constant meter coefficient (i.e. a particular functional representation of the metering equation is assumed with a discharge coefficient or other calibration constant applied to the function), all data collected during the calibration may be ensembled in order to determine the sample standard deviation S. While the uncertainty in an individual measurement may be related to S, the uncertainty in the meter coefficient (the calibration error) would be better characterized by the standard error of estimate $S_e = S / \sqrt{N}$ with N the number of individual calibration points. This characterization implicitly assumes that there is no data reduction error, i.e. that there is no bias between the actual flow and the functional representation of the calibration equation.

Assuming that estimates can be derived for both bias and precision errors for each of the elemental errors listed above, these can be combined into an overall uncertainty estimate. This is performed by computing the root-sum-square of both the bias and the precision error (e.g. $S^2 = S_1^2 + S_2^2 + S_3^2$ where S is the estimate of the overall precision error and S_i refers to the estimates of the individual components of calibration, data acquisition, and data reduction). Finally, the bias and precision uncertainties can be combined to determine an estimate of the overall uncertainty (ISO 5168). While this procedure is relatively straightforward, its implementation is often impossible due to the inability to define the bias components to the overall uncertainty and due to an insufficient number of measurements to properly characterize the precision uncertainty in the calibration. In light of this, the results presented in the case study that follows implement the spirit of this formal procedure while recognizing that additional information would need to be developed in a complete uncertainty analysis. This approach is typical of many metering applications (Dearden, 1992) in which estimates of uncertainty are to be developed.

Since most meters do not measure discharge directly, it is necessary to infer discharge from the measurement of one or more uncertain variables. In this situation it is necessary to investigate the propagation of uncertainty through the function that is used to compute the discharge:

$$Q = f(x_i) \tag{13.1}$$

where:

x_i = the individual variables that enter into the computation of the discharge.

Under the assumption of small uncertainties, the uncertainty in Q may be estimated by the linear terms in a Taylor's series expansion about a base value:

$$dQ = \frac{\partial f(x_i)}{\partial x_1} dx_1 + \frac{\partial f(x_i)}{\partial x_2} dx_2 + \cdots \quad (13.2)$$

where:

dQ = the uncertainty in the discharge and
dx_i = the uncertainty in each of the measured variables that enter into the determination of Q.

Alternatively, uncertainty estimates may be obtained by simply inserting the ranges of each of the variables into the meter equation $f(x_i)$ and determining the maximum range in Q.

For stormwater metering applications, the uncertainties can be grouped into three main categories:

- meter precision,
- calibration, and
- extrapolation.

Meter precision involves the impacts of the individual components of a discharge measurement (depth, velocity etc.) on the discharge computed from the meter output and can be due to both bias and random error (Granger, 1988). It is generally difficult or impossible to determine systematic errors in measurements due to the lack of an independent method for verifying the discharge being reported. Attempts to remove the bias error may be made by calibrating the meter in-situ by comparing the reported discharge against an independent measurement. However, both methods for recording discharge are still subject to precision uncertainties and discharges cannot be considered to be verified except to within the combined uncertainties of the two measurement methods. This uncertainty can be reduced by calibrating over multiple events and averaging the results. Extrapolation errors could arise from utilizing the meter outside the range of flow conditions for which the calibration was determined if there are bias errors in the methodology used for metering flow. This is important because in-situ calibration is often restricted to dry weather flow conditions which may involve discharges substantially less than those associated with the storm events for which numerical models are typically calibrated.

13.2 Case Study

The Detroit sewer system consists of a combination of combined and separate sewer systems with the central, older portion of the system being primarily combined sewers. Flows from the suburbs are often separated but feed into the central system. Twelve new flow meters were installed in the spring of 1992 for the purpose of calibrating the stormwater flow model for the central city. These meters were installed in sewers ranging from 0.61 to 4.72 m in diameter and were calibrated in-situ over the period April-June, 1992. Details of the meter calibrations are discussed below. Four rainfall events in the spring and summer of 1993 were selected for calibration of the hydraulic model. Difficulties were experienced with the attempts to predict reported discharges at several meter stations. This case study reviews the attempts to quantify the measurement uncertainty in these meter installations.

13.2.1 Description of Metering Systems

The two semi-independent methods of discharge metering involve measurements from a single metering installation. The elements of the meter station include a pressure transducer to determine the flow depth and an acoustic Doppler meter, both of which are mounted at the pipe invert. The pressure transducer output was converted directly into depth from which the sectional characteristics such as the area and hydraulic radius could be computed from the known pipe diameter. The acoustic Doppler meter sends a signal of known frequency through the flow which is subject to a Doppler shift after interaction with particles carried with the flow. This Doppler shift is proportional to the particle velocity, but since there are velocity variations throughout the flow depth, a range of frequencies is received by the meter. The meter functions by analyzing the frequency spectrum and determining the peak frequency which in turn relates to the peak velocity in the flow section. A separate calibration was utilized to relate the average velocity to the peak value and this is subsequently converted into discharge by multiplying by the flow area determined from the depth sensor. This type of metering procedure is distinct from other acoustic Doppler meters (acoustic Doppler current profilers or ADCPs) in which the discharge is estimated by measuring the velocities at a number of locations within the flow cross-section and summing the contributions of the products of local velocities and sub-areas (e.g. Gordon, 1989).

All meters were calibrated in-situ to determine meter coefficients. A number of calibration events (of the order of 6 to 10 per meter) were performed during dry weather flow conditions during April-June, 1992. Generally the calibration flows were at substantially smaller flow depths than experienced during the storm events for which the hydraulic model was calibrated. For example, the calibration conditions for a particular meter (HUF1) involved flow depths in the range of 0.45 to 0.58 m in a 4.11 m-diameter sewer; peak storm discharges sometimes approached the full flow condition.

During the calibration event, the flow depth at the center of the channel was measured with a staff gauge. The installed meter was functional during the calibration so that the pressure transducer output could be directly compared to the staff gauge reading.

Velocity calibrations were performed using the three point method described by ISO 748 (1994) in which the local velocities were measured at 0.2, 0.6 and 0.8 of the flow depth by a Marsh McBirney velocity meter. These velocities were measured at three locations, one at the center of the channel and the other two at roughly the quarter points. The average velocity flow velocity was computed by taking the arithmetic average of these nine velocities except that the 0.6 depth velocity was weighted twice according to ISO 748. The discharge was then computed from the product of the average velocity and the flow area determined from the measured depth and the pipe diameter. The peak velocity was also estimated by measuring velocities in a variety of locations throughout the flow cross-section and recording the maximum value. This peak velocity can be directly compared to the acoustic Doppler meter output while the average velocity was used to establish a conversion from the peak to average velocity.

13.2.2 Discharge Metering Methods

The first method estimates the discharge from the measured depth (from the pressure transducer) and an application of the Manning equation; this discharge is hereafter referred to as $Q_{Manning}$. This method for metering flow is dependent upon an assumption of uniform flow which may not be the case for a number of reasons. There may be backwater effects at the meter station due to restrictions further downstream. In addition, there may be nonuniform flow due to the rising and falling hydrographs during a storm event as the water surface slope may exceed the bottom slope during a rising hydrograph and be less than it for a falling hydrograph.

The specific procedure for the determination of discharge is to calibrate a station-dependent hydraulic coefficient HC which is essentially equal to $S_0^{1/2}/n$ with S_0 the sewer slope and n the Manning resistance coefficient. This is used in the following equation to compute discharge:

$$Q_{Manning} = \frac{1.486 A R_h^{2/3} HC}{F_n} \tag{13.3}$$

where:

- 1.486 = the units factor for US customary units,
- A = the flow area (determined from the depth recorded by the pressure transducer and the known pipe diameter),
- R_h = the hydraulic radius, and
- F_n = a resistance correction factor for partially full flow in a circular conduit that is based on the variation in Manning n suggested by Camp (1946). F_n varies from 1.0 (zero and maximum relative depths) to 1.22 (at a relative depth of 0.3).

The hydraulic coefficient HC is solved from Equation 13.3 for each calibration event using the calibration discharge and the area and hydraulic radius terms which are computed from the measured depth. HC is assumed to be a constant for a particular meter station. Uncertainties during the calibration events include both the precision uncertainties associated with the determination of average velocity and the depth and are considered to be reflected directly by the variation in HC among calibration events. During a real metering application, uncertainties in discharge measurement using this method include the precision uncertainty in the depth measurement (as it affects the area and hydraulic radius terms), the calibration uncertainty in HC, and any possible extrapolation uncertainties that may exist. Extrapolation uncertainties may be associated with backwater or other nonuniform flow effects at the particular meter station, the lack of validity of the specific F_n correction, or other unknown effects.

The second discharge metering method is referred to as $Q_{Continuity}$ and is simply computed as the product of the flow area and average velocity. The depth was measured by the pressure transducer and the flow area computed from the depth and the known pipe diameter. The mean velocity is estimated from the peak velocity registered by the acoustic Doppler meter as discussed previously and converted to the average velocity by utilizing an average to peak velocity ratio $R = V_{average}/V_{peak}$ which was determined by calibration. R was assumed to be constant (independent of depth, etc.) at a particular metering station based on experience with the meters in previous installations but the average values determined for the different metering stations varied from 0.75 to 0.93. In addition, the calibration procedure involved the adjustment of the acoustic Doppler meter gain so that the peak velocities measured with the Doppler meter during the calibration event agree with those measured by the Marsh McBirney meter. Uncertainties in this method involved precision uncertainties in the computation of flow area (through the measured depth) as well as the peak velocity. Calibration uncertainties are due to several factors including the uncertainty in the R value, the uncertainty in the flow area and the uncertainty in the peak velocity measurement. Extrapolation uncertainties could be introduced by systematic variations in R with depth of flow or other unknown effects.

13.2.3 Estimates of Meter Uncertainty

Data recorded during the calibration events were reviewed to obtain estimates of the uncertainties associated with discharge measurements. This data included both the records from the installed meter as well as the independent measurements of depth and velocity collected during the calibration events. The installed meter recorded both depth and peak velocity while the calibration reported depth, peak velocity, and average velocity (as determined by the procedure for averaging the nine point velocities). Deviations among these data were analyzed in order to obtain estimates of meter uncertainty. Independent estimates of meter uncertainty for several of the flow meters were estimated by the vendor that installed the meters (Camp, Dresser & McKee, 1993b). Some aspects of these estimates are similar to those presented below but rely more on the precision estimates recorded by field personnel who were involved in the calibration measurements. Uncertainties estimated for dry weather flows generally ranged from 5 to 15% and independent uncertainties (with no formal basis) about twice the dry weather values were assigned to wet weather events. This range of uncertainties can be compared to a precision error estimate of 0.4% and bias errors of the order of 5% estimated for river discharge measurements with an acoustic Doppler current profiler (Gordon, 1989).

Precision Uncertainty

Estimates of the precision uncertainty in the depth measurements were obtained by comparing the two depths recorded (one by the meter and the other by a staff gauge) during a calibration event. The standard deviation (over all calibration events) of the difference between the depths was taken to be an estimate of the uncertainty in depth measurement. These standard deviations were typically of the order of about 2% of the measured depth for most meter installations. Results for specific meters are presented in Tables 13.1 and 13.2. This range is consistent with the precisions recorded on the calibration sheets by the field personnel regarding their estimates of the measurement precision associated with the measurement of depth from the staff gauge. These standard deviations in depth are used to estimate uncertainties in area and hydraulic radius which ultimately influence discharge measurement.

Table 13.1 Estimates of uncertainty in discharge measurement by $Q_{Manning}$ for three flow meters.

Type of Uncertainty	Meter		
	HUF1	SOF1	NWF2
HC uncertainty	1.7 %	2.2 %	3.5 %
Depth Measurement Area Error	2.5% 5.0%	2.5% 5.0%	0.9 % 1.8%
Extrapolation Uncertainty	34	34	40%
$Q_{Manning}$ Uncertainty	≈41%	≈41%	≈45 %

Note : Estimate in HC uncertainty obtained from standard deviation divided by square root of the number of calibration events. The $Q_{Manning}$ uncertainty is the sum of all components.

The process for estimating the precision uncertainty for the velocity measurements is more convoluted because of the procedures employed. Each calibration sheet recorded an estimated precision of the velocity measurement from the Marsh McBirney meter and this was used directly as an estimate of the precision uncertainty of that meter. In addition, both the acoustic Doppler and the Marsh McBirney meters registered a peak velocity at the time of the calibration. The standard deviation of differences between these (over all calibration events) was computed and used as an estimate in the uncertainty in the peak velocity recorded by the acoustic Doppler meter. This standard deviation was always greater than the reported precision of the Marsh McBirney meter and was typically of the order of 10% of the peak velocity. Values are reported for specific meter installations in Table 13.2. The uncertainty in the velocity conversion factor is the standard deviation of the R (average to peak velocity ratio) values.

Table 13.2 Estimates of uncertainty in discharge measurement by $Q_{continuity}$ for three flow meters.

Type of Uncertainty		Meter	
	HUF1	SOF1	NWF2
Depth Measurement	2.5%	2.5%	0.9 %
Area Error	5.0%	5.0%	1.8%
Velocity Precision	4%	7%	4.5%
Velocity Measurement	6%	10.5%	8.6 %
Velocity Conversion Factor	5%	3.5%	7.4 %
$Q_{Continuity}$	≈15%	≈19%	≈13 %

Note: $Q_{Continuity}$ error is estimated as the sum of the standard deviations in the individual components of the calibration uncertainty divided by the square root of the number of calibration events plus the sum of the standard deviations of the velocity and area precision which are related to the metering event.

Calibration Uncertainty

$Q_{Manning}$: The calibration consists only of the computation of the average HC value over all calibration events and the standard deviations of the individual HC values was computed. In order to reflect the fact that the average was used to compute the calibration factor, the standard error of estimate was used to characterize the calibration uncertainty. The standard error of estimate was computed as $\sigma_{HC} / \sqrt{N}$ in which σ_{HC} is the standard deviation of the HC values while N is the number of calibration events.

$Q_{Continuity}$: Uncertainty in the meter calibration was assumed to be associated with two individual components:

1. The estimate of the uncertainty in the peak velocity which was computed from the standard deviation of the difference between the peak velocities recorded by the two meters; and
2. Uncertainty in the conversion from average to peak velocity which is computed from the standard deviation of the individual R values. Again, since the calibration factors (meter gain to adjust the peak velocity and the R value) represent an average over all individual calibration events, the standard error of estimate was used to compute the calibration uncertainties associated with these variables.

Extrapolation Uncertainty

Since the calibration events were for dry weather conditions and at substantially lower depths than the data for calibrating the hydraulic model, the possibility of systematic deviations of the meter calibrations with depth admits the possibility of considerable error in the metered flows. There are a number of factors that could introduce a systematic variation between the metered and actual discharges including:

1 A systematic variation in the average to peak velocity ratio with relative depth. It was assumed that the R value is meter station-specific but constant over the depth range and the basis for this assumption may be questioned. The calibration data for individual events were examined as a function of relative depth in order to determine whether or not any systematic trends can be observed. Data for two individual meters with the largest change in relative depth over the range of calibration flows are presented in Figure 13.1 while the calibration data for all twelve meters is included in Figure 13.2. Although the R values appear to increase with relative depth for both meters in Figure 13.1, the opposite trend is indicated for other meters in Figure 13.2 and the variations for an individual meter are generally within the uncertainty in the calibration measurements and no trend in the data is apparent.

2. Lack of uniform flow due to backwater or other effects, thus invalidating the assumption upon which $Q_{Manning}$ is based. Any systematic deviations might be station-specific but in general, an increasing backwater effect with discharge should result in a systematic over-estimation of discharge with increasing relative depth due to the fact that the water surface slope would be less than the pipe slope, i.e. the HC value should decrease with relative depth. Figures 13.3 and 13.4 present the variations in HC with relative depth

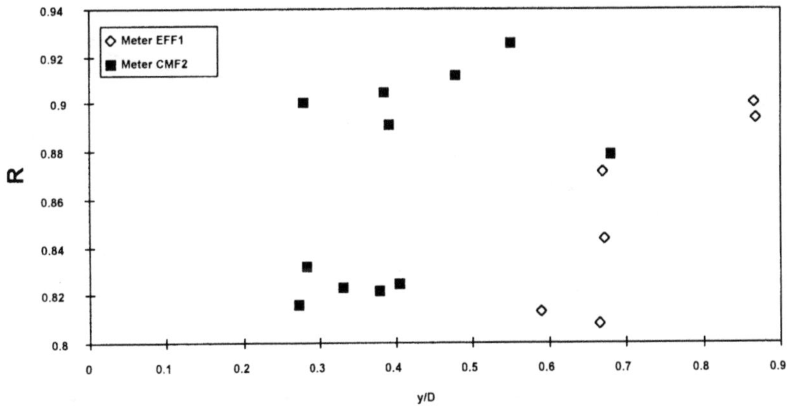

Figure 13.1 Variation in average to peak velocity with relative depth for selected meters.

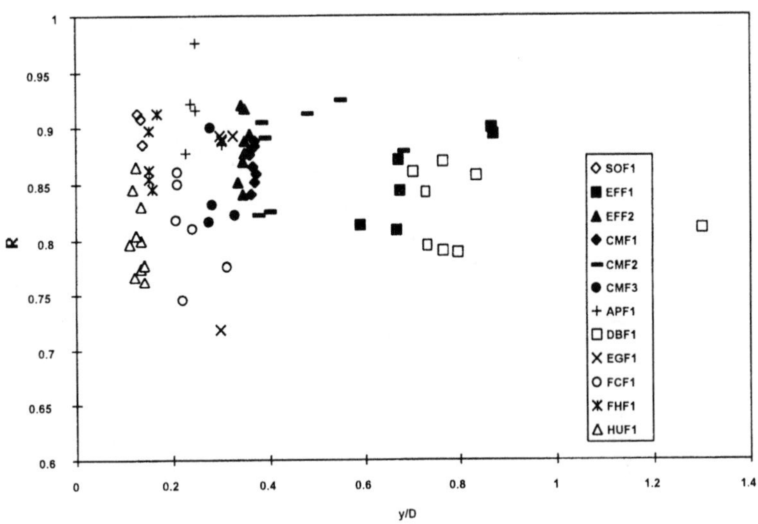

Figure 13.2 Variation in average to peak velocity with relative depth for all meters.

for two meter stations which show an opposite trend of increasing HC with depth; this trend was observed for most of the other meter installations.

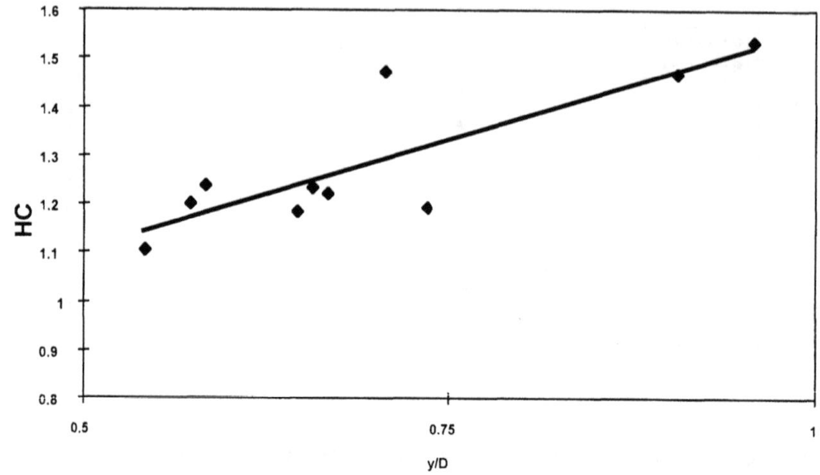

Figure 13.3 Variation in HC with relative depth for the meter NWF2.

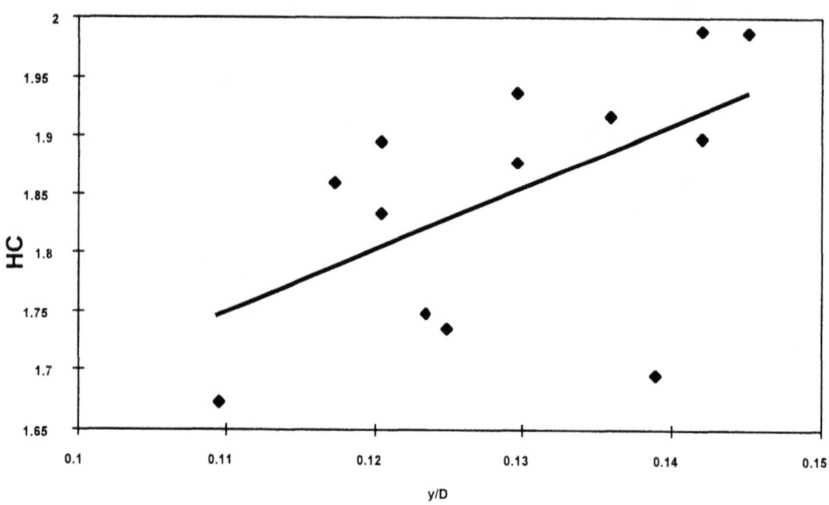

Figure 13.4 Variation in HC with relative depth for the meter HUF1.

Overall Uncertainty

The overall uncertainty in the measurement of discharge may be estimated by summing the individual contributions in each of the three areas noted above. This was done for three representative meter stations and the results are presented in Tables 13.1 and 13.2, for $Q_{Manning}$ and $Q_{Continuity}$, respectively. As noted above, the standard error of estimate was used to estimate the uncertainty in calibration coefficients where these coefficients are obtained from the averages of a set of calibration data. Although recognizing the possibility of an extrapolation error on the average to peak velocity ratio, this was not included in Table 13.2 due to the inability to detect clear trends in the calibration data for the meters in question. The extrapolation uncertainty in HC was estimated by fitting a trend line to the HC variation with relative depth and extrapolating to the full pipe condition. It is seen from Table 13.1 that this is a major contributor to the meter uncertainty for the estimate of $Q_{Manning}$ and cautions against the use of this type of meter in an application which is far outside the range of calibrated depths. On the other hand, the large number of individual components in the $Q_{Continuity}$ determination contribute to a larger overall meter uncertainty when the extrapolation uncertainty is neglected with typical uncertainties of the order of 15 to 20% and somewhat higher than estimated by the meter vendor for these stations.

13.2.4 Evaluation of Metering Records

It was subsequently decided to make a more detailed investigation at one meter site (HUF1) in order to gain some sense of the validity of the uncertainty estimates presented in Tables 13.1 and 13.2. This particular meter is in a 4.11 m-diameter pipe which was calibrated at depths less than 0.6 m and therefore potentially subject to a significant extrapolation error. An additional acoustic Doppler meter was installed at approximately the pipe mid-depth so that the velocity indicated by each Doppler meter could be checked at higher flow rates. The purpose of this action was to determine whether the transmission of the signal through a sediment laden flow might affect the velocity measurement. In addition, a plan to estimate the flow rate by the dye dilution method as an independent method for discharge determination was developed. Although the intention was to implement these procedures for high discharge conditions, there were very few rainfall events over the time period allotted for this investigation. Only one event, on January 28, 1994 resulted in a significant record where both velocity meters were operational for an extended period of time. Dye dilution studies during high flow events have not yet been conducted.

Comparison Between Two Doppler Velocity Meters

The second Doppler velocity meter was activated whenever the flow depth at the meter station was greater than about 1.4 m. This occurred for the six dates. One problem noted in the velocity records for the second meter (identified as MO2) was that whenever the water depth was at the level at which it was initially activated, an erroneously low velocity was registered. In these situations the recorded velocity may be half or less of the velocity recorded by the bottom mounted meter. In order to avoid errors associated with this occurrence, data was discarded at the beginning and ends of the meter records for MO2 whenever it appeared that this effect was present. Consequently, the entire record for one of the six dates was discarded. A total of 44 different velocity comparisons (recordings were made at ten minute intervals) over the five remaining dates were judged to be valid.

The MO2 meter was not verified in place due to the difficulty of calibrating at high discharges. Therefore, the possibility exists of a velocity offset between the two meters. The mean and standard deviation of the differences between the two meter readings were computed; these were 0.070 and 0.108 m/s respectively. The difference of 0.070 m/s (the bottom velocity meter registered greater velocity on average) represents a difference of approximately 4% of the velocity measured (ranging from 1.43 to 1.83 m/s) and thus the difference between the two meters is relatively small. An examination of the records indicates that MO2 registers greater than the bottom velocity meter as the water level first rises high enough to activate it, but then falls below it later in the record. This trend is not sufficiently clear and the number of measurement events are too small to draw definite conclusions, and the effect is small in any case. The standard deviation of 0.108 m/s represents 6.4% of the average velocity for the bottom velocity meter. In the previous analysis, the precision with which velocity could be measured with the Doppler meter was estimated to be approximately 4% for station HUF1. Assuming that the precision is the same for the additional meter MO2, the 6.4% standard deviation falls within the uncertainty in the difference between the two meters. Other components of the uncertainty in discharge determination (such as depth measurement or the peak to average velocity ratio) cannot be assessed from this data set.

Results of Dye Dilution Study

Dye dilution studies were performed on two dates, October 25 and 28, 1993 and involved approximately 45 minutes of dye sampling for which discharges were recorded at five minute intervals. Discharge records with both the stage-discharge calibration ($Q_{Manning}$) and the Doppler meter ($Q_{Continuity}$) were available for this same time period. The discharge computed by the dye dilution procedure for the October 25 event varied from 0.55 to 0.58 m^3/s with an average value of 0.564 m^3/s. The variation of the discharge over the sampling period appears to be fairly random with a slight drop in flow rate over the course of the measurement. The metered discharges for the same period also indicate that the flow rate was relatively constant during the same time period. The record for $Q_{Manning}$ indicates a discharge of approximately 0.67 m^3/s while $Q_{Continuity}$ indicates a discharge of about 0.71 m^3/s with a trend of slightly decreasing discharge over the time interval for which the dye dilution study was performed. The discharge computed by the dye dilution method for the October 28 event varied from 0.48 to 0.55 m^3/s with an average value of 0.50 m^3/s. The first three discharge values are greater than subsequent readings; however discounting these as erroneous, due to some sort of measurement error, only drops the average flow to 0.49 m^3/s. The metered discharges for the same period indicate that the flow rate decreased slightly during the same time period. The records for both $Q_{Manning}$ and $Q_{Continuity}$ indicate a discharge of about 0.67 m^3/s over the time interval for which the dye dilution study was performed.

The discrepancies between the discharge determined from the dye dilution study and the other methods are greater than can be explained on the basis of estimated uncertainties in discharge measurement. In Tables 13.1 and 13.2, the estimated uncertainty in discharge measurement for meter HUF1 was about ± 9% for $Q_{Manning}$ and ± 15% for $Q_{Continuity}$. Observed deviations range from 15.3% for $Q_{Manning}$ on October 25 to around 26% for both $Q_{Manning}$ and $Q_{Continuity}$ on October 28.

Large Discharge Metering

In order to investigate the nature of any possible extrapolation errors, The velocity records for meter station HUF1 were examined to determine the differences between $Q_{Manning}$ and $Q_{Continuity}$ recorded at higher flow rates. Flow data was provided for the period between April 6 and July 17, 1992, for October 10 through December 18, 1993 and for the events which activated the meter M02 in 1994. This data is in the form of discharges at ten minute intervals for both $Q_{Manning}$ and $Q_{Continuity}$. Rainfall events that triggered an increase in discharge at station HUF1 were identified from the data records. The estimate of $Q_{Manning}$ should be incorrect on both the rapidly rising and rapidly falling portions of the hydrograph. The rising portion should be associated with an underestimate of the discharge as the friction slope (the flow acceleration will contribute to this as well) should be greater than the channel slope as the flood wave propagates down the pipe. On the falling leg of the hydrograph, the opposite should be true due to a deceleration of the flow. A detailed inspection of several rainfall events verified this relative trend. A representative plot of the two discharge records for a typical event is presented in Figure 13.5 although it is obscured by the fact that the magnitudes of the discharge peaks from the two meters do not correspond. It was decided to use data only at the peak flow rate for an individual event; this also simplified the data manipulation. A ratio of $Q_{Continuity}$ to $Q_{Manning}$ was computed at each local peak in the depth hydrograph for HUF1 for which the depth exceeded about 0.76 m and for which both discharge measurements were recorded. The ratio of these is presented in Figure 13.6. The data are broken into two sets depending upon whether the measurements were taken before or after June 6, 1992. The Doppler meter was replaced at HUF1 on June 6, 1992 when it was interpreted as providing a faulty signal. Figure 13.6 shows a distinct

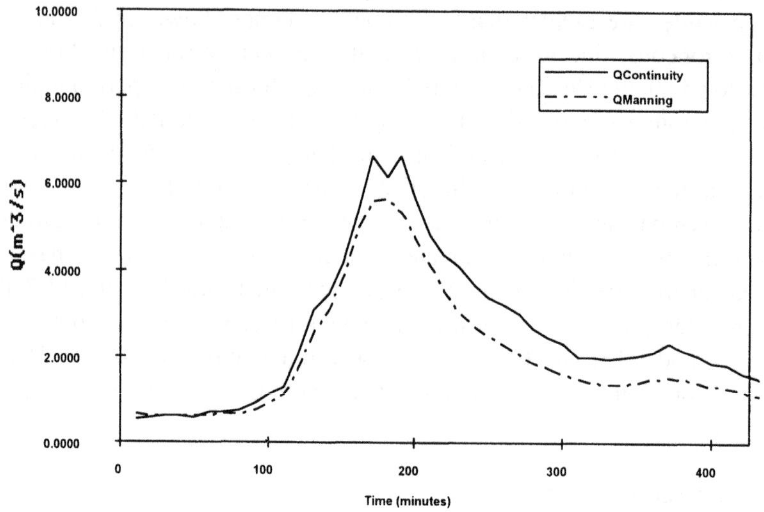

Figure 13.5 Comparison of discharge hydrographs for $Q_{Manning}$ and $Q_{Continuity}$ for a typical storm event.

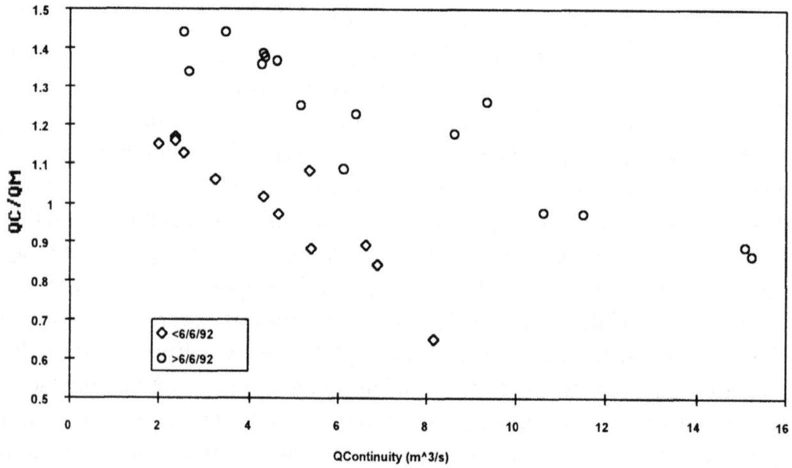

Figure 13.6 Ratio of $Q_{Continuity}$ to $Q_{Manning}$ determined at discharge peaks for storm events.

difference in the magnitude of the ratios for the period prior to June 6 and all data collected thereafter. However, the trends in the ratios are very similar both before and after the meter replacement with an offset in the magnitudes. The meters at station HUF1 were calibrated at discharges between about 0.54 and 1.10 m^3/s

(prior to the June 6 replacement date) and so the discharge ratio should be close to 1.0 in that range. This is apparent in the dry weather flow plots for late 1993 and the fact that the average value of the ratio for a four day dry weather period (6/13/92-6/17/92) is 0.98. This range of discharges is not presented in Figure 13.6 since only higher flow rates associated with storm peaks are included. As the discharge ($Q_{Continuity}$) increases from about 0.71 m^3/s, the ratio $Q_{Continuity}/Q_{Manning}$ increases to about 1.5 at a discharge of about 2.1 m^3/s. This ratio decreases until the two recorded discharges are nearly equal at the highest flows, which have a flow depth about two-thirds the pipe diameter.

In Figure 13.4, it is indicated that the hydraulic coefficient tends to increase with relative depth over the range of the calibrations. A linear extrapolation of this trend to a full pipe condition would result in about a 40% increase in the hydraulic coefficient above the average value determined in the calibrations and used to determine $Q_{Manning}$. However if the hydraulic coefficient did in fact increase, then the discharge computed using the average low flow coefficient would be too low and therefore the ratio $Q_{Continuity}/Q_{Manning}$ would be too large. Although the ratio is greater than one, this cannot be the explanation for the decreasing ratio indicated in the figure, since the discrepancy becomes less with increasing discharge rather than greater. Therefore, it is concluded that there must be some other explanation for the variation noted. It is not known which of the two discharge determinations are correct due to the lack of an independent way of verifying the flow rate, but it is clear that there is an extrapolation error in at least one of the discharge measurements that is of the order of 50% for this metering station. An educated guess would be that the error is in $Q_{Manning}$, because the two different Doppler meters (prior and subsequent to June 6, 1992) show similar trends.

13.3 Conclusions

An analysis was performed in order to estimate and quantify sources of uncertainty in the metering of discharge by two types of low loss meter that were installed to meter flow in a sewer system. These particular meters were calibrated in-situ during dry weather flow conditions in which discharges were substantially less than the range for which results were to be obtained for purposes of calibrating a numerical hydraulic model. In addition to issues associated with meter uncertainty, this calibration approach admits the possibility of considerable extrapolation error if the assumptions upon which the meter calibration is founded are not satisfied. An estimate of the relative contributions of precision, calibration, and extrapolation uncertainty indicate that extrapolation uncertainty may be a controlling factor in overall meter precision. Even when the possibility of extrapolation uncertainty is discounted, the meter uncertainty for the acoustic

Doppler meter was estimated to be of the order of 15 to 20% which is at the limit of the stated calibration goals for the development of the hydraulic model. Given the estimated levels of uncertainty for these meters, these calibration goals were probably unrealistic.

There is clear evidence of a major extrapolation error in at least one of the discharge measurement methods as the two disagree by about 50% for certain ranges of discharges. The ratio of the two flow rates when plotted against system discharge quite clearly indicates a systematic variation, indicating that the calibration method for at least one of the flow rates is inadequate. Based on the limited information, it is suggested that the $Q_{Manning}$ measurement is in error, but this cannot be verified without additional independent discharge measurements at higher flow rates. This finding indicates that all discharges which have been estimated by extrapolation far outside the range for which the meters were calibrated should be viewed with caution unless additional verification is obtained.

Notation

A	Flow area
F_n	Manning n adjustment factor for partially full flow in circular pipe
HC	Hydraulic coefficient used to determine $Q_{Manning}$
n	Manning n
$Q_{Continuity}$	Discharge determined with acoustic Doppler meter
$Q_{Manning}$	Discharge determined on basis of uniform flow assumption
R	Ratio of average to peak velocity
R_h	Hydraulic radius
S	Sample standard deviation
S_e	Standard error of estimate
S_o	Pipe slope
$V_{average}$	Average flow velocity at meter station
V_{peak}	Maximum flow velocity in flow cross-section
σ_{HC}	Standard deviation of HC values

References

Camp, T. R. (1946) "Design of Sewers to Facilitate Flow," Sewage Works Journal, Vol. 18, pp. 3-16.

Camp, Dresser and McKee (1993a) "Model Calibration and Validation Procedure," Technical Memo 2, DWSD Combined Sewer Overflow Study.

Camp, Dresser and McKee (1993b) "Flow Metering Error Analysis," Technical Memo 18, DWSD Combined Sewer Overflow Study.

Dearden, H.T. (1992) "Achieving Flow Accuracy," Intech, Vol. 39 , pp. 37-37.

Gordon, R.L. (1989) "Acoustic Measurement of River Discharge," Journal of Hydraulic Engineering, Vol. 115, No. 7, pp. 925-936.

Granger, R.A. (1988) "Experiments in Fluid Mechanics," Holt, Rhinehart, and Winston, Inc., New York.

ISO 5168-1968 "Measurement of Fluid Flow - Estimation of Uncertainty in a Flow-Rate Measurement," International Organization for Standardization.

ISO-748-1979(E) "Liquid Flow Measurements in Open Channels - Velocity-area Method," International Organization for Standardization, 2nd ed., 1994.

Parshall, R.L. (1926) "The Improved Venturi Flume," Transactions ASCE, Vol. 89, pp. 841-851.

Pomroy, J. (1996) "Selecting the Right Flowmeter," Chemical Engineering, pp. 94-102.

Chapter 14

City of Toronto Experience: the Process of Environmental Approval for the Western Beaches Storage Tunnel

Wayne Green

The City of Toronto's sewer system, parts of which date back to the mid 1800's was originally developed as a combined system, carrying both sanitary and storm flows. As the City rapidly expanded in the early half of the century, the capacity of the combined sewer system was quickly exceeded and basement flooding became a recurring problem.

To solve this problem, the City constructed approximately 680 km of storm sewers in the mid 1960's which separated approximately 70% of the road drainage from the combined sewer system. The combined sewer system which exists throughout the City today, however, continues to pick up sanitary flow and storm runoff from homes and buildings constructed prior to 1965, as well as storm drainage from the remaining 30% of the roadways which have not been separated.

Naturally, during periods of heavy rainfall, the capacity of the combined sewers and interceptor sewers are exceeded, resulting in the excess combined flows discharging directly to the receiving rivers and Lake Ontario.

The Ontario Ministry of Environment investigated the impact of these combined sewer overflows (CSO) and storm outfalls in a 1990 Wet Weather Outfall Study for the Toronto waterfront and determined that the 24 combined sewer overflows and the 29 stormwater outfalls (Figure 14.1) discharge approximately five million cubic metres of flow annually into the City's waterfront from May to October and contribute approximately 1,700 metric tonnes of total

© *Advances in Modeling the Management of Stormwater Impacts - Vol. 5* W. James, Ed.
Pub. by CHI, Guelph, Canada 1997. ISBN 0-9697422-7-4. Fax: +519 767-2770

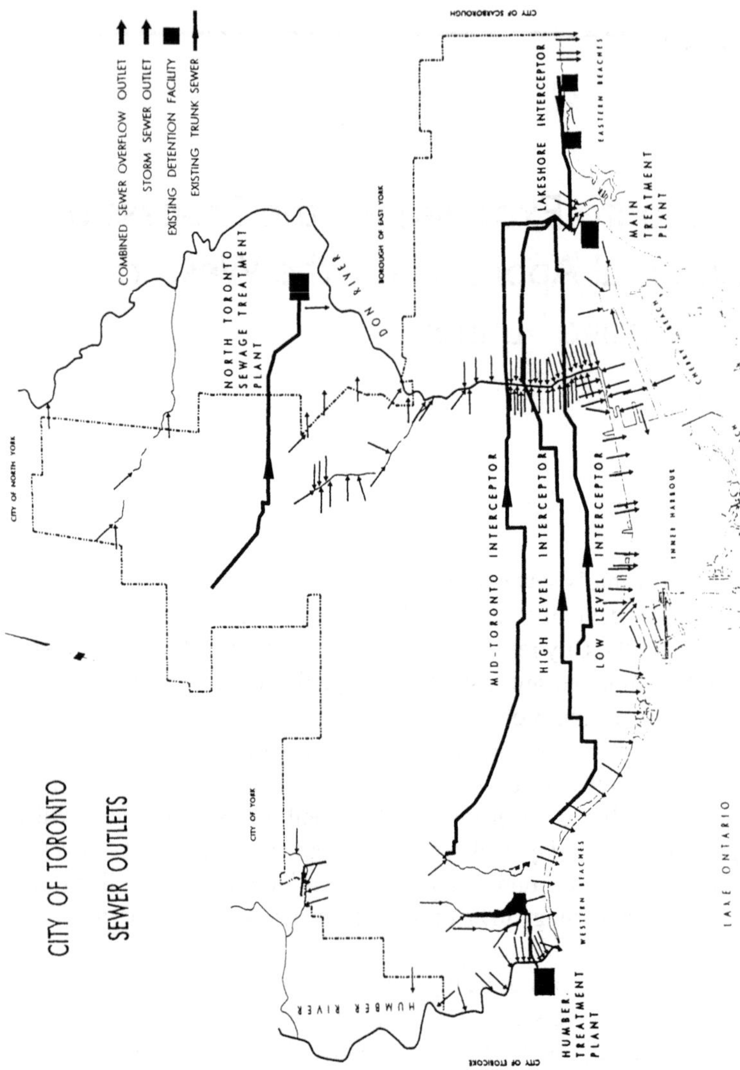

Figure 14.1 City of Toronto sewer outlets.

suspended solids and 6,650 metric tonnes of chemical oxygen demand to the near shore waters, along with concentrations of phosphorous, lead, zinc, aluminium and iron (Metropolitan Toronto Waterfront Wet Weather Outfall Study Phase II, August 1995).

In total, these outfalls severely impact the recreational use of the Toronto waterfront and result in frequent *unsuitable for swimming* beach posting by the Medical Officer of Health.

To solve this problem, the City undertook a Sewer System Master Plan (SSMP) in 1990 with the objective of virtually eliminating the pollutant loadings associated with combined sewer overflows and the control and treatment of stormwater runoff where required.

Virtual elimination of CSO in the context of the SSMP was defined as reducing CSO to an average of one overflow or less per year for the Western Beaches or, alternatively, reducing the average total annual volume of CSO discharging to the Western Beaches by more than 90%. The City chose this criteria which exceeds the Provincial Guidelines for CSO discharges due to the extensive parkland system and swimming beaches along this section of the shoreline.

14.1 Recommended Scheme

The SSMP examined several alternative solutions to solve the CSO and pollutant loading problems, including: the complete separation of storm and sanitary systems on private property and on the remaining 30% of the roads not previously separated; a combination of end of pipe storage tanks and tunnels; and a solution involving a number of near surface storage tanks distributed throughout the sewer system.

The recommended solution under the SSMP includes a combination of sedimentation ponds, tanks and a continuous storage tunnel. The essential elements recommended (see Figure 14.2) are as follows:

- two CSO storage tanks for the Eastern Beaches area;
- marshland habitats and settling ponds in the Grenadier Pond and Humber River areas;
- several swirl concentrators at specific outfall locations along the Don River and Humber River;
- a 5.5 m to 8.0 m diameter (end of pipe) storage tunnel extending 10 km from Parkside Drive in the Western Beaches area to Coxwell Avenue; and
- a second 3 km long storage tunnel is also recommended to extend north along the Don River to the City limits.

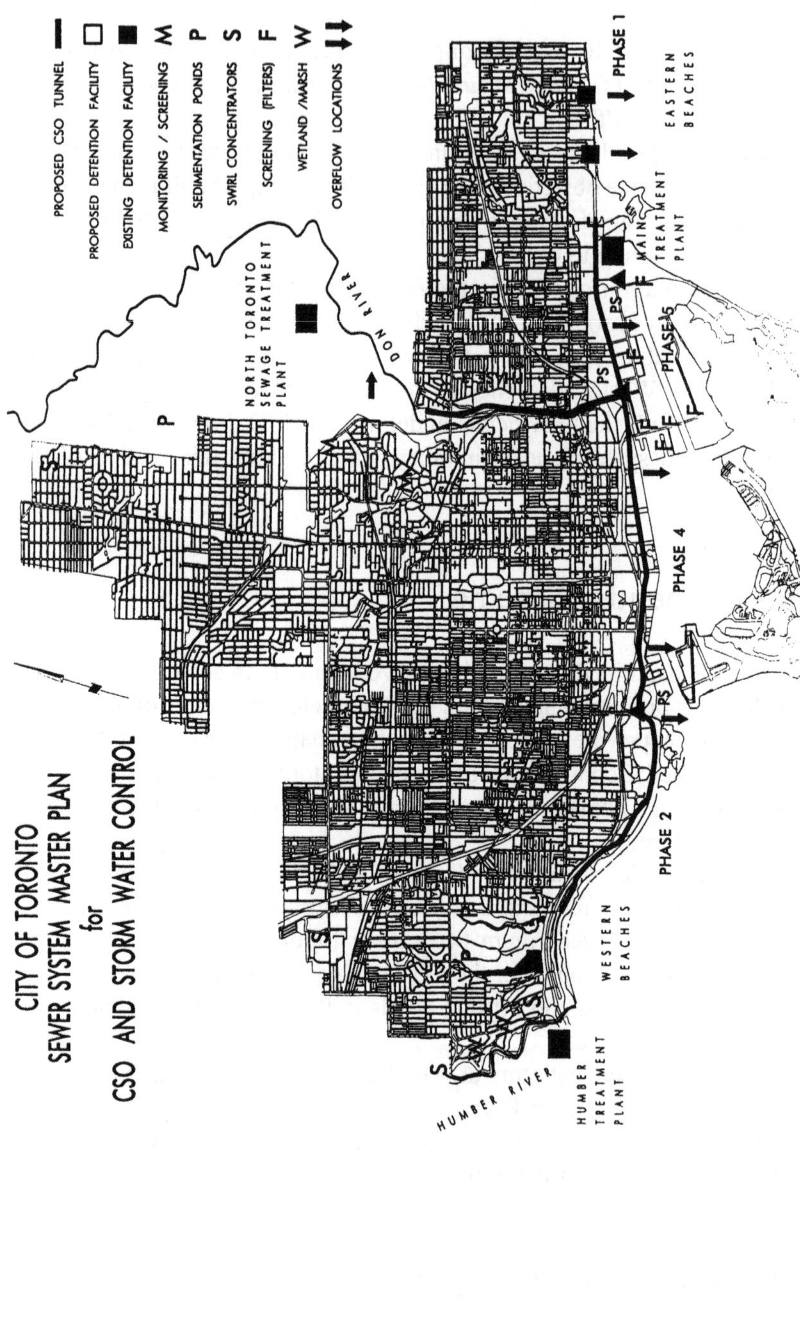

Figure 14.2 City of Toronto sewer system master plan.

A total of 92 storm and 47 combined sewer outfalls are planned to be intercepted by these storage/treatment facilities.

The SSMP improvements are broken down into approximately fifteen component projects prioritized over a 25-year implementation period and have a total cost of approximately $370 million (1996 Cdn$).

Thus far, the first priority projects under the SSMP have been completed including two detention tanks to intercept two CSO and six stormwater outfalls which previously discharged directly into the Eastern Beaches swimming area. These tanks provide approximately 10,300 m^3 of storage and have significantly improved the near-shore water quality for these beaches to meet or exceed the Ministry Guidelines. Further, the construction of a 2,400 m^3 open sedimentation pond for stormwater treatment entering Grenadier Pond commenced in February of 1996 and will be completed by mid-summer of 1996.

In addition to the SSMP recommendations, City Council have approved several non-structural alternatives that are complementary to the SSMP initiatives. These include a voluntary downspout disconnection program, the use of porous pavements, a pilot program for the use of rain barrels, and soak pits. Council requested that these methods be evaluated and incorporated into the SSMP to the greatest extent possible (Figure 14.3).

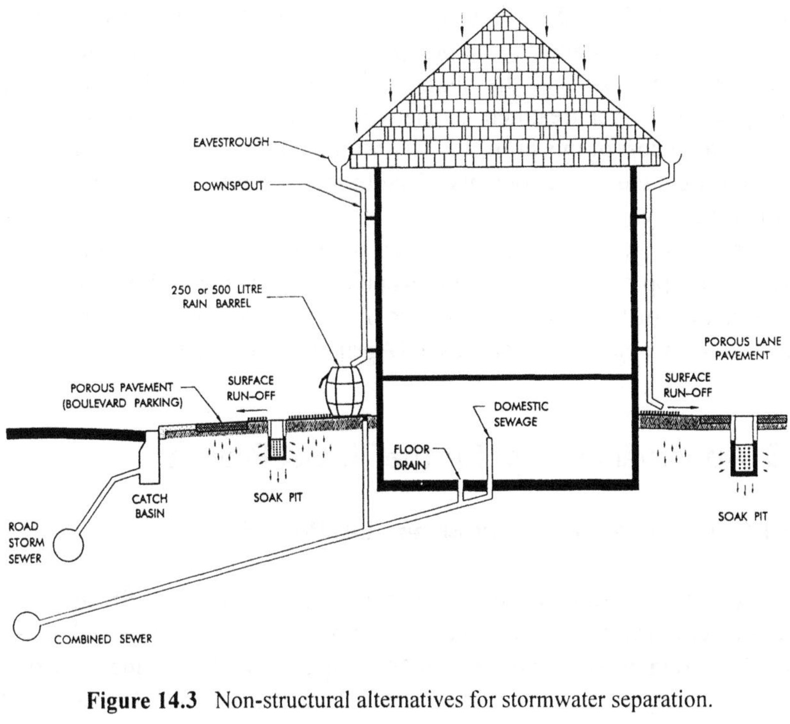

Figure 14.3 Non-structural alternatives for stormwater separation.

14.2 Western Beaches Tunnel

Although work is well underway on using non-structural alternatives and several stormwater diversion and treatment methods, it is important to realize that due to surface and soil conditions as well as the voluntary participation rates by the public, these initiatives are limited in the amount of stormwater which can be removed from the sewer systems. The City's stormwater model predicts that the current City's non-structural programs will reduce the total annual volume of CSO and storm runoff by approximately 15%, however, the remaining 85% of CSO and stormwater which the City's stormwater model estimates to be 2.7 million m^3 annually will continue to impact the near-shore waters along the Western Beaches. Clearly, therefore, the Western Beaches Tunnel is needed in order to meet the objective of one discharge per year or a reduction of 90% of the annual CSO volume.

The first phase of the Western Beaches tunnel system will extend along Lake Shore Boulevard 4 km from Parkside Drive on the west to Strachan Avenue on the east (Figure 14.4). This section of tunnel will be 5.5 m diameter, have a storage capacity of 95,000 m^3 and intercept and store the flows from eight combined sewer outfalls and two storm outfalls which currently discharge into the Western Beaches. The initial operation of the tunnel will result in flows being stored in the tunnel for approximately 24 hours to allow the sludge to settle, after which the clarified water will be pumped through an ultra violet disinfection chamber to the Lake and the sludge will be pumped to the mid-Toronto Interceptor Sewer for treatment at the Main Sewage Treatment Plant. The estimated total cost of this project is $57 million (1996 Cdn$). Ultimately, it is planned to treat the entire contents of the tunnel at a future Metro wet weather treatment facility.

The tunnel was designed to reduce the CSO overflow volume to Lake Ontario from approximately 2.7 million cubic meters to 0.2 million cubic meters or a reduction of 91% and the number of overflows from an average of 80 occasions to two occasions per season (April to November).

14.3 Approval Process for the Tunnel Project

14.3.1 Class Environmental Assessment (EA) Process

The City decided to process each component project of the SSMP through a Class Environmental Study process. This process which is referred to as the "Class Environmental Assessment for Municipal Road, Water and Wastewater Projects", was approved in June 1993 under the Environmental Assessment Act for various types of municipal infrastructure projects. The Western Beaches

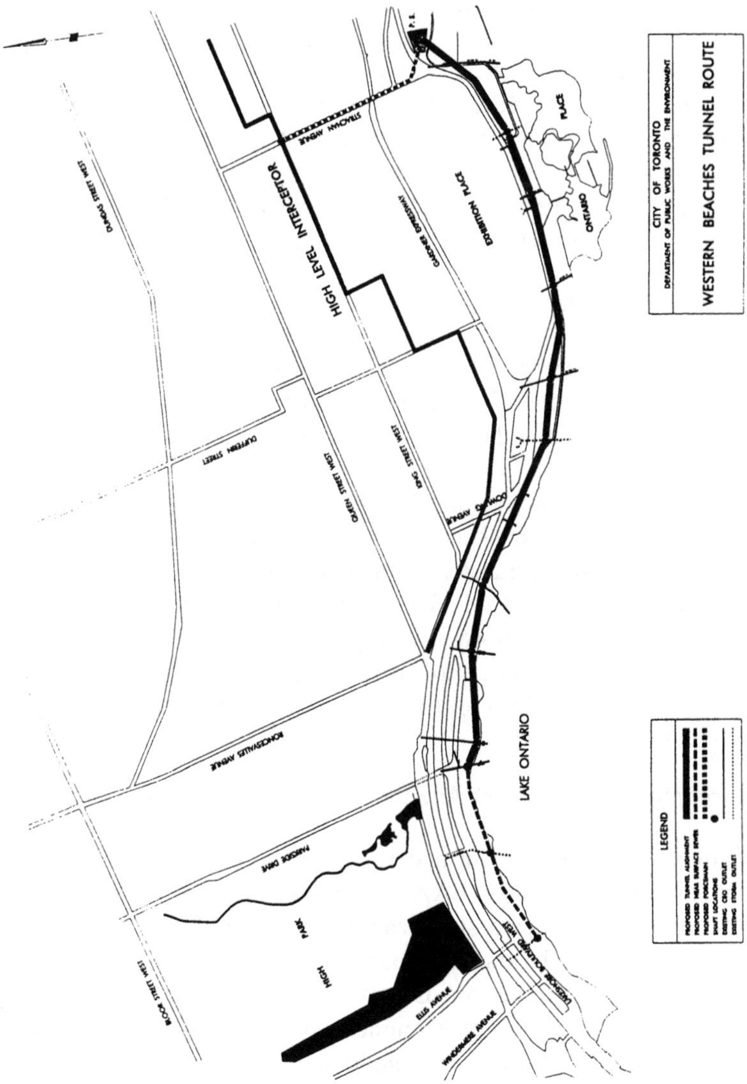

Figure 14.4 Western Beaches tunnel route (4 km).

project was considered as a Schedule 'C' project under the Class process. The Class Environmental Study process requires the proponent (municipality) to define the problem(s), investigate in consultation with the public alternative solutions to the problem(s), and determine the preferred solution(s).

The Class Environmental Assessment process combined with the comprehensive review of alternatives carried out under the SSMP reconfirmed the end-of-pipe tunnel as the preferred solution for control and treatment of water pollutants discharging to the Western Beaches. The results of the Class EA process was summarized in an Environmental Study Report and filed with the City Clerk, local library, the Ministry of the Environment and Energy and local citizens groups for the mandatory 45-day appeal period.

14.3.2 Bump-up Requests

Notwithstanding the approximately three years of the SSMP process and Class EA study process, a total of fourteen objections were filed with the previous Minister of the Environment and Energy requesting a bump-up to a full Environmental Assessment.

The major objections to the tunnel solution related to the approval of the tunnel in advance of the approval and siting of a wet weather treatment facility by Metropolitan Toronto, the transfer of sludge collected in the tunnel from the west side of the City to the east side for treatment and the desire on the part of interest groups to use natural non-structural systems exclusively to deal with the CSO problems, e.g. downspout disconnections, stormwater infiltration, water conservation, rain barrels, cleaning of street surfaces and catch basins, best management practices, etc.

14.3.3 Environmental Assessment Advisory Committee Review

The Minister of Environment and Energy appointed an Environmental Assessment Advisory Committee (EAAC), to review the objections and advise the Ministry on the environmental significance of the unresolved issues between the objectors and the City. To assess the unresolved issues, EAAC conducted a series of meetings with the objectors, City staff, project consultants and Metro staff. EAAC's report to the Minister recommended a conditional approval of the project. However, notwithstanding the commitment on the part of the City and Metro to meet most of the conditions, the Minister chose to grant the bump-up request for a full environmental assessment.

14.3.4 Exemption Order

The current Ministry of the Environment and Energy, following a request from City Council, subsequently reassessed the environmental benefits of the tunnel project and recognizing Metro's support for the project and the City's ongoing commitment to divert the maximum amount of stormwater possible through such programs as the downspout disconnection and porous pavement programs, have drafted an Exemption Order for Cabinet's consideration, who on December 20, 1995 approved the project. Preliminary engineering work is now underway and a Design/Build proposal call is planned for early in 1997 with construction to commence in mid 1997.

14.4 Conclusions

Experience on this project has shown that, notwithstanding the preparation and filing of a complete and well-documented Environmental Study report and the efforts and research undertaken throughout the Sewer System Master Planning process, additional efforts may have proven beneficial. Involvement of interested community groups and citizens both during the planning process and throughout the Class EA Study, should be more aggressively sought.

This may involve establishing community liaison groups and possibly the use of community bulletin boards or newsletters to explain and inform the public of the objectives and merits of the alternative solutions and ultimately the preferred tunnel solution.

The possible additional costs and time associated with this approach may be offset in reducing the number of, or expeditiously resolving, bump-up requests if they occur in the final stages of the EA process.

Also, once community liaison groups are established at the Class EA stage, these same groups can prove to be effective construction liaison groups to mitigate construction impacts.

This suggested broad-based, aggressive community involvement not only ensures the project concerns and issues are dealt with early in the process but also may bring forward community groups and individuals who are in support of the project objectives, who might otherwise be silent observers. If bump-up requests occur later in the EA process, this support may demonstrate a better community balance for the project.

Finally the cost and timing of the planning and Class EA process for a project of this type should not be underestimated. For example, the Class EA process on this project began in October 1991 and the cost, thus far, including the initial SSMP process has exceeded Cdn $1 million.

Chapter 15

A Study of the Impacts and Control of Wet Weather Sources of Pollution on Large Rivers

Christine Hill, John Lyons and Mike Hulley

The Ohio River Sanitation Commission (ORSANCO) has undertaken a study of the sources of wet weather pollution in large rivers. The study is to be completed within two years with work in the first year concentrating on defining wet weather sources of pollution while work in the second year will concentrate on evaluating effective wet weather controls. The project includes both extensive modeling and monitoring of water quality. At this point (March 1996), year one is nearing completion.

As part of this study, a complete XP-SWMM model of the drainage system within the Greater Cincinnati area was developed. This model included interceptor sewers, trunk sewers, all combined sewer overflows (CSOs), significant sanitary sewer overflows (SSOs), treatment plants, and receiving streams and rivers. Both flow and water quality data were modeled for the entire system. The model was calibrated using existing stream tributary flow and concentration data, and interceptor flow and concentration data.

The purpose of the modeling is to generate pollutographs from direct sources to the Ohio River (a significant number of CSOs and three wastewater treatment plants) and tributary streams. Tributary streams within the study area receive CSO and SSO discharges during wet weather. The results of the model are to be interfaced with a WASP model of the Ohio River, currently being developed by another member of the study team.

© *Advances in Modeling the Management of Stormwater Impacts - Vol. 5* W. James, Ed.
Pub. by CHI, Guelph, Canada 1997. ISBN 0-9697422-7-4. Fax: +519 767-2770

The model was used to generate pollutographs for fecal coliform and suspended solids for a total of four wet weather events during the fall of 1995. These four events were selected due to the availability of event mean concentration data from water quality monitoring. Statistical analyses on collected data were completed to define event mean concentrations for CSO discharges.

Model results were compared against data collected during an extensive water quality monitoring program. Water quality was monitored in tributary streams. A comparison of modeled results with measured results indicated good agreement between predicted concentrations of fecal coliform and measured concentrations.

In the next year, the developed model will be utilized to evaluate alternatives for controlling wet weather sources of pollution into the Ohio River

15.1 Introduction

A detailed study of the wet weather sources of pollution has been undertaken for the Ohio River within the states of Ohio and Kentucky by ORSANCO. The overall study consists of water quality modeling of the Ohio River and its tributaries, continuous flow monitoring at CSO and stream locations, and event water quality monitoring using grab samples at CSO locations, stream and river locations. Figure 15.1 presents the overall framework of the study and shows the interaction between the modeling and monitoring phases of the study. This chapter details the modeling approach for the direct discharges into the Ohio River from the Greater Cincinnati Area.

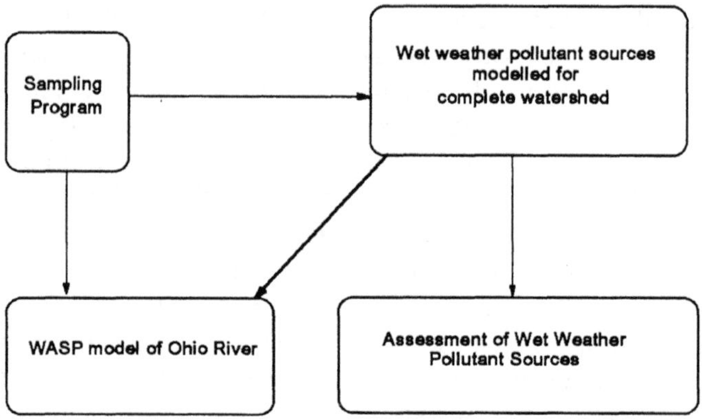

Figure 15.1 Model framework.

15.2 Description of the Study Area

The Greater Cincinnati Area is located within Hamilton County and encompasses all or a portion of 50 municipalities. The study area consists of three large drainage areas: Muddy Creek drainage area; Mill Creek drainage area; and Little Miami River drainage area. Each of these drainage areas is serviced by its own collection system and wastewater treatment plant which discharge directly to the Ohio River. Terrain within the study area is generally rolling with elevations ranging from 955 feet (291 m) in the northern portions of the study area to 455 feet (138 m) at the Ohio River. Table 15.1, 15.2 and 15.3 presents the general characteristics of the Little Miami River, Mill Creek and Muddy Creek drainage areas, respectively.

Table 15.1 Little Miami River drainage area characteristics.

Characteristic	Description
Major Water Courses	• Duck Creek: Drainage area is 27.5 square miles. Of the 18 sub-watersheds contributing to Duck Creek, 14 drain to combined sewers which transport flow to the Little Miami Wastewater Treatment Plant. • Lower Little Miami River: Drainage area of 51.9 square miles. • Upper Little Miami River: Drainage area of 1591 square miles.
Land Use	• Duck Creek: Residential or commerical land uses occupy 71% of the drainage area. • Lower Little Miami River: 53% of the drainage area is developed. About 18% of the area is considered undevelopable due to slope and floodplain restrictions. • Upper Little Miami River: Mostly rural and agricultural land uses.
Sewers	• Duck Creek: Approximately 82% of the drainage area is sewered. About 60% of the sewered area served by combined sewers while remaining 40% of area is served by separate sewers. • Lower Little Miami River: About 34% of the drainage area is sewered with less than 1 % of the sewered area served by combined sewers.
Pump Stations	• There area a total of 27 wastewater pumping stations in the Little Miami Creek drainage area. Most of the pumping stations are in the East Little Miami drainage area in areas not served by gravity sewers.
SSOs	• There are 19 SSOs which discharge directly to adjacent storm sewers or receiving streams.
CSOs	• Of the 56 CSOs in the Little Miami drainage area, 10 are mechanical regulators, 33 are drop gates, and 13 are diversion dams. A total of 11 CSOs overflow as a result of high stage in the Ohio River.

Table 15.2 Mill Creek drainage area characteristics.

Characteristic	Description
Major Water Courses	• South Mill Creek: The drainage area is 63 square miles or about 38% of the total drainage area. • West Branch Mill Creek: The drainage area is 41 square miles or about 25% of the total drainage area. • East Branch Mill Creek: The drainage area is 62 square miles or about 37% of the total drainage area.
Land Use	• South Mill Creek: About 85% of the drainage area is urban land use (residential, commerical or industrial). Only 15% is open space or undevelopable lands. • West Branch Mill Creek: As of 1990, almost 83% of the drainage area was urban land use. • East Branch Mill Creek: About 70% of the drainage area is urban land use.
Sewers	• South Mill Creek: About 75% of the drainage area is served by a combined sewer system. The remaining areas are served by separate sanitary sewers or is unsewered. • West Branch Mill Creek: Only 7% of the drainage area is served by combined sewers. Most of the centre and northern portion of the drainage basin is unsewered. • East Branch Mill Creek: This drainage area is primarily served by a separate sanitary sewer system.
CSOs	• Of the 158 regulating structures in the Mill Creek Drainage Area, 47 are mechanical regulators, 57 are drop gates and 54 are diversion dams.

Table 15.3 Muddy Creek drainage area characteristics.

Characteristic	Description
Major Water Courses	• Muddy Creek : The drainage area is 17 square miles. • Rapid Run Creek: The drainage area is 7 square miles. • River Road: A narrow band of land along each side of River Road provides an additional 6 square miles of drainage area.
Land Use	• Abount 70% of drainage area is urban land use (residential, commerical of industrial).
Sewers	• About 41% of the drainage area is serviced by a combined sewer system.
CSOs	• There are 20 CSO in the drainage area. The CSOs are either diversion dams or drop gate structures.

Overall, a total of 227 CSO locations has been identified within the Greater Cincinnati Area, with a total of 20 CSOs located within the Muddy Creek drainage area, a total of 151 CSOs located within the Mill Creek drainage area, and a total of 56 CSOs located within the Little Miami River drainage area. In addition, a number of SSO locations have been identified. The most significant of the SSO locations are within the Muddy Creek drainage area. Figure 15.2 presents an overall plan of the study area, including the locations of all CSOs.

15.3 Model Development

Computer models were developed to provide predictive tools for the estimation of wet weather pollutant loadings. The models were used to generate estimates of pollutant loadings at specific locations as a function of, among other things, rainfall.

As part of an earlier Facilities Planning Study undertaken for the Greater Cincinnati Area (BBS, 1995), XP-SWMM models of the existing collection systems in the three drainage areas were constructed. These models included all areas tributary to the wastewater treatment plants, all CSO locations, a number of significant SSO locations, all major pumping stations, and all interceptor pipes. These models were calibrated using flow data obtained from six permanent flow monitoring stations located in interceptors and trunk sewers. Rainfall and flow data from January to June 1993 were used to calibrate and verify the models. The calibration process produced reasonable agreement between computed and actual flows (BBS, 1994).

A detailed hydraulic capacity analysis was performed to determine the hydraulic capacity of specific elements, such as individual CSOs, under dynamic flow conditions. The results of this analysis were applied to construct a continuous XP-SWMM model. Using a typical rainfall year, average CSO loadings were estimated. Table 15.4 presents the computed average annual CSO volumes for the three drainage areas and the Ohio River.

The current study involved the extension of the existing models to include the following:

- addition of four watercourses (Muddy Creek, Rapid Run Creek, Mill Creek and the Little Miami River) tributary to the Ohio River complete with boundary locations for Mill Creek and the Little Miami River;
- addition of three wastewater treatment plants (WWTPs) including the Muddy Creek WWTP, the Mill Creek WWTP, and the Little Miami WWTP; and
- water quality modeling explicitly to estimate pollutant loadings into the Ohio River from tributary watercourses and from direct sources such as CSOs and wastewater treatment plants.

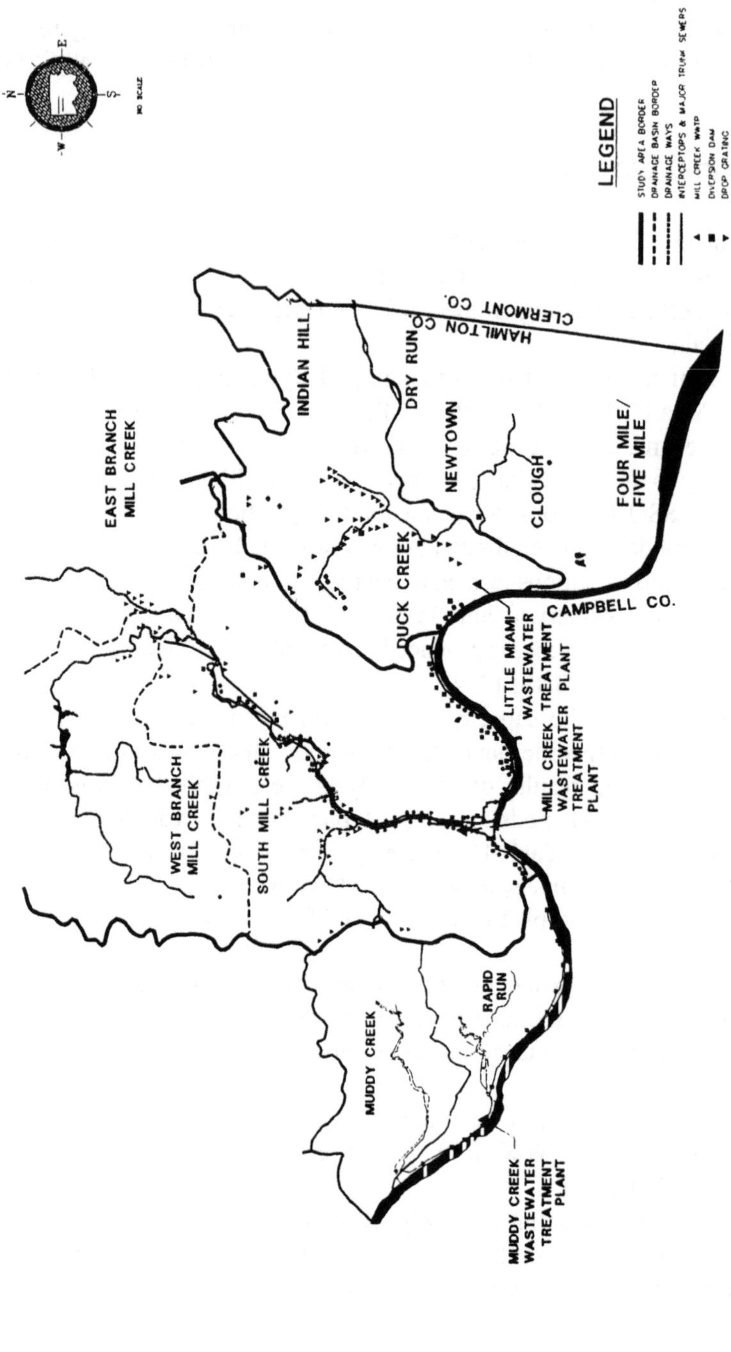

Figure X.2 Study area.

Table 15.4 Average annual CSO loadings.

Source		Average Annual CSO Volume (MG)
Tributary Sources	Little Miami River Drainage Area (including Duck Creek, Little Duck Creek, Clough Creek, and the Little Miami River)	1008
	Mill Creek Drainage Area	3084
	Muddy Creek Drainage Area (including Muddy Creek and Rapid Run Creek)	586
Direct Sources	Ohio River (Muddy Creek Drainage Area)	661
	Ohio River (Little Miami River Drainage Area)	338
	Ohio River (Mill Creek Drainage Area)	1050

Tributary watercourses were modeled using cross sectional data obtained from the U.S. Geological Survey (USGS). Boundary locations were demarcated within the Mill Creek and Little Miami River Drainage Areas. Continuous flow monitoring data was also obtained from permanent USGS stations. For Mill Creek, the upper boundary was selected at Carthage, while the upper boundary for the Little Miami River was placed at Millford. These sites were selected because of the availability of both flow and water quality data.

The models were also extended to include the three wastewater treatment plants within the Greater Cincinnati Area which discharge directly into the Ohio River. Table 15.5 presents a summary of the treatment plant capacities.

Table 15.5 Summary of WWTP treatment capacities.

Location	Capacity (MGD)				
	Raw Sewage Pumping Capacity	Hydraulic Limitations	Primary Treatment Capacity	Secondary Treatment Capacity	Disinfection Treatment Capacity
Muddy Creek WWTP	22	N/A	22	15	30
Mill Creek WWTP	520	N/A	420	240	455
Little Miami WWTP	260	55	118	100	100

Water quality was modeled explicitly using the build-up and washoff routines in XP-SWMM. Event mean concentrations were used to generate pollutographs for direct and indirect sources of pollutants to the Ohio River. Only suspended solids and fecal coliform were modeled explicitly with pollutographs for BOD5, ammonia-nitrogen (NH3-N), organic nitrogen, nitrate nitrogen (NO3-N), total phosphorous, copper, lead, and zinc derived as a constant fraction of suspended solids. The event mean concentrations used for suspended solids and fecal coliform are summarized in Table 15.6. These event mean concentrations were developed from statistical analyses of the results of a number of sampling programs, including the extensive monitoring program undertaken as part of this study. This monitoring program will continue in the next year and it is hoped that data collected during the two-year study will allow a relationship to be developed between land use and event mean concentrations.

Table 15.6 Event mean concentrations.

Source	Suspended Solids (mg/L)	Fecal Coliform (#/100mL)
CSO/SSO	155	1,000,000
Stormwater	93	53,000
WWTP Effluent	12	10
Little Miami River Boundary	149	3400
Mill Creek Boundary	114	7430

The updated model was used to generate pollutographs for direct and indirect sources of pollution into the Ohio River. Figure 15.3 presents a schematic of the updated XP-SWMM model.

15.4 Model Calibration

Data for model calibration and verification was obtained from a number of sources including the following:

- Permanent flow monitors operated by the USGS, located in Mill Creek and the Little Miami River. In particular, flow data at the upper boundaries of Mill Creek and the Little Miami River were used to calibrate a RUNOFF node.
- Data from the extensive water quality sampling program undertaken as part of the overall project was used to calibrate both direct

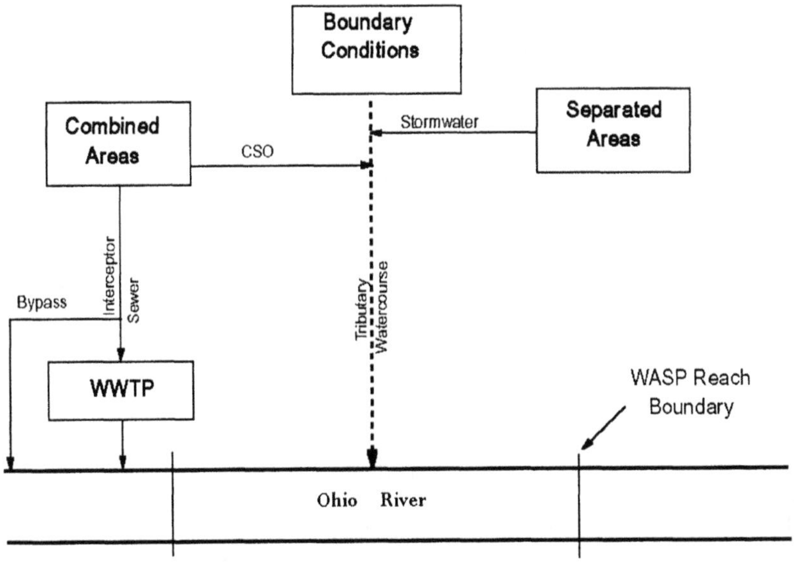

Figure 15.3 Conceptual XP-SWMM model setup.

and indirect pollutant sources. Pollutant concentrations from both CSO locations and tributary watercourses were calibrated and verified using the data collected during the water quality sampling program.

The water quality sampling program consisted of dry and wet weather sampling in the Ohio River and its tributaries, CSO sampling, and Ohio River biomonitoring. During four selected wet weather events, the monitoring program provided a complete dataset of pollutant concentrations from a number of CSOs, the Ohio River at a number of stations, and the tributary watercourses at a number of stations. Monitoring in the Ohio River consisted of samples taken at the left and right banks as well as at the centre of the river.

The XP-SWMM models were calibrated and verified to achieve good agreement between measured and modeled flows and concentrations. Figures 15.4 and 15.5 present the model calibrations of upstream boundary conditions for Wet Weather Event No. 3 of Mill Creek and the Little Miami River, respectively. Figures 15.6 and 15.7 present the model calibrations of fecal coliform concentrations for Wet Weather Event No. 3 in Mill Creek and the Little Miami River, respectively.

The XP-SWMM model output for the four selected wet weather events was used in the calibration of the Ohio River WASP model.

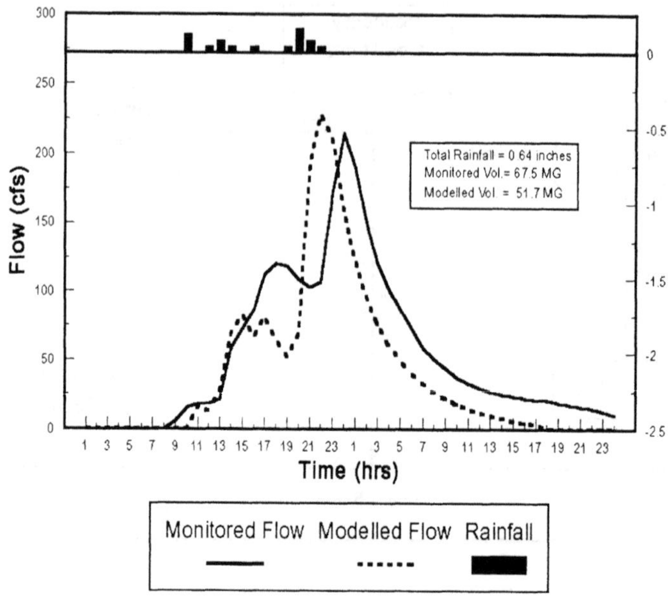

Figure 15.4 Calibration of upstream boundary flows, Mill Creek - wet weather event No. 3.

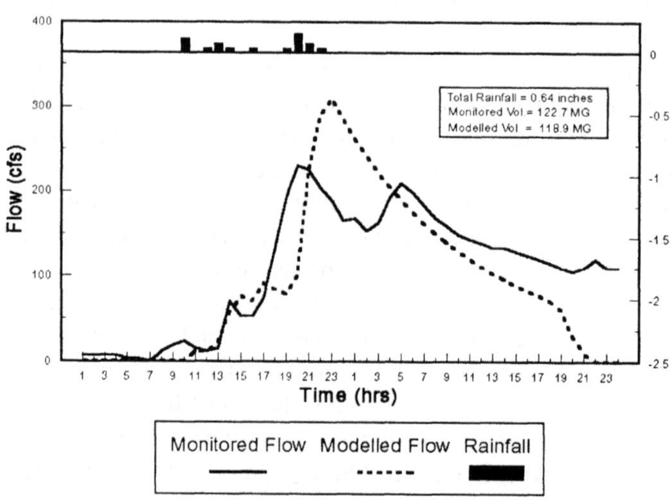

Figure 15.5 Calibration of upstream boundary flows, Little Miami River - wet weather event No. 3.

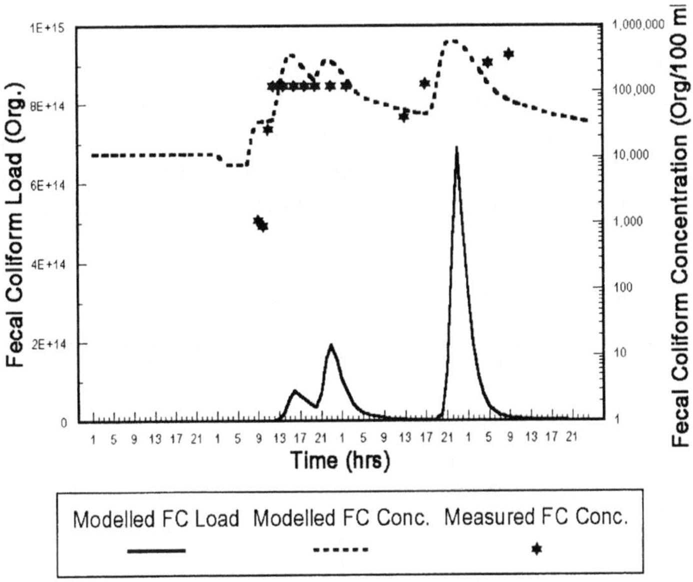

Figure 15.6 Calibration of fecal coliform concentration, Mill Creek at the Ohio River - wet weather event No. 3.

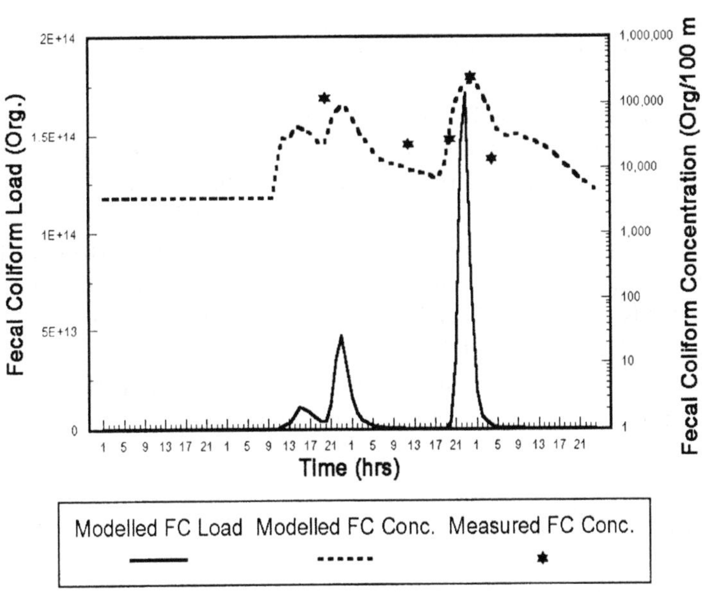

Figure 15.7 Calibration of fecal coliform concentration, Little Miami River at the Ohio River - wet weather event No. 3.

15.5 Model Results

As part of year one of the ORSANCO study, model results were generated for a total of four wet weather events. Model results consisted of total fecal coliform and suspended solids loadings entering the Ohio River from several sources within the study area, including tributary watercourses, wastewater treatment plants, CSOs and stormwater. Table 15.7 presents a summary of the model results.

Table 15.7 Computed suspended solids and fecal coliform loadings to the Ohio River.

Event	Percentage of Total Suspended Solids Load				Percentage of Total Fecal Coliform Load			
	CSO	WWTP	Trib. Water-courses	Storm-water	CSO	WWTP	Trib. Water-courses	Storm-water
Event #1	0.5	13	87	0.5	15	10	73	2
Event #2	0	13	87	0	1	0.2	98	0.8
Event #3	5	14	79	2	24	5	70	1
Event #4	4	10	85	1	33	3	63	1

Examination of Table 15.7 shows that tributary watercourses are computed to be the major sources of both suspended solids and fecal coliforms to the Ohio River from the study area on an event basis contributing 79% to 87% of computed suspended solids loads and 63% to 98% of computed fecal coliform loads. On an event basis, wastewater treatment plants contributed 10% to 13% of computed total suspended solids loads and 0.2% to 10% of computed fecal coliform loads. CSOs contributed 0% to 5% of computed suspended solids loads and 1% to 33% of computed fecal coliform loads on an event basis. Based on these results, stormwater is computed to be an insignificant contributor of both suspended solids and fecal coliforms to the Ohio River.

15.6 Future Program

The second year of the ORSANCO study will concentrate on enhancing modeling tools. An extensive monitoring program will be implemented in order to refine input model parameters such as event mean concentrations. Other model improvements will include the addition of sanitary sewer overflows (SSOs), regulator dry weather bypassing due to high stages in the Ohio River, improved tributary watercourse cross-sectional information to improve the routing of flows, improved boundary conditions, and improved treatment plant hydraulics and treatment efficiencies. In addition, continuous modeling will be undertaken to study the typical, wet and dry annual loadings to the Ohio River from the study area. As in the first year, model results generated will be interfaced with a calibrated WASP model of the Ohio River.

Continuous modeling will also be used to study a number of control alternatives including the removal of all CSOs, the removal of all SSOs, and the removal of all regulator dry weather bypassing. It is hoped that the effects of pollutants entering the river from different sources will be better understood from the results of the model and that this understanding will lead to the development of a control strategy from improving water quality within the Ohio River.

As in the first year, an extensive monitoring program is planned to provide additional data for model calibration and verification purposes.

References

BBS, 1994, Corporation Study Team, Combined Sewer Overflow Strategy Development and Facilities Planning, Technical Memorandum No. 2, Model Development, Prepared for the Metropolitan Sewer District of Greater Cincinnati, p8-16, October 5, 1994.

BBS, 1995, Corporation Study Team, Combined Sewer Overflow Strategy Development and Facilities Planning, Phase III - CSO Implementation Plan, Prepared for the Metropolitan Sewer District of Greater Cincinnati, March, 1996.

Chapter 16

Cost Effectiveness of Urban Runoff and Combined Sewer Control Options

Christine Hill and Barry J. Adams

The negative long term impacts of urban runoff discharges and combined sewer overflows (CSOs) have been widely documented over the past twenty years. Short term and long term impacts can have a negative impact on public water supply, recreational water use and aquatic species. In order to reduce negative impacts, control options can be applied to reduce pollutant mass into receiving water bodies. To achieve maximum effectiveness, a series of source and "end-of-pipe" control options can be applied.

A methodology was developed to assess the effectiveness of a variety of control options. The methodology employed a statistical modeling approach. The statistical approach is meant to be used in preliminary and planning level studies. The approach utilized involved the definition of the statistical properties of rainfall. Closed-form equations have been developed in the past to determine the long term effectiveness of extended detention ponds, wet ponds, and underground storage tanks in removing suspended solids (Guo, 1992). These models were extended to predict the long term effectiveness of wet detention ponds, underground storage tanks, infiltration basins, porous pavements, and outfall treatment in removing both particulate and soluble pollutants (Moroz, 1994).

The least cost for a particular level of service for a control option was determined from available cost data. Capital, operating and maintenance, and land costs were all included in the tabulation of the total cost of a particular option. Cost-effectiveness curves were then generated for each option and level of service.

© *Advances in Modeling the Management of Stormwater Impacts - Vol 5.* W. James, Ed. Pub. by CHI, Guelph, Canada 1997. ISBN 0-9697422-7-4. Fax: +519 767-2770

The results of the analyses on urban drainage control options determined that infiltration ponds with sand base and wet ponds were the most cost-effective options for reducing pollutant loadings from urban drainage discharges. Both options were found to be effective in removing suspended solids, BOD_5, nutrients, heavy metals, and fecal coliform loadings from urban drainage.

The results of the analyses on combined sewer control options determined that outfall treatment with roof disconnection would be the most cost-effective option in removing suspended solids, BOD_5, and fecal coliforms. However, sensitivity analyses showed that, under some circumstances, other options were the most cost-effective.

16.1 Background

Over the last 20 years, the negative impacts of urban runoff and CSOs on water bodies have been studied intensively. Short term impacts, associated largely with bacteria and dissolved oxygen, negatively impact public water supply, recreational use (Hvitved - Jacobsen, 1986) and aquatic species diversity. Long term impacts are typically associated with suspended solids, nutrients, and heavy metals. The presence of these pollutants in receiving waters may result in gill damage to some fish species, excessive algae growth and reductions in aquatic species due to heavy metals ingestion.

In order to reduce these negative impacts, numerous control options have been proposed to reduce urban runoff and CSOs. Control options can be divided into two categories: source controls and downstream controls. Often a comprehensive control strategy will involve both types of control measures. Source control typically involves reducing urban runoff volumes entering a drainage system and includes such options as roof leader disconnection. Downstream controls involve "in pipe" or "end-of-pipe" controls and concentrate on the treatment of discharges entering receiving waters.

16.2 Methodology

In order to assess the effectiveness of a variety of control options, a statistical modeling approach was utilized. Statistical models were developed in order to describe the efficiency of control options in removing suspended solids, BOD_5 , nutrients, bacteria, and heavy metals. These models are derived from the probability density functions of the following rainfall event characteristics (Adams, et al., 1986):

$$f_\upsilon(\upsilon) = \zeta e^{-\zeta\upsilon}, \zeta = \frac{1}{\upsilon} \quad (16.1)$$

$$f_B(b) = \psi e^{-\psi b}, \psi = \frac{1}{b} \quad (16.2)$$

$$f_T(t) = \lambda e^{-\lambda t}, \lambda = \frac{1}{t} \quad (16.3)$$

where:

υ = the rainfall event volume (mm),
b = the interevent time (hrs), and
t = the event duration (hrs).

The values of the parameters used in the present study are $\zeta = 0.193$ /mm, $\psi = 0.0142$ /hr, and $\lambda = 0.288$ /hr and were derived from the Toronto Bloor Street Meterological station. Through the application of derived probability theory, Guo (1992) developed the following equation to predict the long term effectiveness of extended detention ponds and underground storage tanks in removing suspended solids:

$$C_p = E_d \Omega\left(\frac{1}{\lambda} - \overline{T_d}\right)\frac{\xi}{\phi C_1} \times 100 \quad (16.4)$$

where:

C_p = the pollution control effectiveness
E_d = the average removal efficiency,
Ω = the outflow rate (mm/hr),
C_1 = a constant dependent upon the value of depression storage,
$\overline{T}_d$ = the average settling time, and
$(1/\lambda - T_d)$ = the time in which dynamic settling occurs.

The final expression for T_d has been derived by (Guo, 1992) as follows:

$$\overline{T_d} = \frac{S}{C_1^2 C_4^2}[(\Omega + C_5)C_2C_3 - C_5](1 - C_1C_3) + \left(\frac{1}{\lambda} - \frac{1}{\lambda}C_1C_4^2\right)(1 - C_1C_4) - \frac{S}{\Omega}C_1C_2C_3C_4 \quad (16.5)$$

where:

$$C_1 = e^{-C_1 S_d} \quad (16.6)$$

$$C_2 = e^{-\psi \frac{S}{\Omega}} \tag{16.7}$$

$$C_3 = e^{-\lambda \frac{S}{\phi}} \tag{16.8}$$

$$C_4 = \frac{\lambda\phi}{\lambda\phi + \zeta\Omega} \tag{16.9}$$

$$C_5 = \frac{\phi}{\psi\phi + \zeta\Omega} \tag{16.10}$$

and:

ϕ = the runoff coefficient,
S = the storage volume in mm, and
S_d = the depression storage in mm.

Using the same methodology, a series of equations were developed to predict the long term effectiveness of wet detention ponds, underground storage tanks, infiltration basins, and porous pavements in removing both particulate and soluble pollutants from stormwater (Moroz, 1994).

In the study of CSOs, analytical models previously derived to determine pollutant control efficiencies of storage and interception capacity (Li, 1991) were extended to predict the efficiency of outfall treatment in removing particulate and soluble pollutants (Moroz, 1994).

Tables 16.1 and 16.2 present the suspended solids removal equations derived for both urban runoff control and CSO control options, respectively.

16.3 Cost of Control Options

In order to determine cost-effectiveness, the capital, operating and maintenance and land costs were estimated for each control option. Data from recent studies conducted as part of the Remedial Action Plan Program in Ontario (CH_2M HILL, 1992) were used to estimate capital and operating and maintenance costs. Table 16.3 presents the cost equations used to estimate capital and operating and maintenance costs for each control option.

It was recognized that ponds and underground storage tanks would typically require a significant land area to construct. Therefore, land costs were calculated for ponds and underground storage tanks. An equation was developed relating land costs to runoff coefficient and storage volumes based on a regression analysis of existing data (Moroz, 1994). The following equation was derived:

$$C_{LC} = 29.816(\phi S)^{2.11} \tag{16.11}$$

where:

C_{LC} = the land cost in 1995 dollars,
ϕ = the runoff coefficient, and
S = the storage volume of the pond in m^3.

Table 16.1 Suspended solids removal for urban runoff control options.

Control Option	Removal Equation	Source
Extended Detention Ponds, Underground Storage Tanks	$E_d\Omega(\frac{1}{\lambda} - \overline{T_d})\frac{\zeta}{\phi C_1} x100$	Guo (1992)
Infiltration Ponds, Porous Pavements	$1-\dfrac{\frac{\lambda}{\Omega}\frac{\psi}{\Omega}+\frac{\zeta}{\phi}e^{-(\frac{\psi}{\Omega}+\frac{\zeta}{\phi})S}}{\frac{\lambda}{\Omega}+\frac{\zeta}{\phi}\frac{\psi}{\Omega}+\frac{\zeta}{\phi}}$	Adams et. al. (1986)
Wet Detention Ponds	$\dfrac{e^{-\zeta(\frac{S}{\phi}+Sd)}\frac{SV_sA}{S+V_s}+(1-e^{-\zeta(\frac{S}{\phi}+Sd)S})}{\frac{\phi}{\zeta}e^{-\zeta Sd}} \int_{b=0}^{b=\infty}\int_{\upsilon=0}^{\upsilon=\frac{S}{\phi}+Sd} f(\upsilon,b)d\upsilon db$	Moroz (1994)

Table 16.2 Suspended solids removal by CSO control options.

Control Option	Removal Equation	Source
Downstream Storage, Increased Interceptor Capacity	$1-\dfrac{\frac{\lambda}{\Omega}\frac{\psi}{\Omega}+\frac{\zeta}{\phi}e^{-(\frac{\psi}{\Omega}+\frac{\zeta}{\phi})S}}{\frac{\lambda}{\Omega}+\frac{\zeta}{\phi}\frac{\psi}{\Omega}+\frac{\zeta}{\phi}}$	Adams et al (1986)
Outfall Treatment	$n_s\dfrac{(R-Pu_1)}{R}+n_o\dfrac{(Pu_1-Pu_t)}{Pu_1}$	Moroz (1994)

Table 16.3 Cost of control options (1995 Canadian dollars).

Extended Detention Ponds	$857V^{0.5461}$, V is the storage volume in m^3.
Wet Detention Ponds	$1500V^{0.5461}$, V is the storage volume in m^3
Underground Storage Tanks	$4048V^{0.79}$, V is the storage volume in m^3
Roof Leader Disconnection	0.827AR, AR is the disconnected roof area in ha.
Porous Pavement	$1157SA_p+180.75A_p$, A_p is the pavement area in ha. and S is the depth of stone in mm.
Increased Interceptor Capacity	$504{,}800Q^{0.67}$, Q is the interceptor capacity in mm/hr
Sewer Separation	$176{,}400A_S$, A_S is the area separated in ha.
Outfall Treatment	$17{,}493Q^{0.82}$, Q is the outfall treatment rate in mm/hr

16.4 Cost Effectiveness Analysis Methodology

The purpose of the cost-effectiveness methodology is to determine the least-cost option for a particular level of pollution control performance. Therefore, the methodology involves minimizing the cost of each option considered, subject to performance constraints. For two inputs, the cost-effectivness curve can be developed from the expansion path by plotting the level of service versus the cost at each tangency point where tangency points represent the minimum cost to achieve the level of pollution control performance. For optimization problems with two or more variables, this procedure can also be applied. The cost minimization problem can be expressed mathematically as follows:

Minimize $$C_T = f_c[(X_i - X_{io}),...(X_n - X_{no})]; i = 1,...n \qquad (16.12)$$

subject to: $$Y_k(X_i,...X_n) = Y_{ko}; k = 1,...m \qquad (16.13)$$

$$X_i \geq X_{io}, i = 1,...n \qquad (16.14)$$

where:

C_T = the total cost of the control option,
$f_c[(X_i - X_{io}),...]$ = the total cost of the control option dependent upon n variables,
n = the number of variables,
$Y_k(X_i,...X_n)$ = the kth performance constraint,
X_i = the total value of the ith variable,

X_{io} = the existing value of the ith variable, and
m = the total number of performance constraints.

By employing the Lagrange multiplier method, the following Lagrange function can be formed by combining the objective function with the constraint equation:

$$Z = f_c[X_i - X_{io},...(X_n - X_{no})] + E_k(Y_k(X_i,...X_n) - Y_{ko}) \quad (16.15)$$

where:

E_k = the kth Lagrange multiplier or the marginal cost of an additional unit of Y_{ko}.

In order to optimize the performance, the objective function was differentiated with respect to variables X_i,...X_n and multipler E_k and the derivatives set to zero. The resulting set of equations was then solved to determine the optimal mix of measures. The solution to this set of equations is the tangency point between the isoquant and isocost lines. Li (1991) developed this methodology for water quality analysis.

16.5 Cost Effectiveness Analysis Results

Cost-effectiveness curves were developed for each control options for each pollutant of interest. The following sections summarize the findings of the completed analyses.

16.5.1 Urban Drainage Control Options

Figures 16.1 and 16.2 present the cost-effectiveness curves for urban drainage control options in removing suspended solids and fecal coliforms respectively. The results of the analysis showed infiltration ponds with sand base and wet ponds were the most cost-effective options. Both of these control options are capable of removing suspended solids, BOD_5, nutrients, heavy metals, and fecal coliform from urban drainage. These control options probably could not both be utilized in a specific area, since infiltration ponds require high hydraulic conductivity soils in order to encourage infiltration while wet ponds normally require a low permeability clay lining in order to maintain a permanent pool of water. The results of analyses for other pollutants were found to be similar.

The analyses determined that storage tanks had limited effectiveness in removing fecal coliforms (maximum of 40%) due to the short residence time of these tanks. In addition, storage tanks had high costs. Furthermore, the analyses showed that even with a roof leader disconnection program in place, storage tanks still had a poor effectiveness in removing fecal coliforms.

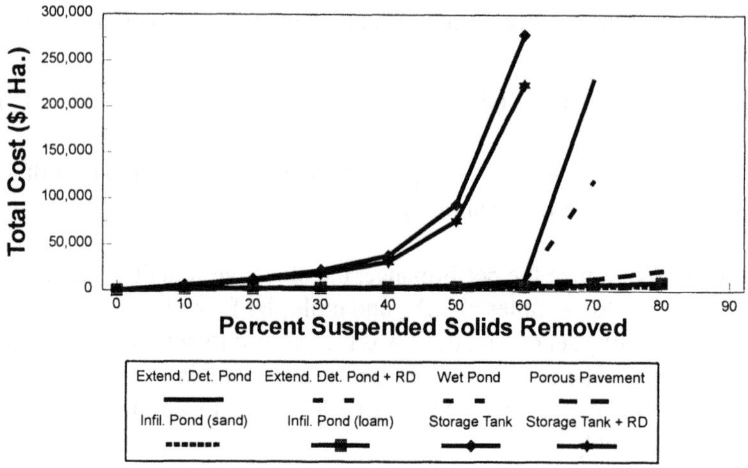

Figure 16.1 Urban runoff control - cost effectiveness of suspended solids removal.

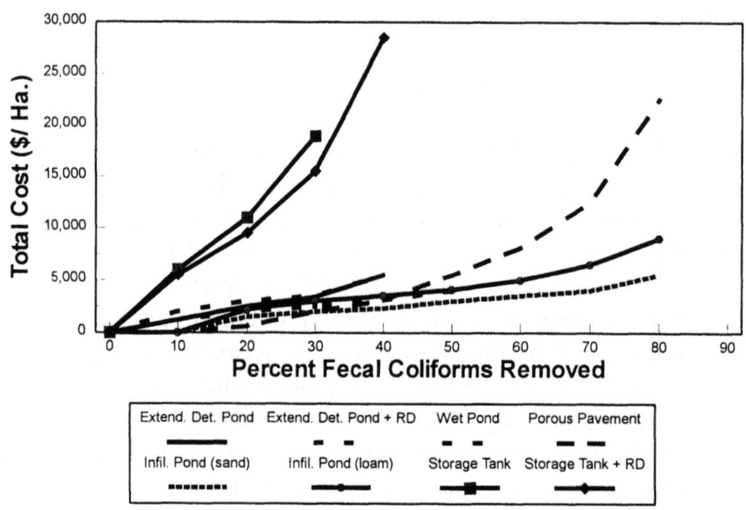

Figure 16.2 Urban runoff control - cost effectiveness of fecal coliform removal.

16.5.2 Combined Sewer Rehabilitation Control Options.

Figures 16.3 and 16.4 present the cost-effectiveness curves developed for the removal of suspended solids and fecal coliforms; respectively. The results of the analyses showed that outfall treatment with a roof leader disconnection program would be the most cost-effective option in removing a high percentage of suspended solids, BOD_5 and fecal coliforms. If lower removal rates are required (i.e. < 80%), downstream storage combined with increased interception capacity and roof leader disconnection was found to cost effective until the removal rate reached 80%. If removals above 80% were required, the cost effectiveness curve of this option became exponential.

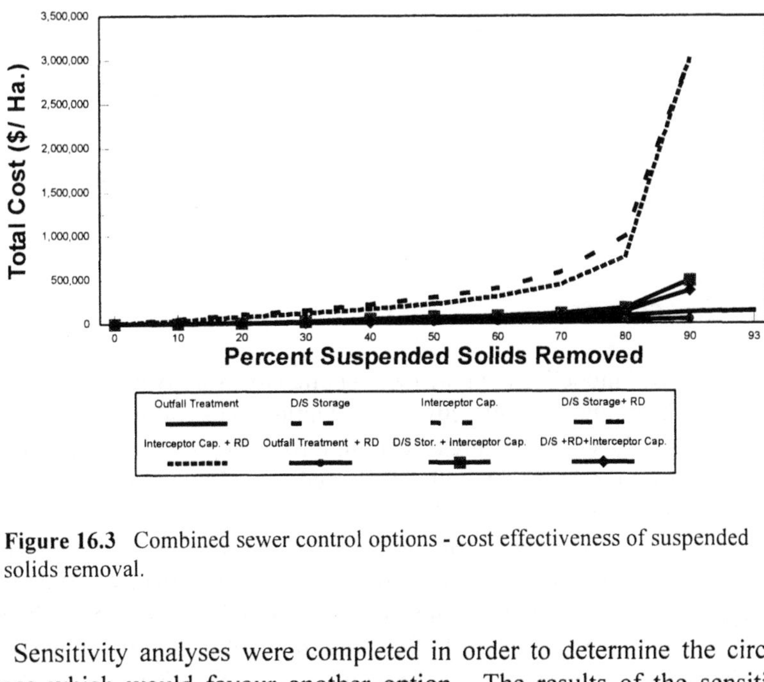

Figure 16.3 Combined sewer control options - cost effectiveness of suspended solids removal.

Sensitivity analyses were completed in order to determine the circumstances which would favour another option. The results of the sensitivity analyses showed that only under extreme circumstances would outfall treatment with a roof leader disconnection program not be the most cost-effective option. These extreme circumstances included an existing interception capacity in excess of 80 times dry weather flow, a high cost to implement a roof leader disconnection program (i.e. $100,000/ha), and a low efficiency for outfall treatment units. If the existing interception capacity was in excess of 80 times dry weather flow, downstream storage was found to be optimal. Increased interception capacity was found to be optimal if the efficiency of outfall treatment was below 23%.

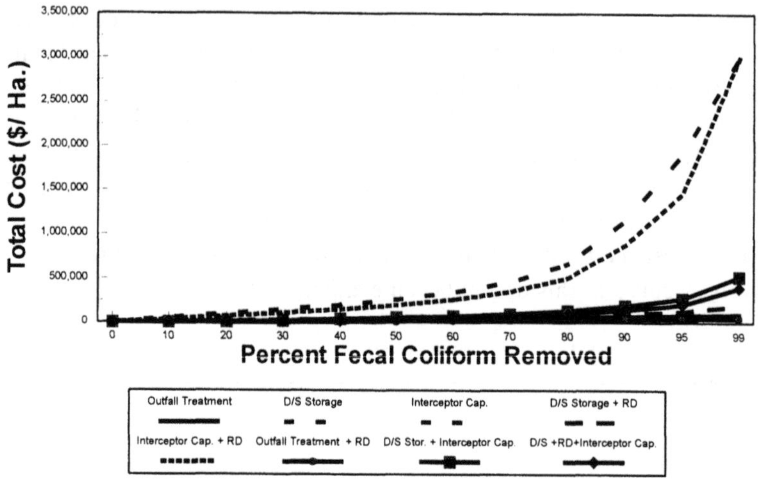

Figure 16.4 Combined sewer control options - cost effectiveness of fecal coliform removal.

16.6 Conclusions

A statistical modeling approach was applied to a variety of control options in order to determine their effectiveness. Optimal cost information for each option was developed by solving a cost minimization problem. The statistical approach allows many options to be evaluated. In addition, a comprehensive sensitivity analysis was performed.

The results of the applied methodology indicated that in some cases, a combination of control options may be the most cost-effective. For urban drainage systems, it was found that wet ponds and infiltration ponds with sand base were found to be the most cost-effective options. Both of these options could not be implemented due to local soil conditions. For combined sewer rehabilitation, outfall treatment combined with a roof leader disconnection program was found to be the most cost-effective option under most circumstances. A series of sensitivity analyses determined the circumstances under which other control options could be considered most cost-effective.

References

Adams, B.J., Bontje, J.B., "Microcomputer Applications of Analytical Models for Urban Stormwater Management", Proceedings of an Engineering Foundation Conference on Emerging Techniques in Stormwater and Flood Management, Niagara-on-the-Lake, Ontario, 1983.

Adams, B.J., Fraser, H.G., Howard, C.D.D., and Hanafy, M.S., "Meterological Data Analysis for Drainage System Design", Journal of Environmental Engineering, ASCE, Vol. 112, No. 5, October 1986.

CH2M HILL Engineering Ltd., "Combined Sewer Overflow Control Costs at Ontario RAP Sites", Summary Report, Environmental Canada and the Ontario Ministry of the Environment, 1992.

Guo, Y., "Long Term Performance of Stormwater Quality Ponds", MASc Thesis, Department of Civil Engineering, University of Toronto, Toronto, Canada, 1992.

Hvitved-Jacobsen, T., "Conventional Pollutant Impacts on Receiving Waters", in Urban Runoff Pollution, Springer-Verlag, Berlin, Germany, 1986.

Li, J.Y., "Comprehensive Urban Runoff Control Planning", PhD. Thesis, Department of Civil Engineering, University of Toronto, Toronto, Canada, 1991.

Moroz, C.L. "Cost Effectivness of Urban Runoff Options and Combined Sewer System Control Options", M.Eng. Design Project, Department of Civil Engineering, University of Toronto, Toronto, Canada, 1994.

Nix, S.J., "Graphical Design and Optimization", Civil Engineering for Practicing and Design Engineers, Vol. 3, 1984.

Chapter 17

Modeling Methodology for Determining Pollutant Concentrations and Loadings for Combined Sewer Overflows: A Simplified CSO Model

Young-Yun Rhee, Yinlun Huang, and Ralph H. Kummler

The Detroit Metropolitan area occupies much of the basin of the Detroit River - the basin as a whole is home to about four million people. The Detroit River provides important habitat for fish and birds. Atmospheric transport and deposition of chemical contaminants over the entire basin, leakage from sites contaminated according to Michigan's Act 307, and industrial effluents to the system provide sources for a variety of toxic chemicals to the Detroit River through the combined sewer overflow (CSO) process (Arimoto, 1989; Michigan Department of Natural Resources, 1993; Michigan Department of Natural Resources and Ontario Ministry of Environment, 1995). The Detroit River receives treated and untreated waste water from the City of Detroit and suburban districts, industries, runoff from urban and agricultural lands, and effluent from combined sewer overflows (Lin, 1994; Roginski, 1981; USEPA and EC, 1988; Rhee, 1995). The city is entirely served by combined sewers. As might be expected, CSOs have been demonstrated to be a major source of conventional and toxic contamination to the Detroit River, for example by the 1979-1980 major monitoring and modeling work on the section 201 Final Facilities Plan

© *Advances in Modeling the Management of Stormwater Impacts - Vol 5.* W. James, Ed. Pub. by CHI, Guelph, Canada 1997. ISBN 0-9697422-7-4. Fax: +519 767-2770

conducted by the Joint Venture (Giffels et al., 1980), to characterize the Detroit River CSO loadings and the Detroit River Remedial Action Plans (Stage I & II) (Michigan Department of Natural Resources and the Ontario Ministry of the Environment, 1991; Michigan Department of Natural Resources and Ontario Ministry of Environment, 1995).

A simplified CSO model has been developed as a rapid and inexpensive way to estimate and assess potential environmental hazards, as opposed to standard cause-and-effect modeling which typically costs $2,000,000 to $15,000,000 per study for Detroit. This model is a large-scale multiple-input multiple-output (MIMO) system. The modeling task can be eased by applying the decomposition-coordination strategy of the large-scale system theory (Huang and Fan, 1993). The goal of the present work is to use the field data generated by the current Southeast Michigan Council of Governments, U. S. Geological Survey, and Detroit Water and Sewerage Department (SEMCOG/USGS/DWSD CSO; Perry, 1996) study, and other resources to build a simplified CSO model capable of correlating and predicting the CSO concentrations and loadings to the Detroit River. Three groups of input variables to the model have been considered: the runoff concentrations of pollutants by land use; the runoff and CSO volumes; and the dry weather flow rates and concentrations of pollutants. This study focuses on three constituents, cadmium, copper, and lead.

The *first* input data, the runoff concentrations of cadmium, copper, and lead as a function of land use, were obtained from the stormwater pollutant loading factors presented in the Rouge River National Wet Weather Demonstration Project Report (Cave, et al., 1994). These runoff concentrations by land use already incorporate whatever leakage from Michigan's Act 307 contaminated sites and atmospheric deposition occur through runoff during the antecedent period. The *second* input data, the runoff and CSO volumes, were obtained from SEMCOG's database system (daily precipitation) and the CDM report (Camp Dresser and McKee, 1993). The *third* input data, the dry weather flow rates and concentrations of cadmium, copper, and lead, have been obtained from the DWSD Industrial Waste Control (IWC) database (1993-1994) and residential sources (Salley and Kummler, 1986; Garakani et al., 1991).

Because actual case studies involved complicated geometry and source functions, the modeling was performed in two model scenarios. First, daily average based estimation of CSOs concentrations was conducted using 1980 cadmium data from the City of Detroit's Section 201 Final Facilities Plan report for whole Detroit River basin (Kummler, 1983; Salley and Kummler, 1987). Second, a storm event based estimation of CSO concentrations was performed for two sampling sites (Conner Creek and Fischer sites). Four CSOs discharging to the Detroit River were monitored in 1994-1995 to characterize storm-related water quantity and quality to calculate their respective pollutant loads.

17.1 Theoretical Modeling Background

A simplified CSO model is presented as a rapid and inexpensive way to predict the CSO loadings and concentrations of contaminants. Figure 17.1 shows a simplified sewerage/CSO system.

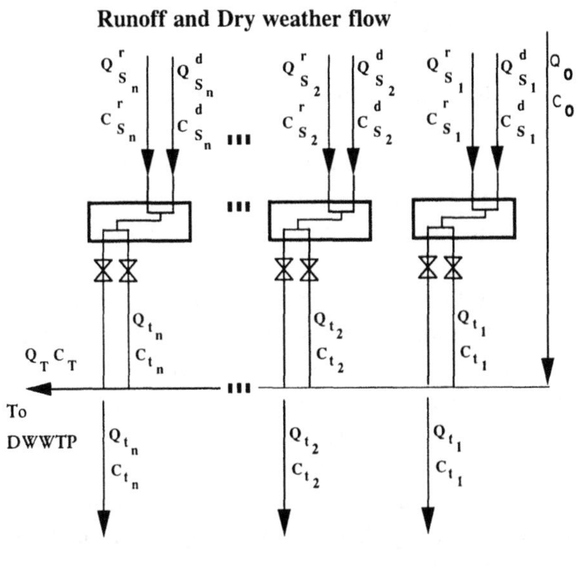

Figure 17.1 Simplified sewerage/CSO system.

The system has (n+1) inputs and (n+1) outputs. Each input represents a sewer line. Variables Q_{S_i} and C_{S_i} (i=1,2, . . . , n) are designated, respectively, the volumetric flow rate of the i-th stream and the concentration of a pollutant in it. These variables are further divided into variables $Q^r_{S_i}$ and $C^r_{S_i}$ for the runoff, and $Q^d_{S_i}$ and $C^d_{S_i}$ for the dry weather flow of conservative contaminants.

$$Q_{S_i} = Q^r_{S_i} + Q^d_{S_i} \tag{17.1}$$

$$C_{S_i} = \frac{Q^r_{S_i}}{Q_{S_i}} C^r_{S_i} + \frac{Q^d_{S_i}}{Q_{S_i}} C^d_{S_i} \tag{17.2}$$

Q_0 and C_0 characterize a wastewater stream from all other sources that connect directly to the main sewer interceptor without being part of CSOs. Corresponding to each sewer line, there is a CSO outfall to the river. Variables Q_{C_i} and C_{C_i} (i=1,2, . . ., n) are designated, respectively, the volumetric flow rate of the CSO stream and the concentration of the pollutant in it. The other output is the main stream going to the Detroit Wastewater Treatment Plant (DWWTP). This main stream is characterized by variables Q_{t_i} and C_{t_i} representing the volumetric flow rate of the stream and the concentration of the pollutant in it. Variables Q_T and C_T are designated, respectively, the total volumetric flow rate and concentration to the DWWTP.

17.1.1 Sub-system Model for a Daily-Average-Based Estimation

The structure of each of the n sub-systems is illustrated in Figure 17.2 for a detailed analysis. Here we neglect hydraulic interactions between adjacent sub-systems. Each sub-system is still modeled as a mixer-splitter, with the following assumptions:

1. As a mixer, the flow rate is the sum of the flow rates of all input streams. In the mixer, perfect mixing is assumed. Hence, the concentration of a pollutant is the ratio of the total input of the pollutant to the total volume flow into the sub-system.
2. As a splitter, the concentration of the pollutant in each split branch is the same as that before splitting; the sum of the flow rates in all branches is the same as the flow rate before splitting.
3. Any wastewater leakage from the mixer-splitter is negligible.

In each sub-system i, there exist two inputs, a dry weather flow and a runoff. Each input is characterized by the flow rate of a stream and the concentration of a pollutant in the stream. The two inputs are well mixed in the mixer. The mixed stream is characterized by the flow rate, Q'_{S_i}, and the concentration of a pollutant, C'_{S_i}. The mixed stream is then split into two branches. One branch as a CSO goes directly to the Detroit River. The flow rate of this stream, Q_{C_i}, is a portion of the runoff into the same sub-system. This portion is characterized by the factor, f'_i. Thus:

$$Q_{C_i} = f'_i Q^r_{S_i} \tag{17.3}$$

According to the CDM report (Camp, Dresser, and McKee, 1993), this factor averages 0.51 for the entire DWSD system; this value is used throughout this study, but can be changed if site-specific information is available. The concentration of the CSO, C_{C_i}, is the same as that of the mixed stream, C'_{S_i}, as the splitting only affects the flow rate.

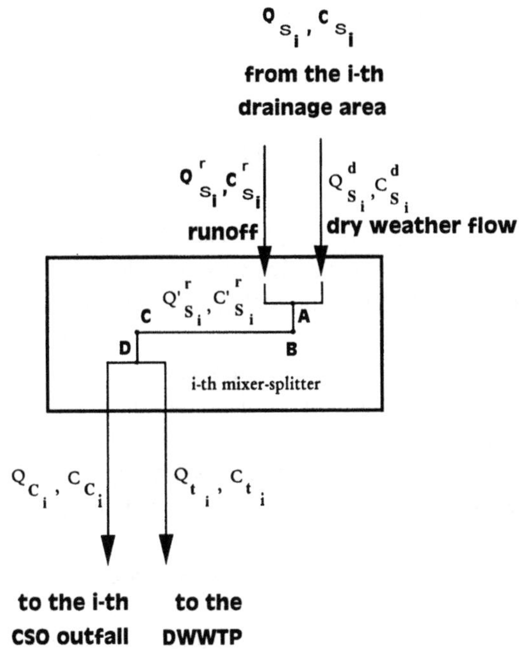

Figure 17.2 Structure of a sub-system.

The other branch joins other down-flow streams from the sub-systems ahead of it and eventually goes to the DWWTP. The flow rate of this branch, Q_{t_i}, is:

$$Q_{t_i} = Q'_{s_i} - f_i Q^r_{s_i} \tag{17.4}$$

From Figures 17.1 and 17.2, we can find that each of the n sub-systems is independent in estimating the concentration and flow rate of the CSO to the Detroit River. In other words, the concentration and the loading of any pollutant of the CSO are solely determined by the runoff and dry weather base flow into the same sub-system (neglecting infiltration and inflow). The concentration of a particular pollutant in the CSO in the sub-system i is:

$$C_{C_i} = C'_{s_i} = \frac{Q^r_{s_i} C^r_{s_i} + Q^d_{s_i} C^d_{s_i}}{Q^r_{s_i} + Q^d_{s_i}} \tag{17.5}$$

The loading of the pollutant in the CSO, L_i, is:

$$L_i = Q_{C_i} C_{C_i}$$

$$= f'_i Q_{R_i} \frac{Q^r_{S_i} C^r_{S_i} + Q^d_{S_i} C^d_{S_i}}{Q^r_{S_i} + Q^d_{S_i}} \qquad (17.6)$$

The values of $Q^r_{S_i}$, $C^r_{S_i}$, $Q^d_{S_i}$, $C^d_{S_i}$, and f'_i can be obtained or estimated from the CDM report, the Technical Memorandum No. 34 of the Rouge River national wet weather demonstration project report, and the DWSD Industrial Waste Control (IWC) database.

Note that the selection of the number of sub-systems is determined by the data availability, the requirement of prediction precision, etc. The index i for differentiating sub-systems can be individual CSOs (45 CSO locations along the Detroit River in the U.S. side) or can be aggregated CSOs for comparison to Treatment Plant or globally averaged CSO statistics.

17.1.2 Sub-system Model for an Event-Based Estimation

The model given in Equations 17.5 and 17.6 is a steady-state model and thus can not be directly used for an event-based estimation. However, this model can be easily modified as a pseudo-dynamic model with an acceptable prediction error, if the following assumptions hold:

1. The dry weather flow data inputs are constant. This implies that $Q^d_{S_i}$ and $C^d_{S_i}$ do not change during wet weather events.
2. The runoff concentration, $C^r_{S_i}$, is independent of time during the event. The runoff concentration, $C^r_{S_i}$, can be obtained from the stormwater pollutant loading factors presented in the Rouge River National Wet Weather Demonstration Project Report (Cave et al., 1994).

The SWMM RUNOFF model calculates the amount and timing of overland discharge $Q^r_{S_i}$ from defined combined sewered areas for specific rainfall events. This SWMM model can simulate the runoff response with a variety of different rainfall hyetographs imposed on different portions of the system.

With the above two assumptions, we can derive a pseudo-dynamic model based on the steady-state model in Equations 17.5 and 17.6. The concentration of a pollutant in the CSO at the time t, i.e. $C_{C_i}(t)$, can be evaluated:

$$C_{C_i}(t) = \frac{Q^r_{S_i}(t) C^r_{S_i} + Q^d_{S_i} C^d_{S_i}}{Q^r_{S_i}(t) + Q^d_{S_i}} \qquad (17.7)$$

The loading of the pollutant in the CSO to the Detroit River at time t, i.e. $L_i(t)$, is:

$$L_i(t) = Q_{C_i}(t)C_{C_i}$$
$$= f_i\, Q^r_{s_i}(t)\frac{Q^r_{s_i}(t)C^r_{s_i} + Q^d_{s_i}C^d_{s_i}}{Q^r_{s_i}(t) + Q^d_{s_i}} \qquad (17.8)$$

17.2 Modeling Methodology

A simplified CSO model has been developed for the prediction of CSO loadings and concentrations to the Detroit River. We have considered three groups of input variables to the model: the runoff loadings and concentrations of pollutants by land use; the runoff and CSO volumes; and the dry weather loadings and concentrations of pollutants. In this study we focused on three constituents, cadmium, copper, and lead.

The first input data, the runoff concentrations by land use of cadmium, copper, and lead were obtained from the stormwater pollutant loading factors presented in the Rouge River National Wet Weather Demonstration Project Report (Cave et al., 1994). These runoff concentrations, which are a function of locally derived land use empirical values, automatically incorporate the average leakage from Michigan's Act 307 contaminated sites and atmospheric deposition through runoff.

The second input data, the runoff and CSO volumes, were obtained from SEMCOG's database system (daily precipitation) and the CDM report (Camp Dresser and McKee, 1993). CDM developed a complete hydraulic model of the major sewer interceptors to estimate response to rainfall events and assess potential combined sewer overflow control measures within the region tributary to the DWWTP in 1993. They selected the period of record of precipitation for January 1, 1982 through December 31, 1992 and calculated the annual average rainfall to be 34.25 inches (870 mm). However we require the precipitation record for the specific year. The Michigan Department of Agriculture (MDA) manages the SEMCOG rain gage network data; they supplied the daily average precipitation summary of the 72 stations in the SEMCOG Network for the years 1982 through 1994 in a digital format (the MDA/Climatology and MSU/ Agricultural Weather Service Program's jointly operated Bulletin Board System). We selected ten rain gage stations over the Detroit River basin and calculated the arithmetic (equivalent to area averaged) average of gaged quantities for any specific year.

The third input data, the dry weather flow rates and concentrations of cadmium, copper, and lead, were obtained from the DWSD IWC database (1993-1994) and residential sources (Garakani et al., 1991; Kuplicki, 1995).

The schematic diagram of simplified CSO modeling is shown in Figure 17.3. Because actual case studies involved complicated geometry and source functions, the modeling was performed in two model scenarios. First, daily-average-based estimation of CSOs concentrations was conducted using the cadmium data in 1980 from Section 201 Final Facilities Plan report for the whole Detroit River basin. Second, storm-event-based estimation of CSOs concentrations was performed for two sampling sites (Conner Creek and Fischer sites).

The U.S. Geological Survey (USGS) monitored four CSOs discharging (Conner Creek, Fischer, Schroeder, and Rosa Parks sites) to the Detroit River in 1994-1995 to characterize storm-related water quantity and quality and to calculate their respective pollutant loads. Flow measuring stations are located as near as practicable to the outlet of each CSO. Water-level, velocity, discharge, and precipitation were measured continuously. Samples at all sites were collected at discrete times during each storm event. Thus estimates of variability of pollutant concentrations during a single event can be made (these could be stochastically distributed in a next version).

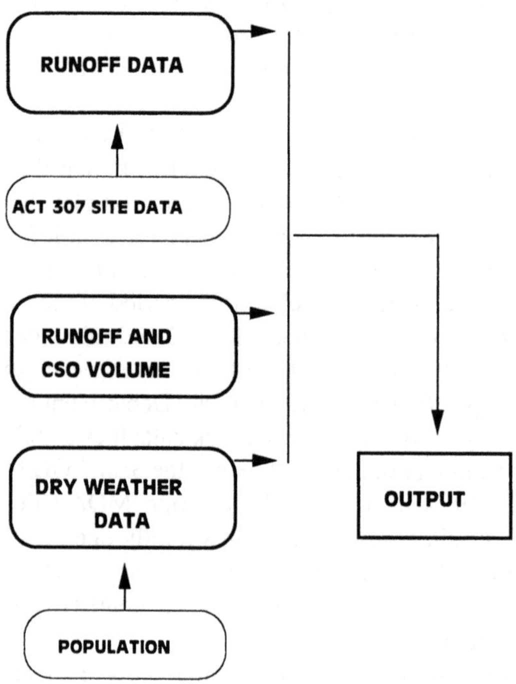

Figure 17.3 Schematic diagram of simplified CSO model.

17.2.1 Daily-Average-Based Estimation: Model Scenario I

In daily-average-based estimation we used the flow rates and concentrations data of cadmium in 1980 from Section 201 Final Facilities Plan Report (Giffels et al., 1980). The dry weather quality and industrial waste loadings were available for the Detroit Water and Sewerage Department's service area at various locations.

17.2.2 Data Input for Model Scenario I

Three input variables are used for this Model Scenario 1: the average daily runoff concentrations by land use, the average daily runoff and CSO volumes, and the average daily dry weather flow rates and concentrations of cadmium.

Average Daily Runoff Concentrations by Land Use for Cadmium

We used the recommended concentrations by land use category (Table 17.1) presented in the Rouge River National Wet Weather Demonstration Project report (Cave et al., 1994).

Table 17.1 Average daily runoff concentrations by land use category for four constituents.

Land use Category	Percent Imperv.	Cd (μg/l)	Cu (μg/l)	Pb (μg/l)
Forest/Rural Open	2.0 %	0	0	0
Urban Open	11.0 %	1	0	14
Agricultural/Pasture	2.0 %	0	0	0
Low Density Residential	19.0 %	4	26	57
Medium Density Residential	38.0 %	4	26	57
High Density Residential	51.0 %	3	33	41
Commercial	56.0 %	3	37	49
Industrial	76.0 %	5	58	72
Highways	53.0 %	3	37	49
Water/Wetlands	51.0 %	1	7	11

Average Daily Runoff and CSO volumes

We decomposed the Metro-Detroit basin into four sub-systems and obtained the runoff and CSO volumes (Table 17.2) following the CDM report (Camp Dresser and McKee, 1993).

These concentrations are typically total recoverable metals, and are used because the dissolved fraction is not available. Sedimentation and resuspension values are unknown and in this work that level of detail is not desired.

Table 17.2 Runoff and CSO volumes for sub-systems.

	District	Runoff (MGD)	CSO (MGD)
Sub-system 1	Fox Creek	5.8	3.0
Sub-system 2	East Side	8.6	4.1
Sub-system 3	Central City, S.E. Oakland, Evergreen-Farmington	86.6	46.2
Sub-system 4	West Side, Western Wayne County	40.1	18.6

Dry Weather Flow and Concentrations

The dry weather flow and concentrations of cadmium in 1980 were obtained from Section 201 Final Facilities Plan report (Giffels et al., 1980).

17.2.3 Storm Event-Based-Estimation: Model Scenario 2

In storm-event-based estimation we used the flow rates and concentrations data of cadmium, copper, and lead in 1993 from the DWSD Industrial Waste Control (IWC) database.

Monitoring Sites

Conner Creek District. Conner Creek district drains primarily into the Conner Creek Sewer, which flows in a general north-south direction and discharges through the Conner Creek regulator into the Detroit River Interceptor (DRI) or through the Conner Creek outfall to the Detroit River. The Conner Creek Sewer also transports suburban flow from the City of Centerline.

Fischer District. Fischer district is a small overflow draining area from primarily residential areas in the vicinity of Indian Village. Flow transported in the Fischer Sewer is either pumped into the DRI by the Fischer Pump Station or overflowed to the Detroit River through a triple box outfall.

Data Input for Model Scenario 2

Three input variables are used for storm-event-based estimation (Model Scenario 2): the average runoff concentrations by land use, the average CSO volumes by unit time, and the average dry weather flow rates by unit time and concentrations of cadmium, copper, and lead.

Average Runoff Concentrations by Land Use for Three Constituents. We used the recommended concentrations by land use category (Table 17.1) presented in the Rouge River National Wet Weather Demonstration Project report (Cave, et al., 1994).

Average CSO Volumes per Unit Time. The Storm Water Management Model (SWMM) Runoff block calculates the amount and timing of overland discharge from defined combined sewered areas for specific rainfall events. This model can simulate the runoff response by a variety of different rainfall hyetographs imposed on different portions of the system. This makes the model ideally suited for computing runoff from a number of tributary areas based on a spatially varied rainfall input. Since the objective of this work is to create a simplified model not dependent on SWMM, we arbitrarily used a hypothetical Gaussian storm model.

Dry Weather Flow by Unit Time and Concentrations. The average dry weather flow rates per unit time and concentrations of cadmium, copper, and lead, have been obtained from the DWSD IWC database (1993-1994) and residential sources.

17.3 Modeling Results and Discussion

Two model scenarios were evaluated using a simplified CSO model. First, daily-average-based estimation of CSOs concentrations was conducted using the cadmium data in 1980 from Section 201 Final Facilities Plan report for the whole Detroit River basin. Second, storm-event-based estimation of CSOs concentrations was performed for two sampling sites (Conner Creek and Fischer sites).

17.3.1 Daily-Average-Based Estimation: Model Scenario 1

The estimated average daily CSO concentration of cadmium for overall system in 1980 is 27.2 μg/l, which is 15% lower than the CSO monitoring data (32.0 μg/l in 1979 from Section 201 Final Facilities Plan Report (Giffels et al., 1980). The computed average daily concentration of DWWTP inflow (30.6 μg/l) in 1980 is 11% higher than the estimated average daily CSO concentration (27.2 μg/l). The computed results are illustrated in Figure 17.4.

17.3.2 Storm-Event-Based Estimation: Model Scenario 2

Quantification of Combined Sewer Overflow

We used the hypothetical time-series CSO volume for each district (Table 17.3).

Dry Weather Flow Rates and Concentrations

The Industrial Waste Control (IWC) database includes information only for Significant Industrial Users (SIU) that are required to participate in the Industrial Pretreatment Program. Hence we treated the concentration of the flow contributed

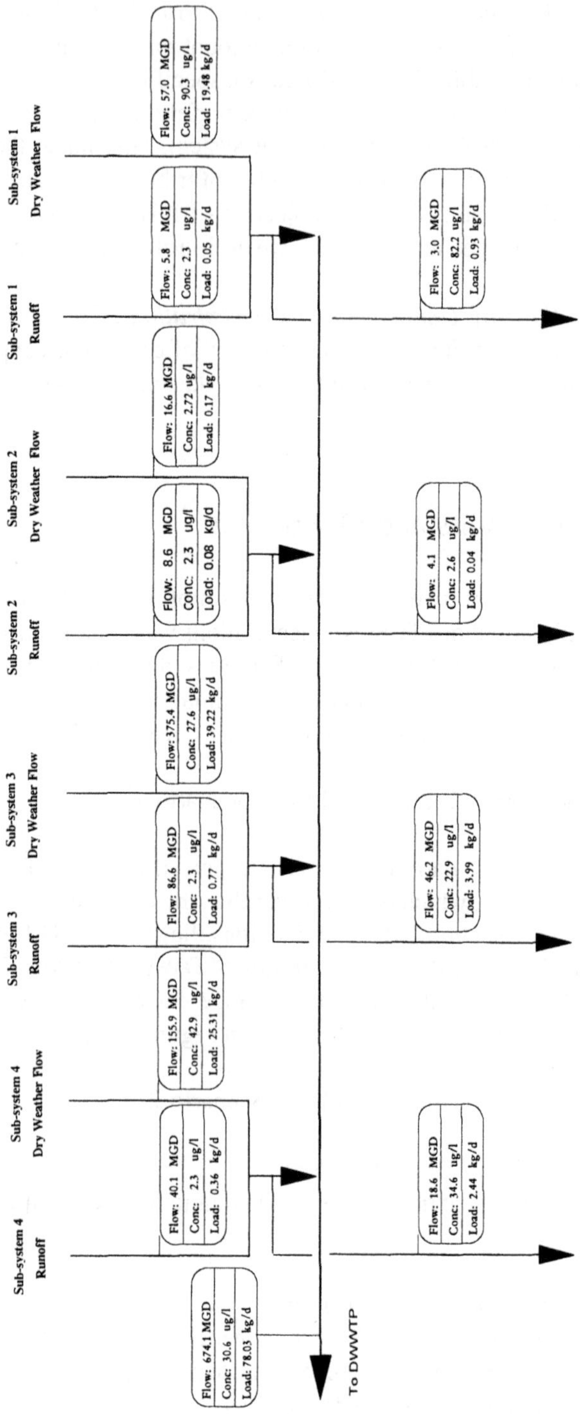

Figure 17.4 Average daily flow rates, concentrations, and loadings of cadmium for DWWTP inflow, CSO runoff, and dry weather flow in 1980.

Table 17.3 Time-series CSO volume (MGH) distribution for typical storm event.

	Conner Creek	Fischer
Time + 1 hr	16.12	0.21
2 hr	37.36	0.50
3 hr	24.75	0.33
4hr	12.00	0.16
5 hr	6.00	0.08
6 hr	2.81	0.04
7 hr	1.35	0.02
8 hr	0.68	0.01
9 hr	0.34	0.005
10 hr	0.15	0.002

from remaining Industrial Users as the same as that of residential flows. The concentrations of residential flow for cadmium, copper, and lead, were obtained from the dry weather flow analysis of domestic/commercial discharges in the DWSD Industrial Waste Control Program Report (McNamee et al., 1995). Table 17.4 summarizes the dry weather flow concentrations for three parameters.

Table 17.4 Dry weather flow concentrations for industrial and residential flows.

	Industrial Effluent (μg/l)	Residential Effluent (μg/l)
Cadmium	13.8	0.8
Copper	209.7	25.3
Lead	111.3	6.2

Predicted Time-series CSO Concentrations

The computed time-series CSO concentrations of three constituents at two sampling sites (Conner Creek and Fischer) are illustrated in Figures 17.5 through 17.7. The averages of computed CSO concentrations for each parameter were compared with the averages of pollutant qualities (Table 17.5) that were sampled and analyzed during 1994-1995 by USGS and DWSD.

As shown in Figure 17.5 for CSO concentration of cadmium for the 10 hour duration, the concentration follows the runoff flow so that the CSO concentration profiles illustrate a decreasing trend after the peak flow (2nd hour) of CSO. This decreasing trend occurs in both Conner Creek district and Fischer district. Figure

17.6 represents the CSO concentration of copper at two sites. The concentration of copper in the Conner Creek district shows dilution by runoff flow that is opposite to the Fischer site. For the Conner Creek district, the industrial effluents from many plants are diluted with runoff flow. For the Fischer district, the CSO volume consists primarily of residential effluents and is concentrated with runoff flow during the storm event. Figure 17.7 illustrates how the lead concentration for two CSO sampling sites increases during the 10 hour period.

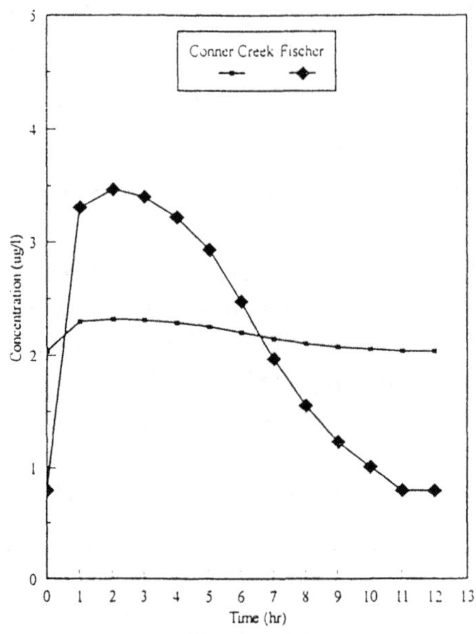

Figure 17.5 Storm-event-based computation of time-series CSO concentration of cadmium in 1993 for two sampling sites.

The USGS has measured the CSO event concentrations at four CSO locations. We can compare the theoretical predictions with the USGS field data if we adjust the USGS data sets for samples with values below the detection limit. We assumed that the concentrations which are recorded as below detection limit (BDL) are equal to 50% of the detection limit.

Based upon the results (Table 17.5) for the storm-event-based estimation for the Conner Creek site, the average of computed CSO concentrations of cadmium during a 10 hr period deviates by 60% from the measured value, copper (Cu) 18%, lead (Pb) 28%. For the Fischer site, the average of computed CSO concentrations of cadmium (Cd) differs by 49%, from the monitored value, copper 58%, lead

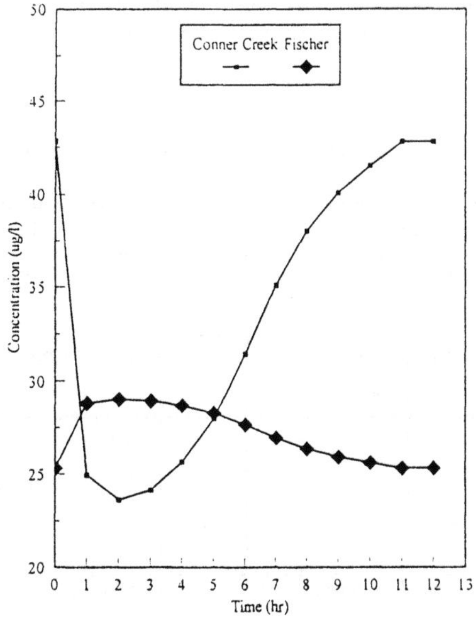

Figure 17.6 Storm-event-based computation of time-series CSO concentration of copper in 1993 for two sampling sites.

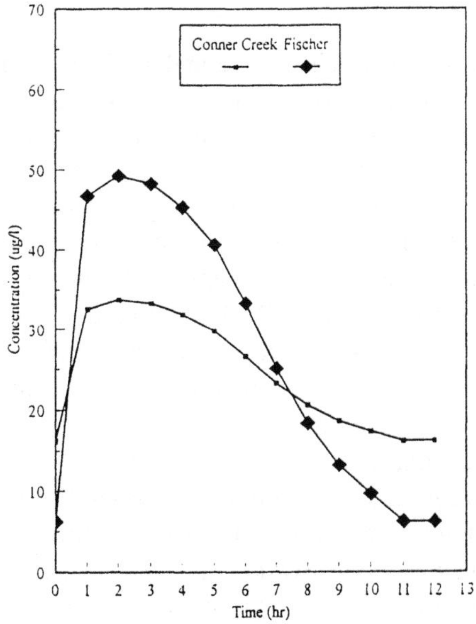

Figure 17.7 Storm-event-based computation of time-series CSO concentration of lead in 1993 for two sampling sites.

44%. Values for Conner Creek are believed to be more accurate because Fischer flow rates are much lower and concentrations are therefore subject to a greater representativeness problem.

Table 17.5 Comparison of averages of computed and monitored concentrations (μg/l).

	Conner Creek					Fischer				
	Computed		Monitored*		Difference (%)	Computed		Monitored*		Difference (%)
	Conc.	Std. Dev.	Conc.	Std. Dev.		Conc.	Std. Dev.	Conc.	Std. Dev.	
Cd	2.21	0.10	5.50	4.63	-60	2.46	0.90	4.84	5.26	-49
Cu	31.2	6.6	39.8	38.2	-18	27.6	1.3	65.0	68.1	-58
Pb	26.8	6.0	37.1	52.9	-28	33.0	14.5	59.1	71.0	-44

* From USGS/DWSD monitored data during 1994-1995 (Conner Creek: 24 samplings, Fischer: 36 samplings).
Assumed that concentrations which are recorded as BDL are equal to 50% of the detection limit.

17.4 Conclusion

Representing a design in which sewage and surface runoff combine and exceed retention and treatment system capacity, CSOs are identifiable as a troubling source of contamination for water bodies that are cumulative recipients of overflows. CSOs collect and transmit contaminants from a variety of other sources. Since 1989, the DWSD has been engaged in discussions with the Michigan Department of Environmental Quality (MDEQ), the National Wildlife Federation (NWF) and the Greater Detroit Chamber of Commerce (CofC) on the activities to be undertaken to minimize PCB and mercury inputs to the sewer system (Detroit Water and Sewerage Department, 1995). Because of the City of Detroit's Industrial Pretreatment Program, the general decline of industrial activity in the area, the conscientious efforts of industries to reduce discharges through pollution prevention, and other factors, the CSO concentrations and loadings of pollutants were substantially reduced during the last decade.

This study is a planning level approach to the CSO concentration prediction at CSO outfalls using a simplified CSO model. The average runoff concentrations of pollutants by land use, the average runoff and CSO volumes, and the average dry weather flow rates and concentrations of pollutants were used as inputs to the model. The modeling was performed for two model scenarios: a daily-average-based estimation and a storm-event-based estimation.

The results for the daily-average-based estimation indicated that the computed average daily CSO concentration of cadmium for the overall system in 1980 (27.2 μg/l) is close to the monitored data (32.0 μg/l) in 1979. The results for the storm-event-based estimation revealed that the average of computed CSO concentrations of cadmium in the Conner Creek site during a 10 hr period deviates by 60% from the measured value, copper 18%, lead 28%. For the Fischer site, the average of computed CSO concentrations of cadmium differ by 49% from the monitored value, copper 58%, lead 44%.

According to the computed CSO concentration of cadmium for the 10 hour duration, the concentration occurs by the runoff flow so that the CSO concentration follows a decreasing trend after the peak flow (2nd hour) of CSO. This decreasing trend occurs in both the Conner Creek district and the Fischer district. The CSO concentration of copper in the Conner Creek shows dilution by runoff flow that is opposite to the Fischer site. For the Conner Creek district, the industrial effluents from many plants are diluted with runoff flow. For the Fischer district, the CSO volume consists primarily of residential effluents and is concentrated with runoff flow during the storm event. For the CSO concentration of lead in two sampling sites, the runoff flow increases the CSO concentrations during a 10 hour period.

The shape of the hyetograph can have a major effect on the runoff characteristics in relatively small, urbanized basins. In order to estimate the more accurate time-series CSO volumes for sampling sites, the SWMM model is a useful tool for the evaluation.

Consequently, the simplified CSO model offers a good procedure to estimate the CSO concentrations for pollutants to assist in remediation decision-making. This model is applicable to any Area of Concern, for example, Rouge River, although the Rouge studies can afford predictive, continuous modeling.

References

Arimoto, R. 1989. Atmospheric Deposition of Chemical Contaminants to the Great Lakes. J. Great Lakes Res., 15(2). pp:339-356.

Camp Dresser & McKee. 1993. Model Development Status Report. Prepared for the Detroit River Water and Sewerage Department. pp:27-61.

Cave, K., Quasebarth, T., and E. Harold. 1994. Technical Memorandum of Selection of Stormwater Pollutant Loading Factors. No. RPO-MOD-TM34.00: Rouge River National Wet Weather Demonstration Project.

Detroit Water and Sewerage Department. 1995. PCB/Hg Minimization Program: Sampling Activities Status Report.

Garakani, S., Salley, S., Kummler, R.H., and A. Rothberger. 1991. Database Management for the Detroit Industrial Pretreatment Program (II) in Computing in Civil Engineering. ASCE, New York. pp:356-365.

Giffels/Black & Veatch. 1980. Quantity and Quality of Combined Sewer Overflows Volume II. Prepared for the Detroit River Water and Sewerage Department.

Huang, Y. And L. T. Fan, 1993. A Fuzzy-Logic -Based Approach to Building Efficient Fuzzy Rule Based Expert Systems. Computers and Chemical Engineering, 17, pp181-192.

Kummler, R.H. 1983. SWMM Modeling for the Detroit 201 Final Facilities Plan: Final Results. Proceedings of the USEPA SWMM Meeting, Ottawa, Canada.

Kuplicki, S. 1995. Private Communication.

Lin, C.C. 1994. Modeling the Detroit River Aquatic and Sediment Systems. Ph.D. dissertation. Dept. of Chemical Engineering, Wayne State University, Detroit, Michigan. pp:1-5.

McNamee, Porter & Seeley, Inc. 1995. Industrial Waste Control Program: Re-evaluation of Local Limitations for Incompatible Substances, Vol. II Development of Local Limits. Prepared for Detroit Water and Sewerage Department.

Michigan Department of Natural Resources. 1993. Proposed List for Michigan Sites of Environmental Contamination.

Michigan Department of Natural Resources and Ontario Ministry of the Environment and Energy. 1995. Remedial Action Plan for Detroit River Area of Concern. Draft Biennial Report.

Michigan Department of Natural Resources and Ontario Ministry of the Environment and Energy. 1991. Remedial Action Plan Stage I for Detroit River Area of Concern.

Perry, S., 1996. Final Report on SEMCOG Detroit River Toxics Study.

Rhee, Y. Y. 1995. Modeling Methodology for Determining Pollutant Concentrations and Loadings for Combined Sewer Overflows: A Simplified CSO Model. Ph.D. dissertation. Dept. of Chemical Engineering, Wayne State University, Detroit, Michigan. pp:1-7.

Roginski, G. T. 1981. A Finite Difference Model of Pollutant Concentrations in the Detroit River from Combined Sewer Overflows. Ph.D. dissertation. Dept. of Chemical and Metallurgical Engineering, Wayne State University, Detroit, Michigan. pp:1-10.

Salley, S. and R.H. Kummler. 1987. Hazardous Waste Management: The Municipal Interest - Cadmium in Management of Hazardous and Toxic Wastes in the Process Industries: Elsevier Applied Science Publishers Ltd., London: edited by S.T. Kolaczkowski and D. B. Crittenden.

Salley, S. and R.H. Kummler. 1986. A Data Management System for the Detroit Industrial Pretreatment Program: Elsevier Applied Science Publishers Ltd., Amsterdam.

U.S. EPA and EC. 1988. Upper Great Lakes Connecting Channels Study. Volume II. Final Report. pp:456-476.

Chapter 18

Modeling Fecal Coliform In Mill Creek

Jennifer D. Xie, Philip Gray, Dante Zettler and Betsy Yingling

In 1995, the Northeast Ohio Regional Sewer District (NEORSD) initiated the Mill Creek Interceptor (MCI) Project, a multi-year undertaking aimed at developing a comprehensive facilities plan for the area. It is envisioned that, once implemented, the plan will alleviate sewer system surcharging and control combined sewer overflows (CSOs) within the Mill Creek drainage basin, at the same time complying with current and potential state and local permitting requirements.

In light of the fact that significant water quality impacts to Mill Creek had been identified as part of previous studies of the area, it was determined that an assessment of Mill Creek be undertaken. As part of the assessment of Mill Creek, an extensive monitoring program was completed, the results of which complemented the considerable data previously collected. The results of the monitoring program, along with the previously collected data, were used to develop a water quality model for Mill Creek. Once developed, the Mill Creek water quality model was used to assist in the formulation of an overall drainage basin plan, as well as in the analysis of a Use Attainability Analysis for the Mill Creek.

This chapter discusses the process undertaken to set up, calibrate and apply a water quality model to support the CSO facility planning process. Specifically, in the Mill Creek drainage area, there was a significant amount of effort spent in defining and quantifying sources of fecal coliform whether from CSOs, stormwater, dry weather seepage, or other sources contributing to Mill Creek that were

© *Advances in Modeling the Management of Stormwater Impacts - Vol. 5* W. James, Ed.
Pub. by CHI, Guelph, Canada 1997. ISBN 0-9697422-7-4. Fax: +519 767-2770

identified in the watershed. Source identification was extremely important to the water quality model in order that the impact of source correction could be reliably predicted. Fecal coliform was modeled by the TRANSPORT module of XP-SWMM in the study. The calibrated water quality model was used in continuous simulations to evaluate the sensitivity of water quality in Mill Creek to various control alternatives being considered and for the evaluation of the overall facility plan.

18.1 Study Area

Mill Creek is a small tributary of the Cuyahoga River that discharges to Lake Erie in the Cleveland area. Mill Creek is approximately 20 kilometres (12 miles) in length and has a total drainage area of 6,000 hectares (15,000 acres).

The MCI drainage basin comprises the south-eastern portion of the Mill Creek basin and is tributary to the Southerly Wastewater Treatment Plant (SWTP). The MCI drainage basin covers an area of approximately 6,880 hectares (17,000 acres) and includes all or part of eleven communities. The sewerage system tributary to the Interceptor is comprised of separate (53%), combined (26%), and dual (21%) sewers.

Land use within the drainage basin is primarily urban, most of it zoned for single and multiple family dwellings. Commercial and industrial areas are generally adjacent to main streets, with open space limited to small parks, one large park (Garfield Heights) on the south side of the Creek, cemeteries, a golf course and a race track.

There are 175 outfalls documented as having outlets that discharge directly into Mill Creek and its tributaries from areas serviced by combined sewer, separate sewers and dual sewers. The creek also receives effluent discharges from septic tanks and semi-public disposal systems. Mill Creek is, as well, subject to both municipal and industrial spills and inputs of leachate from three landfill sites (one active and two closed) in its lower reaches through bank seepage.

Wet weather discharges to Mill Creek are mainly from combined sewer overflows, stormwater, and overflows from the dual sewer areas. There are two dual sewer systems in the Mill Creek drainage basin. The first has been designated a common trench - dividing wall (CTDW) system. The CTDW system consists of separate storm and sanitary pipes constructed in a common trench and a partial wall in common manhole structures that separate sanitary and storm flows. During wet weather periods, it is possible for sanitary flows to overflow into the storm system and/or for the stormwater to overflow into the sanitary system, resulting in both cases in combined sewage flow. The second type of dual sewer has been designated a common trench - separate (CTS) system. The CTS system is characterized as having both sanitary and stormwater pipes within the same trench, similar to the CTDW system. However, the systems do not share common manhole structures. Nevertheless, cross-mixing of flows can occur as a result of leakage from one conveyance system infiltrating the other.

18.2 Model Set-Up

18.2.1 Overview

The TRANSPORT module of XP-SWMM was used to simulate fecal coliform levels in the Mill Creek water quality study. Figure 18.1 shows the schematic of the Mill Creek water quality model. The Mill Creek model includes about 9.3 miles (15 km) of stream and it is divided into 24 segments averaging a half mile (0.8 km) each.

For the water quality model, the number of outfalls was reduced through grouping common pollutant source types. In total, 62 pollutant loading points that contribute both flow and pollutants into Mill Creek are used in the model.

18.2.2 Bacterial Decay

Fecal coliform (FC) modeling involves the use of a first-order decay expression to describe bacterial die-off. The coliform levels are a function of initial loading and the disappearance rate. The disappearance rate is a function of:

- time or distance of travel from the source, and
- environmental factors such as temperature, salinity, and light intensity.

The following formula was used to calculate the decay of fecal coliform in the model:

$$C_t = C_0 e^{-kt} \tag{18.1}$$

where:

C_0 = initial fecal coliform concentration, org/100 ml
C_t = fecal coliform concentration, org/100 ml
k = decay rate constant, day^{-1} or hr^{-1} (accounts for various factors including temperature, salinity, and light intensity)
t = exposure time, days or hours (travel time)

The decay rate in the context of the Mill Creek was not a sensitive parameter during wet weather. The length of Mill Creek and the flow velocity was such that there was minimal decay of fecal coliform levels observed. Calibration of decay rates was undertaken using dry weather data collected at the four instream monitoring locations. The fecal levels at the upstream station were compared to

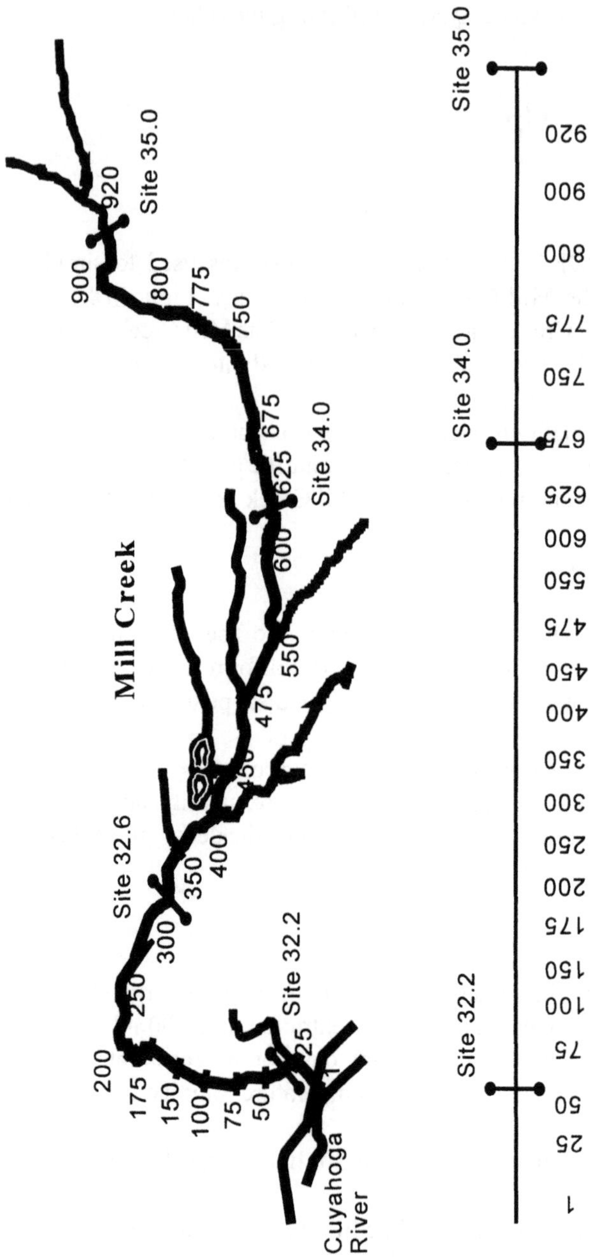

Figure 18.1 Mill Creek model schematic.

the level of the next station downstream and intermediate sources accounted for in determining the overall decay factor for each reach. Overall, a decay factor of 0.3 hr^{-1} was found to be representative for Mill Creek for dry weather conditions. This rate agrees with the decay rate ranges listed in EPA surface water quality modeling report (Tetra Tech, 1985).

18.2.3 Model Inputs

Stream Flows

Modeling of the collection system within the Mill Creek drainage basin was completed by Metcalf & Eddy Inc. using the SWMM RUNOFF and EXTRAN blocks. The collection system SWMM output hydrographs are used as inflow hydrographs to the Mill Creek water quality model. In the water quality model, the TRANSPORT block is used to route the flow hydrographs, and pollutographs, in the stream. This process required the development of suitable tools to reformat the collection system interface files containing the flow hydrographs into importable files for the XP-SWMM water quality model.

Collection system runoff and routing parameters were adjusted to calibrate the stream flows in the water quality model. Fecal coliform calibration was undertaken following the flow quantity calibration and is discussed later in the chapter.

Source Concentrations

Fecal coliform concentrations were input at the same 62 loading points as the flow hydrographs.

Dry weather fecal coliform densities were established using source outfall, boundary, tributary and in-stream water quality and flow data collected as part of the monitoring and sampling program.

An event mean concentration (EMC) method was employed for wet weather fecal coliform inputs. Event mean concentrations of fecal coliform were calculated based on sampling and monitoring data collected at various source outfalls. Specifically, the event mean concentrations were calculated using data collected form five wet weather events at 17 different source sampling sites and four instream monitoring sites during the period May to July 1995.

It was important to identify the source of fecal coliform loadings. To this end, wet weather source flows from the collection system were classified into the following groups based on the sewer service types contained in each of the subcatchments:

- combined sewer overflows (CSO);
- stormwater - common trench dividing wall (CTDW);
- stormwater - common trench separate (CTS);
- stormwater - highway drainage(HWY);

- stormwater - separate sewer (S); and,
- stormwater - open space, unserviced area 9 (i.e. park) (OS).

Fecal coliform EMCs were calculated for each flow source type and applied in the receiving water model.

18.3 Model Calibration

18.3.1 Flow Calibration

The water quality model was calibrated using flow data (stage-discharge relationships) collected at three stream sites (sites 35, 34 and 32.2) between May and July 1995. Four dry weather calibration events were used to establish a typical level of dry weather flow in Mill Creek, representing the summer period. Three wet weather events were used to calibrate and one event was used to verify the model for wet weather conditions.

Figures 18.2 to 18.4 show the results of the model verification using the July 15, 1995 event. These figures show very good agreement between the modeled and measured flows. The event flow volume difference between modeled and measured flows was less than 10% at all three flow monitoring sites for the event. The peak flow rates compared well except for Site 35, the most upstream monitoring location. In reviewing the flow data provided by the United States Geological Survey (USGS) it was found that the site was calibrated for flows of less than 3 feet (0.9 m) in depth, beyond 3 feet, the flow rate is extrapolated. In all calibration events Site 35 shows a significantly higher peak flow which is not reflected downstream. As such, it was determined that if flows were greater than 3 feet the flow rate was not representative. The timing of measured flows versus modeled flows compared well for all events.

18.3.2 Water Quality Calibration

Dry weather fecal coliform densities were established using source outfall, boundary, tributary and in-stream water quality and flow data collected as part of the monitoring and sampling program. The decay rate of fecal coliform was calibrated using the four dry weather events. The decay rate accounts for decay between instream monitoring sites and dry weather inputs from other intermediate sources (i.e. dry weather seepage, leachate).

The three May 1995 wet weather events were used to validate wet-weather bacterial model calibration while the July 15, 1995 event was employed for verification. Fecal coliform densities for the July 15, 1995 verification event are presented for sites 35, 34 and 32.2 in Figures 18.5 to 18.7, respectively. FC loadings for the July 15, 1995 event for sites 35, 34 and 32.2 are presented in

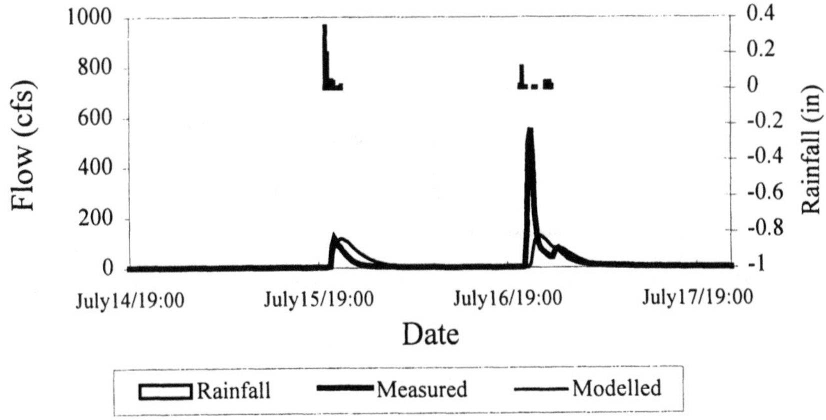

Figure 18.2 Wet weather flow calibration event - July 15, 95 Site 35.

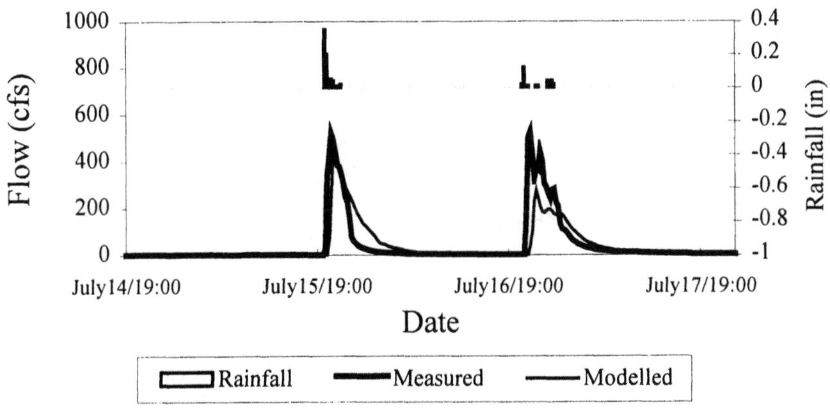

Figure 18.3 Wet weather flow calibration event - July 15, 95 Site 34.

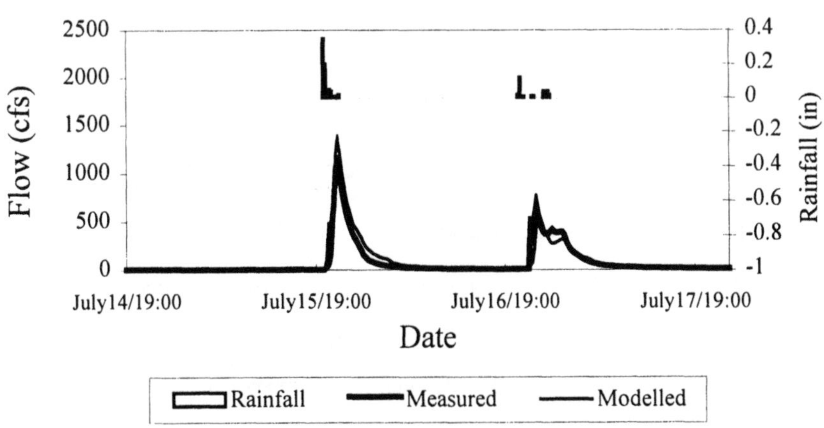

Figure 18.4 Wet weather flow calibration event - July 15, 95 Site 32.2.

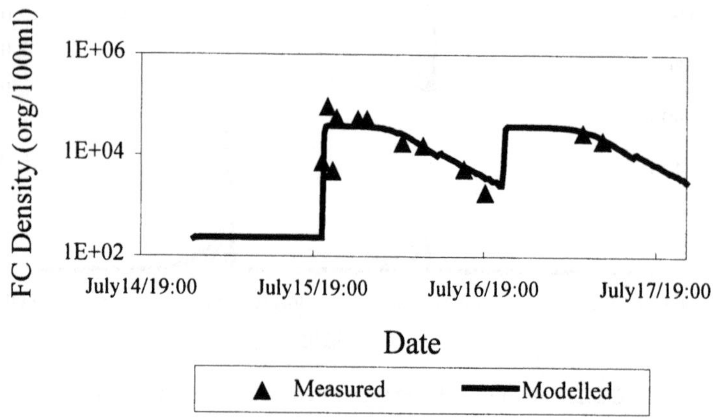

Figure 18.5 Wet weather FC concentration verification event - July 15, 95 Site 35.

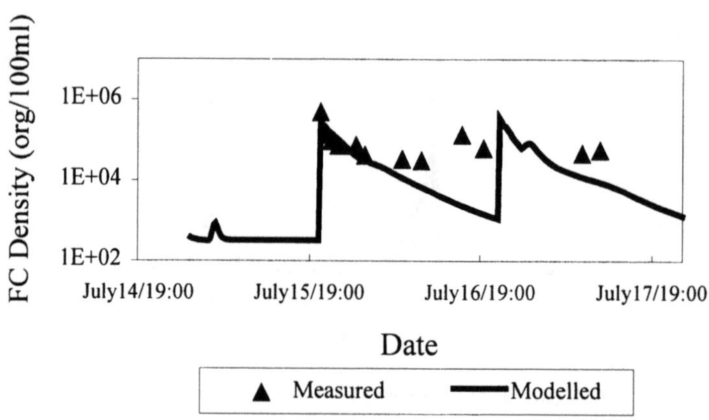

Figure 18.6 Wet weather FC concentration verification event - July 15, 95 Site 34.

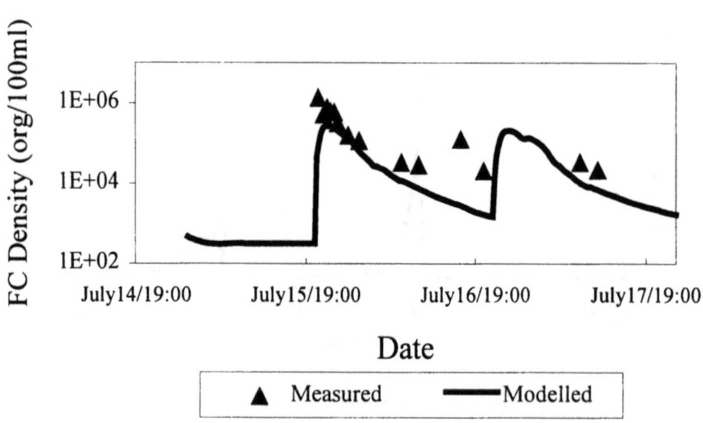

Figure 18.7 Wet weather FC concentration verification event - July 15, 95 Site 32.2.

Figures 18.8 to 18.10, respectively. The verification results of the fecal coliform densities and loadings show very good agreement following the calibration process. [*Editor's note: Figures 18.8-18.10 present the same information as Figures 18.2-18.7.*]

18.4 Water Quality Simulations

Following calibration and verification of the Mill Creek water quality model, continuous simulations were performed to examine the water quality of Mill Creek in support of pollutant source control plans. A total of six alternative cases were simulated for the swimming season from May 1st to October 15th; three cases are discussed in this chapter. The remaining three simulations were variations of the three presented and were found to reflect similar results with regards to fecal coliform concentrations and loadings.

The runoff hydrographs for the simulations were generated from the collection system model with one hour time steps for a typical year of rainfall developed by others. All the simulations were carried out using the defined baseline dry weather water quality stream model.

The continuous simulation results for fecal coliform were compared with the current Ohio State Water Quality Standards (WQS). As part of the WQS, each water body is assigned a Recreational Use Designation. Mill Creek is defined as having a "primary contact" recreational use, meaning that it is suitable for full body contact such as swimming or canoeing. The primary contact recreational criteria require that:

- geometric mean fecal coliform content based on not less than five samples within a 30 day period shall not exceed 1,000 per 100 ml; and
- geometric mean fecal coliform content shall not exceed 2,000 per 100 ml in over 10% of the samples taken during any 30 day period.

The three cases simulated by the receiving water model are for the following conditions:

- Case 1 - Baseline - existing conditions;
- Case 2 - No CSO - assumes that all combined sewer overflows will be collected by a proposed tunnel and removed from Mill Creek;
- Case 3 - Complete separation - assumes that all wet weather flows from combined sewer overflows; sewer overflows from common trench - dividing wall; and sewer overflows from common trench - separate become stormwater. In this scenario, the stream receives all of the wet weather source flows with the same fecal coliform density of stormwater from a separate system. This is in effect a fully separated system.

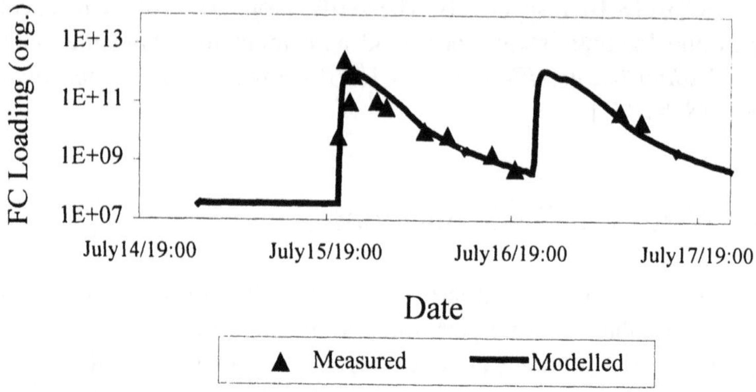

Figure 18.8 Wet weather FC loading verification event July 15, 95 Site 35.

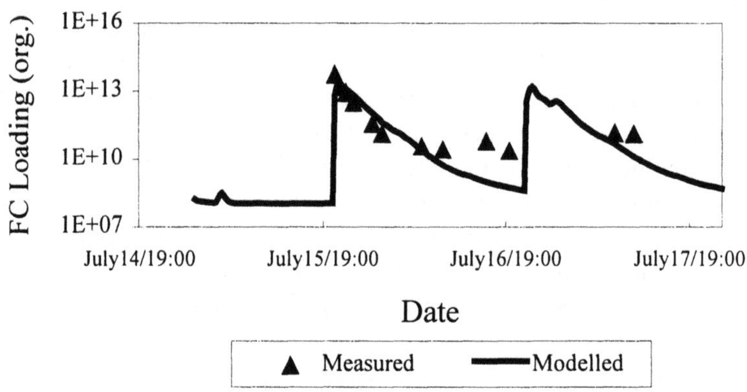

Figure 18.9 Wet weather FC loading verification event July 15, 95 Site 34.

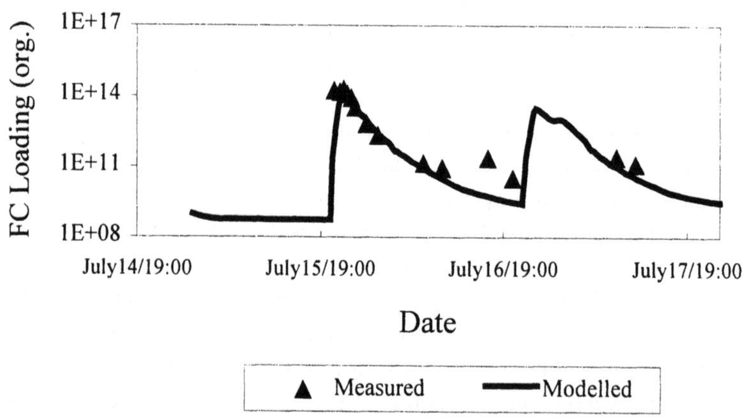

Figure 18.10 Wet weather FC loading verification event July 15, 95 Site 32.2.

Table 18.1 shows the seasonal geometric means of computed fecal coliform densities for the three cases. The computed fecal coliform densities in Case 1 - Baseline exceed the primary contact recreational criteria in the middle and lower reaches. The fecal coliform results of Case 2 - No CSO show that only the middle reaches exceed the criteria. The computed fecal coliform densities in Case 3 present quite similar results: the computed fecal coliform densities exceed the criteria in the middle section of the Creek, and meet the criteria in the upper and lower portions of the stream. Figure 18.11 displays profiles of geometric means calculated for the above three cases over the May 1st to October 15th period.

Table 18.1 Seasonal summary. FC Geometric mean (May 1-Oct. 15). org/100ml.

Stream Node	Baseline	No CSO	Complete Separation
920	580	580	580
900	547	547	524
800	436	436	411
775	500	500	472
750	630	630	589
675	779	773	698
625	727	722	643
600	732	726	641
550	854	849	727
475	732	728	605
450	1188	814	644
400	1911	1521	1164
350	1829	1433	1094
300	1766	1385	1053
250	1811	1411	1071
200	1199	954	750
175	1179	935	730
150	1140	903	704
100	1111	878	683
75	1093	862	669
50	1068	841	651
25	1102	857	667
1	1070	828	644

A similar trend is shown in Table 18.2. In Case 1 - Baseline, the percentages of the time that computed fecal coliform densities exceed the criteria of 2,000 org/100 ml range between 19% to 36%. When the CSO sources are removed from the model, the computed percentages of the time are reduced to a range of between 17% and 20%. Case 3 - complete separation, the percentages of the time that

Table 18.2 Seasonal summary; computed hours with FC density>2000 org/100ml (May 1 - October 15).

Stream Node	Baseline		No CSO		Complete Separation	
	Hours	%	Hours	%	Hours	%
920	784	19%	783	19%	781	19%
900	754	19%	753	19%	742	18%
800	735	18%	734	18%	718	18%
775	732	18%	731	18%	717	18%
750	733	18%	732	18%	713	18%
675	698	17%	697	17%	677	17%
625	703	17%	702	17%	680	17%
600	696	17%	695	17%	671	17%
550	697	17%	696	17%	671	17%
475	713	18%	712	18%	684	17%
450	1439	36%	780	19%	699	17%
400	1326	33%	826	20%	679	17%
350	1303	32%	815	20%	676	17%
300	1297	32%	815	20%	681	17%
250	1311	33%	826	20%	699	17%
200	912	23%	715	18%	609	15%
175	915	23%	717	18%	610	15%
150	878	22%	720	18%	621	15%
100	877	22%	733	18%	635	16%
75	879	22%	743	18%	642	16%
50	841	21%	741	18%	649	16%
25	850	21%	758	19%	658	16%
1	835	21%	759	19%	662	16%

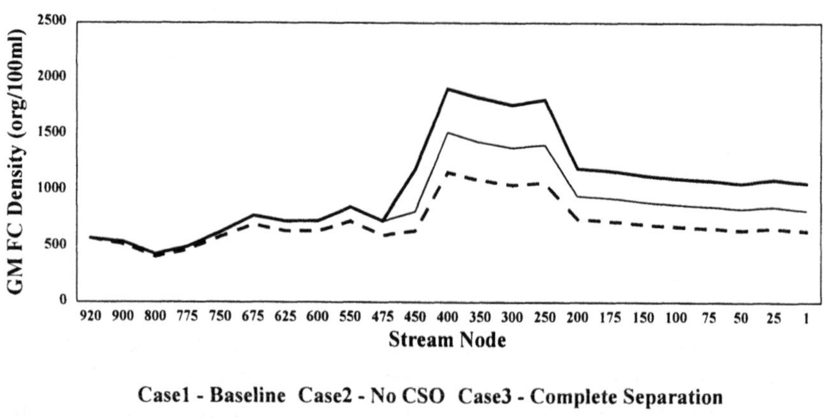

Figure 18.11 FC density profile (May 1-October 15).

computed fecal coliform densities exceed the criteria range between 15% and 19%, which includes no significant improvement when compared with Case 2.

18.5 Conclusions

The field program undertaken was critical to the success of the water quality model to ensure a reliable and representative model for assessing various control alternatives. Careful evaluation of the monitoring data was imperative to ensure that the data accurately represents the existing condition.

The Event Mean Concentration method which associated a fecal coliform EMC with a specific service area type (i.e. CTDW, storm, etc.) combined with TRANSPORT module of XP-SWMM model has been shown to be a successful tool in fecal coliform density and loading simulations. This methodology allowed the water quality model to be used to simulate the correction of fecal coliform pollutant sources. The sensitivity analysis investigated the removal of CSO and the complete separation of the combined and mixed service areas. The outcome of the simulations indicated computed fecal coliform levels would still exceed primary contact limits despite correction of all sanitary sources. These findings indicate that correction of CSOs alone will not be sufficient to bring the fecal coliform levels into compliance with regulatory requirements.

The sensitivity analysis undertaken using the water quality model was used by others in developing the CSO Facility Plan. To this end, the benefits of the recommended CSO Facility Plan were simulated The simulated fecal coliform levels in Mill Creek, following the implementation of the CSO Facility Plan, were similar to those presented in Case 2, No CSO. The CSO Facility Plan would not be sufficient to control computed fecal coliform levels to meet the primary contact recreational criteria. Additional analysis was undertaken using the water quality model in the form of a stream loadings assessment. The loadings assessment indicated that the CSO Facility Plan would reduce computed pollutant loadings, such as suspended solids, metals, total phosphorus, BOD_5 by up to 40 to 50%.

The water quality model developed for the Mill Creek project is a tool that the NEORSD can use in the future to estimate fecal loadings and pollutant loadings as the CSO Facilities Plan is implemented to compare their progress with their objectives.

Reference

Tetra Tech, Inc., 1985, Rates, Constant, and Kinetics Formulations in Surface Water Quality Modeling. For USEPA Environmental Research Lab., Athens, GA., by Tetra Tech Inc, Lafayette, CA. Second Edition. NTIS. Springfield, Virginia. Page 436-437.

Chapter 19

StormTreat System Installation at Elm Street, Kingston, Massachusetts

Scott W. Horsley

In October of 1994, four StormTreat Systems (STS) tanks were installed at Elm Street in Kingston, Massachusetts (see Figures 19.1 and 19.2). This site was identified as a significant stormwater pollution site by the Jones River Watershed Association. Prior to the STS installation, stormwater discharged directly to the Jones River from a 850 foot (259 m) length of road surface (Elm Street) and associated parking area at the Kingston Water Department building. This drainage area is estimated to contain 18,700 square feet, or 0.43 acres (1740 m^2), of pavement. The StormTreat System is designed to capture and treat the first 0.5 inches (12.7 mm) of runoff, i.e. the "first flush".

19.1 Procedure and Analysis

The four STS tanks were installed by the Kingston Highway Department, connecting to an existing catch basin with PVC piping. Four individual effluent lines constructed of 1.5 inch (38 mm) PVC were installed, and discharge at an average rate of 0.25 gallons/minute (3.97 Liters/s), into a small rip-rapped area adjacent to the Jones River. Four effluent lines were installed at the site for research purposes (testing alternative conditions in each tank). Normal STS installations utilize one effluent line serving the entire cluster of tanks at each location.

© *Advances in Modeling the Management of Stormwater Impacts - Vol. 5* W. James, Ed.
Pub. by CHI, Guelph, Canada 1997. ISBN 0-9697422-7-4. Fax: +519 767-2770

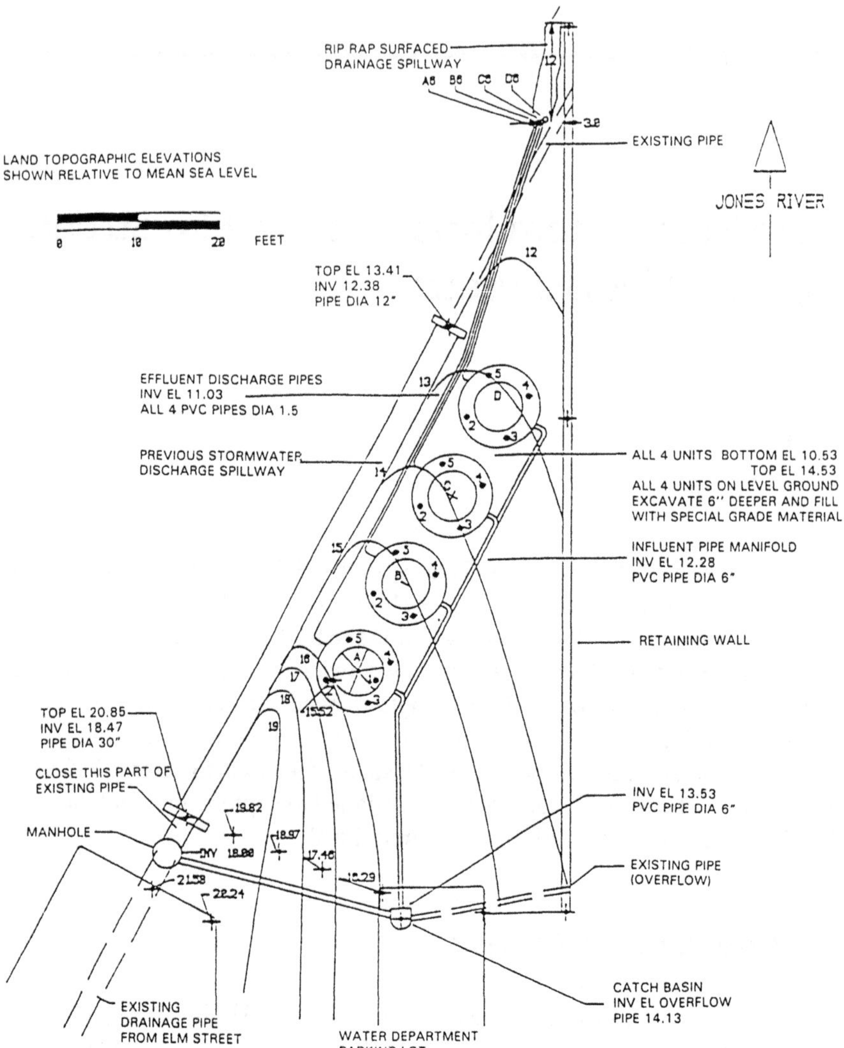
RIP RAP SURFACED
DRAINAGE SPILLWAY
EXISTING PIPE
LAND TOPOGRAPHIC ELEVATIONS
SHOWN RELATIVE TO MEAN SEA LEVEL
FEET
JONES RIVER
TOP EL 13.41
INV 12.38
PIPE DIA 12"
EFFLUENT DISCHARGE PIPES
INV EL 11.03
ALL 4 PVC PIPES DIA 1.5
PREVIOUS STORMWATER
DISCHARGE SPILLWAY
ALL 4 UNITS BOTTOM EL 10.53
TOP EL 14.53
ALL 4 UNITS ON LEVEL GROUND
EXCAVATE 6" DEEPER AND FILL
WITH SPECIAL GRADE MATERIAL
INFLUENT PIPE MANIFOLD
INV EL 12.28
PVC PIPE DIA 6"
RETAINING WALL
TOP EL 20.85
INV EL 18.47
PIPE DIA 30"
CLOSE THIS PART OF
EXISTING PIPE
MANHOLE
INV EL 13.53
PVC PIPE DIA 6"
EXISTING PIPE
(OVERFLOW)
CATCH BASIN
INV EL OVERFLOW
PIPE 14.13
EXISTING
DRAINAGE PIPE
FROM ELM STREET
WATER DEPARTMENT
PARKING LOT

Figure 19.1 Elm Street StormTreat system site plan.

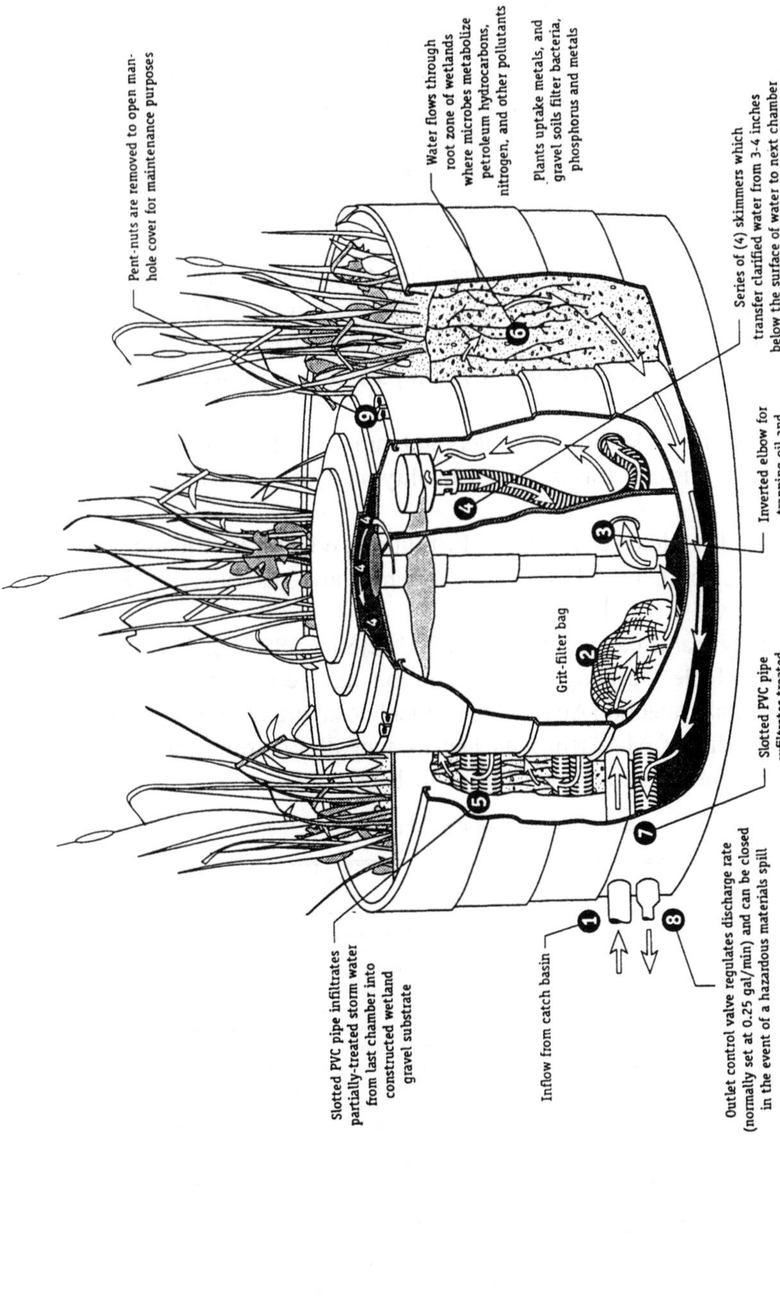
Pent-nuts are removed to open man-hole cover for maintenance purposes
Water flows through root zone of wetlands where microbes metabolize petroleum hydrocarbons, nitrogen, and other pollutants
Plants uptake metals, and gravel soils filter bacteria, phosphorus and metals
Series of (4) skimmers which transfer clarified water from 3-4 inches below the surface of water to next chamber
Inverted elbow for trapping oil and other floatables
Slotted PVC pipe exfiltrates treated water to outlet
Grit-filter bag
Slotted PVC pipe infiltrates partially-treated storm water from last chamber into constructed wetland gravel substrate
Inflow from catch basin
Outlet control valve regulates discharge rate (normally set at 0.25 gal/min) and can be closed in the event of a hazardous materials spill
1
2
3
4
5
6
7
8
9

Figure 19.2 Storm Treat Systems tank.

Water quality sampling of six independent storm events has been conducted over the past 24 months. A total of 643 analyses have been completed on 41 samples. Sampling was conducted during and following storm events by members of the Jones River Watershed Association under the supervision of its Executive Director. According to our quality assurance plan (QAP), the Field Captain determines the likelihood of a one-half inch event using local weather forecasts and notifies samplers to report to the Elm Street location. No sampling is conducted without a minimum of 3 days of preceding dry weather.

First flush stormwater samples are taken at the entry point to the STS tanks by opening the manhole cover. Effluent samples are taken during the 5 days following the storm event. They are obtained at the sampling ports (A6, B6, C6 and D6) where the effluent pipes discharge.

Water samples are obtained using laboratory-prepared sampling bottles by taking grab samples and measuring flow rates. Samples for dissolved nutrients are filtered in the field using 0.45-micron filter paper and filtering syringe. Following the storm event the first flush samples are composited immediately (by flow-weighting method).

Samples are packed in iced coolers and shipped to the analytical laboratories with chain of custody forms. Water quality analyses are conducted by three laboratories:

1. Woods Hole Oceanographic Institution [WHOI], (nutrients)
2. Barnstable County Health & Environmental Department [BCHED], (bacteria, TSS, metals, COD)
3. Analytical Balance Corp. [ABC], (bacteria, TSS, metals, COD)

Woods Hole Oceanographic Institution has been certified by the U.S. Environmental Protection Agency. The other two laboratories are certified by the Commonwealth of Massachusetts Department of Environmental Protection. Specific laboratory methods are shown in Table 19.1.

Table 19.1 Laboratory analytical methods.

Parameter	Analysis
TN-part	Elemental analysis
TN-diss	Persulfate digestion
NO3, N	Cadmium reduction
NH3-N	Indophenol
TP	Persulfate digestion
O-P	Molydbenum blue
FC	Membrane filter procedure
TSS	Standard methods 2540D
Cr, Pb, Ni, Zn	Atomic absorption spectrophotometry

Our quality assurance plan requires a minimum of 5% of samples as duplicates. This is accomplished by splitting samples between two certified laboratories and by providing them with blind duplicates.

19.2 Results

To date we have successfully sampled five stormwater events at the Elm Street installation. The results are summarized below in Table 19.2 and Figure 19.3. At the time of printing, not all of the analytical results had been received from the laboratories.

Table 19.2 - Summary of water quality monitoring results of the StormTreat system - Kingston, Massachusetts.

Pollutant	Stormwater influent	Treated effluent	Percentage removed
FC (orgs/100 ml)	690	20	97
TSS (mg/liter)	93	1.3	99
COD (mg/liter)	95	17	82
TDN (micrograms/liter)	3569	520	77
TP (micrograms/liter)	300	26.5	89
TPH (mg/liter)	3.4	0.34	90
Pb (micrograms/liter)	6.5	1.5	77
Cr (micrograms/liter)	60	1	98
Zn (micrograms/liter)	590	58	90

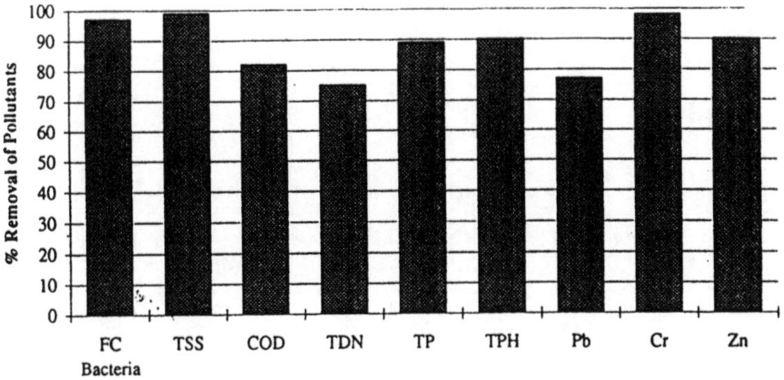

Figure 19.3 Water quality monitoring results of the Storm Treat system - Kingston, Massachusetts (1994-1996)

The laboratory results of six rounds of samples from four stormwater events: 11/11/94, 11/28/94, 12/5/94, 2/17/95, 10/9/95, and 10/1/96 have been averaged. We are evaluating system performance by calculating the percentage removal of each parameter by comparing the influent with the average of the effluent samples. We are not directly comparing the effluent results with those in chamber A1 because we believe that resuspension of sediments within this chamber significantly alters the quality of the incoming stormwater.

These results indicate that an average of 94% of the total coliform bacteria and 97% of the fecal coliform bacteria, 99% of the total suspended solids (TSS) and 90% of total petroleum hydrocarbons is removed. Preliminary nutrient results suggest a removal rate of 77% for total dissolved nitrogen (TDN) and 89% for total phosphorus. We expect that the nitrogen removal rate will improve during the growing season when the wetland plants are more active. Removal rates for metals are as follows: lead -77%, chromium - 98%, and zinc - 90%.

Chapter 20

Issues Regarding the Application of a Mass-Balance Equation to an Urban Creek

Peter Hicks, Ed McBean, Steve Quigley and Bruce Polan

Uniroyal Chemical Ltd. (Uniroyal) operates a chemical manufacturing facility in Elmira, Ontario (hereinafter referred to as "the site"). Contaminated groundwater from beneath some areas of the site discharges to Canagagigue Creek, which flows through the site. As part of a proposal to contain a portion of this groundwater, the Ontario Ministry of Environment and Energy (MOEE) required Uniroyal to prepare a contaminant loading model that would estimate the improvement in surface water quality following implementation of the groundwater containment system. This chapter describes the preparation of this model, which was developed by Conestoga-Rovers & Associates (CRA) on behalf of Uniroyal.

20.1 Background

Uniroyal operates a chemical manufacturing facility at the site as shown on Figure 20.1 Operations at the site began circa 1897, with the production of footwear prôducts, and continued, in summary, as follows:

- 1897-1917 - footwear products;
- 1917-1929 - tennis shoe manufacturing;
- 1929-1941 - closed;

© *Advances in Modeling the Management of Stormwater Impacts - Vol. 5* W. James, Ed.
Pub. by CHI, Guelph, Canada 1997. ISBN 0-9697422-7-4. Fax: +519 767-2770

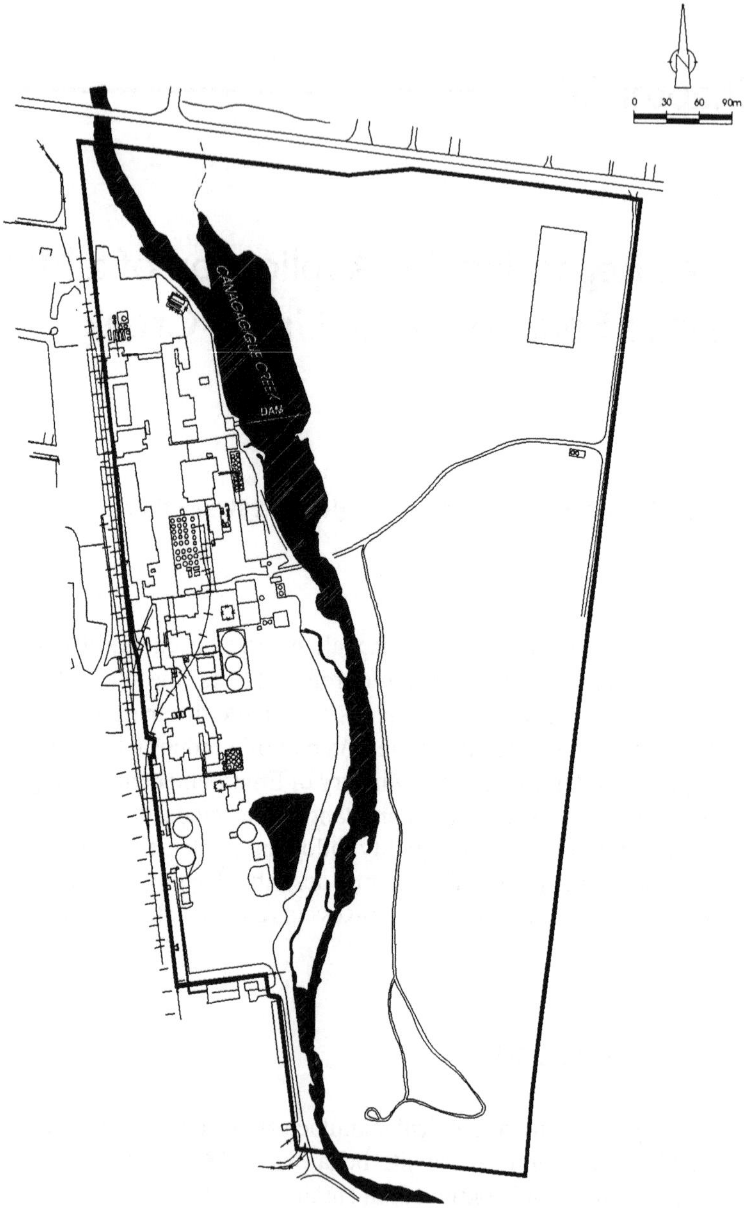

Figure 20.1 Site layout - Uniroyal Chemical Ltd. Elmira, Ontario.

- 1941-late 1940s - aniline and diphenylamine production; and
- late 1940s-present - organic chemical products and intermediates for agricultural, plastics and rubber industries.

Historic waste disposal and manufacturing practices have contaminated the underlying soil and groundwater. The site is underlain by a complex geologic and hydrogeologic environment comprising glacial and fluvial deposits. Canagagigue Creek (Creek), a tributary of the Grand River, flows through the site.

The principal aquifer units beneath the site of interest (in order of depth from the ground surface) are:

- the Upper Aquifer (UA) comprising three subunits UA_1, UA_2, UA_3;
- the Upper Aquitard (UAT);
- the Municipal Aquifer (MA); and
- the Municipal Aquitard (MAT).

Figure 20.2 depicts a typical cross-section of these units and their relationship to the Creek.

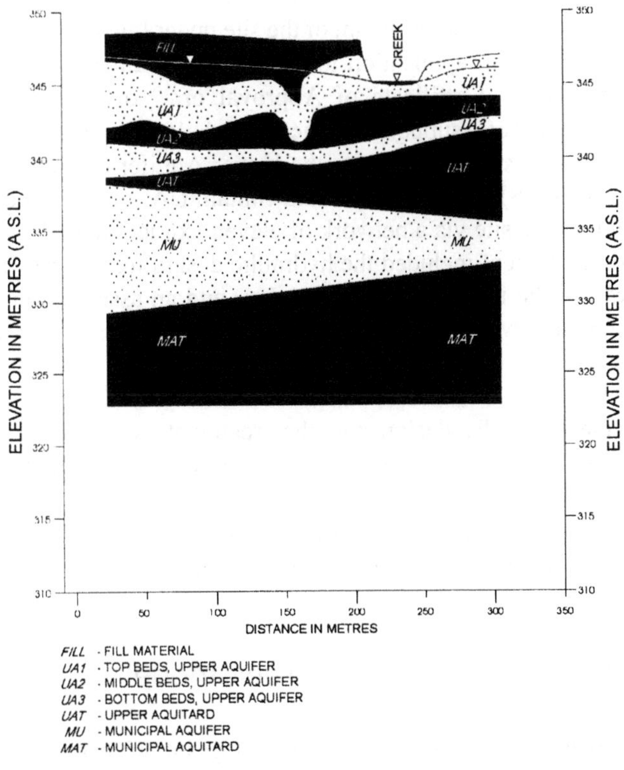

Figure 20.2 Typical hydrogeologic cross-section. Uniroyal Chemical Ltd., Elmira, Ontario.

Following the discovery of n-nitrosodimethylamine (NDMA) in the Elmira water supply system in 1989, the MOEE issued a Control Order (an administrative order, similar to a U.S. EPA Unilateral Administrative Order) to Uniroyal which required Uniroyal to, among other environmental controls, contain contaminated groundwater under the site. Initial groundwater containment efforts were directed at the MA utilizing a network of groundwater extraction wells. The MA contaminant and treatment system, known as the on-site contaminant and treatment systems (OSCTS) was commissioned in 1992 with two extraction wells and supplemented with a third extraction well in 1993.

Surface water quality sampling conducted immediately downstream of the site indicated that contaminated UA_1 groundwater was discharging to the Creek. Water quality concerns with the results of Canagagigue Creek surface water monitoring led to the preparation of the Upper Aquifer Feasibility Study (UAFS) (CRA, 1994a). This report evaluated alternatives for UA groundwater containment that would improve surface water quality in the Creek. The UAFS concluded that containment of groundwater in the southwest portion of the site would result in a 95.5% reduction in the mass loading of contaminants to the Creek, and improvement in surface water quality such that Ontario surface water quality criteria would be met downstream of the site under low flow conditions.

Uniroyal submitted an application to the MOEE to amend its approved treatment system permit (Certificate of Approval or C of A) in a formal application accompanied by a support document (CRA, 1994b). The application requested approval to install and operate the Upper Aquifer containment and treatment system (UACS). The UACS would contain UA_1 groundwater in the southwest portion of the site via a network of eleven extraction wells.

The support document included the hydraulic design for the groundwater containment system, the treatment system design, and an assimilation model. The assimilation model was utilized to predict the impact to the Creek and the associated improvement in surface water quality that would result from the operation of the UACS.

Although the hydraulic design and the treatment system design were significant engineering and technical efforts, the assimilation model was the subject of intense technical review, evaluation and controversy.

The following sections describe how the model was constructed, manipulated and utilized to predict water quality conditions in the Creek.

20.2 Elements of Canagagigue Creek Model

There are several sources of water that discharge into Canagagigue Creek as it passes through the site. In order to create a model of the Creek, it was necessary to first enumerate these sources of water and determine how they would be treated as elements of the model. Each of the elements is discussed below in Sections 20.2.1 to 20.2.6, and illustrated schematically in Figure 20.3.

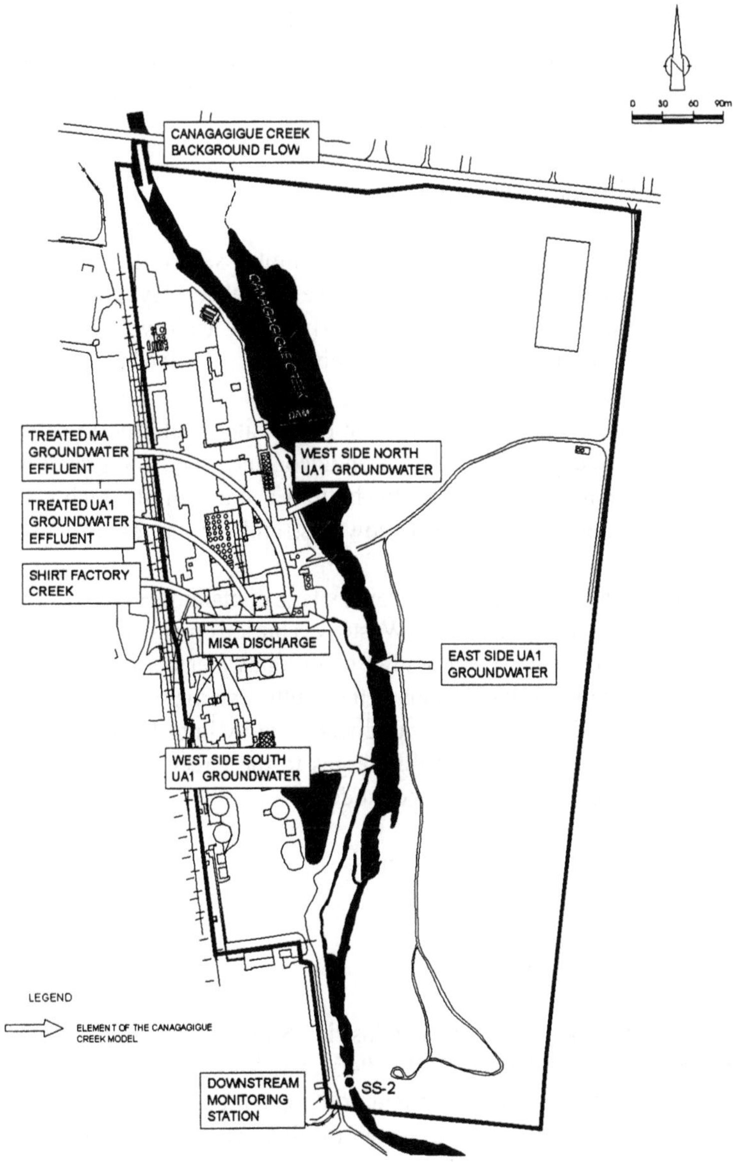

Figure 20.3 Elements of the Canagagigue Creek model, Uniroyal Chemical Ltd. Elmira, Ontario.

20.2.1 Canagagigue Creek Background Flow

The first element of the model is the background flow of the Creek as it enters the site. The MOEE required that the model reflect low flow conditions, so that the model would estimate contaminant levels in the Creek under those infrequent conditions which would result in the maximum contaminant concentrations in the stream. The water in the Creek is not free of contaminants when it enters the site. A former municipal waste landfill is located immediately upstream of the site on the banks of the Creek.

20.2.2 UA_l Groundwater Discharge

The Uniroyal site is underlain by a complex series of glacial and fluvial units, ranging in composition from coarse gravel to clay till. Much of the understanding of the hydrogeology underlying the Uniroyal site, and the surrounding Elmira area, is derived from the work of CRA's Uniroyal project hydrogeologist, Alan Deal.

The surficial aquifer beneath most of the site is referred to as the Upper Aquifer - Top Beds, or UA_l. UA_l is an unconfined sand and gravel formation which is approximately 5 m thick, the bottom 1 m to 2 m of which is saturated. UA_l groundwater generally flows towards, and discharges to, Canagagigue Creek.

Groundwater beneath the Uniroyal site is contaminated due to Uniroyal's past waste management practices. Waste was buried on-site at several locations, however, the most significant source of groundwater contamination was the former unlined wastewater lagoons in the southwestern portion of the site. Due to the significant difference in groundwater quality in different portions of the site, the UA_l groundwater that discharges to the Creek was divided into the following three components for the model:

- West Side-North;
- West Side-South (WSS); and
- East Side.

20.2.3 MISA Discharges

There are three Municipal-Industrial Strategy for Abatement (MISA) discharge points (MISA 0200, MISA 0400, and MISA 0800) to the Creek at the site, which discharge stormwater and non-contact cooling water to the Creek. Although non-contact cooling water discharges at two (MISA 0200 and MISA 0800) of the three MISA discharge points, it was treated as one discharge in the model. This simplification was appropriate because the same water quality (municipal potable water) discharges at both points, and because the purpose of the model was to predict Creek water quality at a point downstream of both the MISA non-contact cooling water discharge points. Stormwater discharges from the MISA outlets were not considered in the model because the assimilation assessment was completed for low flow (drought) conditions, and therefore no stormwater would be present.

20.2.4 Municipal Aquifer Treated Groundwater Effluent

The treated groundwater from the MA treatment system (see Section 20.1) is discharged to the Creek at MISA 0800.

20.2.5 Shirt Factory Creek

Shirt Factory Creek is a small tributary which feeds into Canagagigue Creek on site. Shirt Factory Creek passes beneath the site inside a 200 m long corrugated steel culvert. This culvert discharges to Canagagigue Creek at MISA 0800.

20.2.6 UA_I Treated Groundwater Effluent

The UACS will contain UA_I groundwater in the southwestern portion of the site by extracting UA_I groundwater via a network of eleven extraction wells. The UA_I groundwater will be treated by granular activated carbon, followed by biological treatment. The UA_I groundwater will then be mixed with MA groundwater and treated by ultraviolet oxidation followed by granular activated carbon. The treated UA_I groundwater will then be discharged to the Creek at MISA 0800.

20.2.7 Parameter Selection

CRA developed a list of parameters to be included in the model using the following general steps:

1. An initial list of 204 parameters was created. This list included all parameters that had been analyzed for in 1993 in surface water samples, WSS groundwater, and the influent and effluent of the existing MA treatment system.
2. Parameters that had not been detected in any of these streams were eliminated from further consideration.
3. Parameters that had maximum detections below their respective surface water quality standards (Provincial Water Quality Objectives or PWQOs) were eliminated from further consideration.

4. Parameters that had no PWQOs, but had Ontario Drinking Water Quality Objectives (ODWOs), were eliminated from further consideration if there were no exceedences of the relevant ODWOs.
5. Inorganic parameters that were not indicative of contamination at the Uniroyal site were eliminated from further consideration.

The finalized parameter list for the model is provided in Table 20.1.

Table 20.1 Final parameter list.

General Chemistry		
Ammonia as N	Chloride	Cyanide
Formaldehyde	Sulfate	
Metals		
Cadmium	Iron	Lead
Sodium	Strontium	
Volatile Organic Compounds		
Benzene	Chlorobenzene	Ethylbenzene
m,p- Xylenes	o- Xylene	Toluene
Trichloroethylene		
Base/Neutral/Acid Extractables		
2,3,4-Trichlorophenol		2,4,5-Trichlorophenol
2,4,6-Trichlorophenol		2,4-Dichlorophenol
2,6-Dichlorophenol		2-Chlorophenol
2-Mercaptobenzothiazole		3/4-Methylphenol (m,p-Cresols)
4-Chloro-3-methylphenol (p-Chloro-m-cresol)		Aniline
Benzothiazole		bis(@-Ethylhexyl) phthalate
N-Nitrosodi-N-Butylamine (NBDA)		N-Nitrosodimethylamine (NDMA)
N-Nitrosodiphenylamine/Diphenylamine (NDPA/DPA)		
N-Nitrosomorpholine (NMOR)		Oxathiin (Carboxin)
o-Cresol		Phenol
Herbicides/Pesticides		
2,4-Dichlorophenoxyacetic Acid		Lindane
Dioxins/Furans		
2, 3, 7, 8-Tetrachlorodibenzo-p-dioxin (2, 3, 7, 8-Tetra CDD)		
2, 3, 7, 8-Tetrachlorodibenzofuran (2, 3, 7, 8-Tetra CDF)		

20.2.8 Structure of the Model

The purpose of the model was to estimate parameter concentrations in Canagagigue Creek at SS2, a compliance monitoring point immediately downstream of the site. Parameter concentrations at SS2 were calculated using the mass balance equation:

$$C_{SS2} = \frac{Q_{BACK}C_{BACK} + Q_{MISA}C_{MISA} + Q_{UT}C_{UT} + Q_{MT}C_{MT} + Q_{SFC}C_{SFC} + Q_{UA}C_{UA}}{Q_{TOTAL}}$$

where:

C_{SS2} = concentration in Canagagigue downstream of Uniroyal.
Q_{BACK} = background stream flow (Canagagigue Creek)
C_{BACK} = background stream concentration
Q_{MISA} = flow rate from MISA 0200 and MISA 0800 cooling water
C_{MISA} = concentration in MISA 0200 and MISA 0800 cooling water
Q_{UT} = effluent flowrate of on-site treatment plant for UA groundwater
C_{UT} = concentration in treated UA groundwater effluent
Q_{MT} = effluent flowrate of on-site treatment plant from MA groundwater extracted from PW1, PW3 and PW4
C_{MT} = concentration in treated MA groundwater effulent
Q_{SFC} = flow rate from Shirt Factory Creek
C_{SFC} = concentration in Shirt Factory Creek
Q_{UA} = discharge flowrate of UA groundwater (varies according to scenario of containment/no containment)
C_{UA} = discharge concentration of UA groundwater
Q_{TOTAL} = $Q_{BACK}+Q_{MISA}+Q_{UT}+Q_{MT}+Q_{SFC}+Q_{UA}$

The model did not attempt to account for losses of contaminants via mechanisms such as volatilization, photo-oxidation, or biological uptake/degradation. The model was calibrated with real Creek monitoring data (see Section 20.5) to verify that losses from such mechanisms were not significant.

The model was run as an Excel 4.0 spreadsheet. Each element of the model was represented by a column, and each parameter by a row.

20.3 Determination of Flow Values

The initial step in quantifying flows in the model was the determination of an appropriate background Creek flow that represented drought flow conditions. Typically in assimilation assessments, the published $7Q_{20}$ value is used, which represents the minimum consecutive 7-day average flow that has a probability of occurrence of once every 20 years. These values are published for numerous rivers and creeks in Ontario (Inland Waters Directorate, 1992). In the case of Canagagigue Creek, the published $7Q_{20}$ value was based on almost 30 years of creek flow measurements.

Two significant problems were associated with the published $7Q_{20}$. The first was that the $7Q_{20}$ was based on flow records both prior to, and after construction of the Woolwich Reservoir, a dam on Canagagigue Creek upstream from the site. The purpose of the reservoir is to maintain base flows in the Creek, and particularly to provide a minimum base flow for the Elmira sewage treatment plant (STP), located immediately downstream of Uniroyal.

The published $7Q_{20}$ was an unrealistically low value which did not reflect the significance of flow regulation following construction of the Woolwich Reservoir. CRA undertook an extensive effort to re-evaluate the Creek flow records for the period following construction of the reservoir to determine an appropriate $7Q_{20}$ for use in the assimilation assessment. This re-evaluation has been the subject of intense controversy with the MOEE.

Another difficulty with determining a suitable background Creek flow upstream of Uniroyal was that the Creek flow records used for $7Q_{20}$ calculations were obtained from an Environment Canada monitoring station which is located several kilometres downstream of Uniroyal. However, because of the extensive historical database available, it was possible to quantify flows from the various sources (such as the Elmira STP and flow from a tributary to the Creek) between the downstream gauging location and Uniroyal's site to determine a suitable upstream low flow value. Once this value was determined, assimilation calculations were performed with this background flow rate.

The background flow rate was determined as follows:

$$Q_{Back} = Q_{Total} - Q_{Other}$$

where:

Q_{Back} = background low flow upstream of Uniroyal

Q_{Total} = minimum observed 7-day average flow at downstream station; 290 L/s (the minimum observed 7-day average flow occurred in 1987 and was 3% less than the $7Q_{20}$ of 298 L/s calculated by CRA).

Q_{Other} = flows from other sources that flow into the Creek between the downstream station and Uniroyal (approximately 120 L/s)

Therefore:

$$Q_{Back} = 290 \text{ L/s} - 120 \text{ L/s} = 170 \text{ L/s}$$

Again, this approach was the subject of much debate between the MOEE, Uniroyal, CRA and other interested parties. It is interesting to note that if the published $7Q_{20}$ was used in the above equation, the resultant upstream flow would have resulted in a negative number. This fact supports CRA's approach.

Most sources that discharge to the Creek on Uniroyal's property are point-source discharges from outfalls, and these flows were quantified in the model by determining average flow values from historical databases.

An exception to the point source discharges noted above is the non-point source discharge from UA_1, which discharges to the Creek along its banks. These discharges were quantified using Darcy's Law:

$$Q = KiA = KibL$$

where:

Q = flow (m^3/s)
K = hydraulic conductivity (m/s)
i = horizontal hydraulic gradient (dimensionless)
A = cross-section area (m^2)
b = saturated thickness (m)
L = cross-section length (m)

Detailed calculations of the groundwater flux from UA_1 to the Creek were presented in the UAFS. Extensive hydrogeologic data were available from numerous groundwater monitoring sampling rounds, pumping tests, and single-well response tests performed using the on-site monitoring wells. Figure 20.3 presents a schematic representation of UA_1 discharges to the Creek.

The proposed UACS will contain the UA_1 groundwater flux currently discharging to the Creek along the West Side South (WSS) portion of the property. Therefore, two conditions were modeled, first with the UA_1 WSS uncontained, and second with the UA_1 WSS contained. This was accomplished by setting the UA_1 WSS flux to zero for the contained scenario. In this way, the effects of the UACS on the Creek water quality could be modeled. With the UA_1 WSS contained, flow from the treated UA_1 groundwater was considered to be a source of discharge to the Creek in the model.

Current target flow rates were used to quantify flows from the three operating on-site MA extraction wells. The treated groundwater from these wells was included as a source of flow to the Creek.

Non-contact cooling water is obtained from the municipal water supply and is discharged through MISA 0200 and 0800. Average flow rates for the cooling water were obtained from MISA monitoring and municipal water usage records, and these were used to quantify this source in the model.

20.4 Determination of Water Quality Chemical Profiles

Water quality chemical profiles for the various sources considered in the model were developed using an extensive historical database which included:

1. the Uniroyal database which comprises all groundwater, surface water, and treatment system effluent analytical work conducted by Uniroyal from 1982 to the present;

2. Uniroyal MISA monitoring analytical results for cooling water discharges; and
3. the MOEE, Regional Municipality of Waterloo, and Ontario Clean Water Agency database for selected locations.

The manipulation of these data was not a trivial task. It involved compiling data from the several sources listed above, sorting through several thousand analytical results, and removing duplication.

Parameter levels for the Creek background water quality were calculated by combining historic data from several upstream Creek water sampling locations, including samples obtained by both Uniroyal and the MOEE.

Non-contact cooling water quality at the MISA outfalls was quantified by compiling Uniroyal's MISA analytical results.

Shirt Factory Creek (a tributary to Canagagigue Creek) water quality was determined by reviewing MOEE water quality data.

UA_1 contaminant concentrations were quantified by considering the most recent analysis of each parameter at each monitoring well screened in UA_1.

Treatment system discharges were handled differently than conventional assimilation assessments, where maximum allowable concentrations at maximum discharge rates are typically considered. The available database for the MA treatment system was used to determine representative effluent concentrations for this treatment system. The proposed UACS treatment system will utilize similar (and in some cases, better) treatment technology than the MA treatment system. Therefore, an effluent profile for the UACS was developed based on expected removal efficiencies determined from data obtained for the MA treatment system. This approach was taken because the intent of the assimilation assessment was to determine the impact of the expected discharge concentrations on the Creek, rather than to determine effluent criteria for the UACS. Effluent criteria for the UACS were ultimately determined by the MOEE from treatment technology limitations, not from the assimilative capacity.

A number of technical issues arise in determining a typical water quality profile for a body of water. The estimation of parameters is difficult because the data are frequently skewed (asymmetrical) and the data record usually contains numerous values reported as less than the detection level. CRA utilized two novel approaches to quantify representative concentrations for the various sources in the model. These were:

1. use of the geometric mean (as opposed to an arithmetic mean or 75th percentile); and
2. use of a sliding scale to quantify concentrations reported as being non-detect.

The rationale for these assumptions is discussed below.

CRA accounted for the skewed data by utilizing a log normal distribution to characterize the water quality data for the individual constituents. When data are skewed, the estimation of the arithmetic mean is strongly biased by a very small subset (i.e. the outliers) of the entire data set. As a result, the arithmetic mean is not a very useful indicator of the central tendency of the data set. Consequently, CRA utilized the geometric mean as the indication of the central tendency. The geometric mean is also consistent with the log normal distribution indicated above.

To confirm the appropriateness of the log normal distribution assumption, CRA calculated the means and coefficient of variation (standard deviation divided by the mean) for both the arithmetic and geometric (log-transformed) data, for all inputs into the model. The coefficient of variation test indicated that the lognormal distribution better characterized the data than the normal distribution. Figure 20.4 illustrates the skewing effect of high analytical results on two relevant contaminants (toluene and chlorobenzene) in the UA_1 groundwater, and demonstrates that the geometric mean better represents the central tendency of the data.

An additional issue is how to characterize data reported as non-detect or less than the analytical detection limit. A number of different approaches have been published but the most relevant for this type of situation (for estimating the geometric mean) involve either:

1. assuming values of zero, one-half the detection level, or equal to the detection level, or
2. a sliding scale procedure where the replacement value is selected on the basis of three-quarters of the detection level if one-quarter of the data are less than, half the detection level if one half of the data are less than, and one-quarter of the detection level if three-quarters of the values are less than.

With respect to 1. above, the two end conditions (zero or equal to the detection limit) are extreme situations which do not reasonably reflect the real condition, and the middle condition (assuming one-half the detection level) is somewhat arbitrary.

CRA instead used a sliding scale procedure to quantify non-detects. When a sliding-scale calculation is performed, the percentage of the total number of samples that are non-detect values is determined. In the ensuing calculations, the sliding-scale factor associated with this percentage is multiplied by the detection limit of each non-detect to create a value to be used in calculation.

The sliding scale procedure adjusts the value for assignment of the non-detect data in accordance with the percentage of samples that are reported as non detects. If the number of samples that are reported as non-detects are a relatively large percentage of the total data set, then one expects that the actual concentration of the parameter in the non-detect samples must be close to zero. Alternatively, if the number of non-detect values is a relatively small percentage of the total data set, then one expects that the actual concentration of the parameter in the non-detect samples must be close to the detection limit.

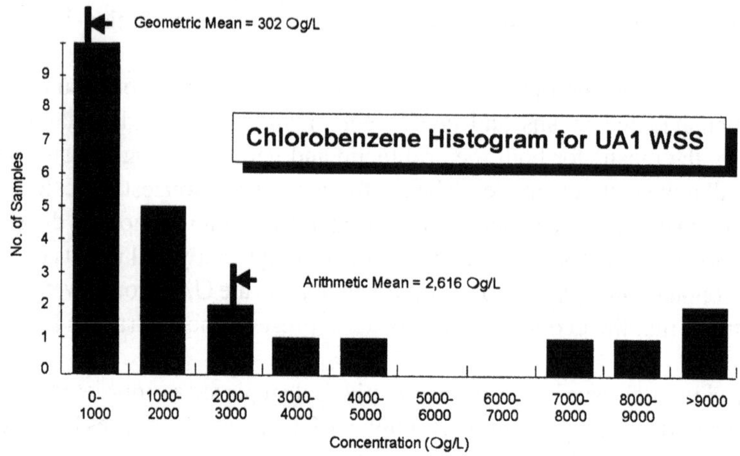

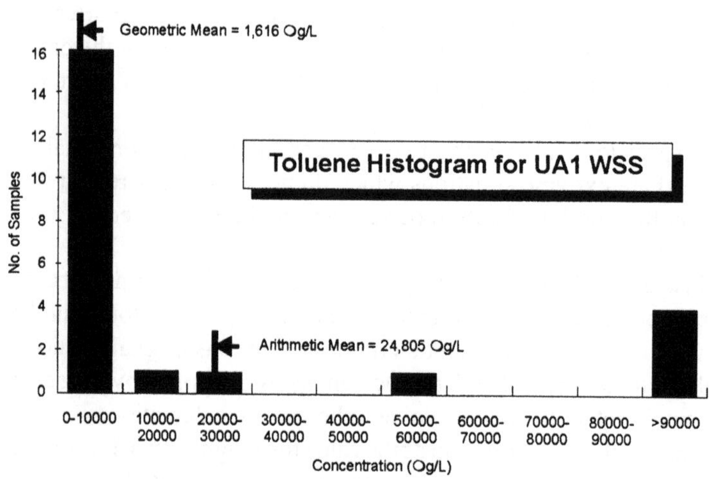

Figure 20.4 Geometric means vs. arithmetic means. Surface water modeling of Canagagigue Creek.

The sliding-scale factors used by CRA in the determination of all groundwater, surface water and effluent water quality profiles are:

Percent of Samples Which are Non-detect	Sliding-scale Factor (Fraction of Detection Limit)
≤ 37.5%	0.75
> 37.5% and ≤ 62.5%	0.5
> 62.5%	0.25

Both the use of the geometric mean, and the use of a sliding scale to quantify non-detects, were the subject of debate between the MOEE and CRA.

20.5 Simulation of Existing Conditions

After setting up the model, it was necessary to determine if the model would provide estimates of Creek water quality downstream of the site that were accurate. Surface water stations upstream and downstream of the site were monitored for water quality and flow during four monitoring rounds in 1993. For each sampling round, CRA input the observed upstream flow and upstream water quality into the model, and then used the model to estimate downstream water quality. Each modeled downstream parameter concentration was divided by the actual parameter concentration (as measured during the sampling round) to provide a comparative ratio. These comparative ratios were thus indicative of the validity of the model.

The MOEE fundamentally disagreed with CRA's decision to calculate average UA_1 groundwater contaminant levels using geometric means. The MOEE felt that using geometric averages would inappropriately mitigate the impact of those monitoring well samples that contained high concentrations of contaminants. However, CRA believes that the use of geometric averages was appropriate, based on the lognormal distribution of the data. To address the MOEE's concerns, comparative ratios were calculated for the model using both geometric means and arithmetic means to quantify UA_1 groundwater quality. These comparative ratios are summarized in Table 20.2

As shown by the comparative ratios in Table 20.2, the model using geometric means estimated downstream parameter concentrations that more closely matched the actual sampling data than the model using arithmetic averages. In fact, the version of the model using geometric means was remarkably accurate in modeling the downstream Creek water quality for the four 1993 surface water sampling events. The comparative ratios ranged from 0.022 to 6.148, and the overall average comparative ratio was 1.284. Given the complexity of the many different elements that were being modeled, the

Table 20.2 Summary of comparative ratios.

	Using Geometric Mean Values For UA$_1$ Groundwater			Using Arithmetic Mean Values For UA$_1$ Groundwater		
Parameter	**Minimum Comparative Ratio**	**Maximum Comparative Ratio**	**Average Comparative Ratio**	**Minimum Comparative Ratio**	**Maximum Comparative Ratio**	**Average Comparative Ratio**
General Chemistry						
Ammonia as N	0.431	1.461	0.940	0.887	2.032	1.473
Volatile Organic Compounds						
Benzene	0.345	1.457	0.975	1.491	6.619	3.539
Chlorobenzene	0.287	5.763	2.343	2.112	51.305	19.269
Ethylbenzene	0.670	1.890	1.267	3.413	10.027	5.620
m,p-Xylenes	1.248	1.925	1.672	1.130	12.105	7.029
o-Xylenes	1.185	1.689	1.500	3.207	9.229	6.975
Toluene	0.109	1.354	1.689	1.357	10.516	6.469
Trichloroethylene	0.022	1.053	1.697	1.050	2.512	1.323
Base/Neutral/AcidExtractions						
2,3,4-Trichlorophenol	0.978	0.994	0.986	1.066	1.246	1.159
2,4,5-Trichlorophenol	0.492	0.995	0.865	0.685	1.407	1.093
2,4,6-Trichlorophenol	0.982	0.998	0.993	1.198	1.673	1.402
2,4-Dichlorophenol	0.230	1.121	0.858	1.310	6.860	3.546
2,6-Dichlorophenol	0.967	1.020	0.998	1.344	2.315	1.932
2-Chlorophenol	1.002	1.006		4.109	12.596	8.522
2-Mercaptobenzothiazole	0.107	3.501	2.361	0.788	22.224	14.436
Aniline	0.693	6.148	2.710	3.747	31.057	13.260
Benzothiazole	0.264	2.900	1.025	3.858	15.607	7.386
Carboxin	0.348	1.494	1.107	2.655	10.212	6.029
m,p-Cresols	1.043	1.217	1.121	1.151	1.584	1.383
NDBA	2.328	3.064	2.696	3.972	5.619	4.796
NDEA	2.020	2.020	2.020	5.736	5.736	5.736
NDMA	0.227	1.840	0.894	3.690	20.972	10.094
NMOR	0.387	1.851	1.117	1.407	9.611	5.509
o-Cresol	0.684	1.010	0.925	0.908	1.503	1.189
p-Chloro-m-Cresol	0.984	0.996	0.990	1.022	1.081	1.053
Phenol	1.013	1.022	1.016	7.474	14.262	14.262
Herbicides/Pesticides						
2,4-D	0.907	1.107	1.012	2.235	5.021	3.165
Lindane	0.993	1.447	1.172	0.993	1.448	1.172
	Range of Comparative Ratios		0.022-6.148			0.050-51.305
	Range of Average Comparative Ratios		0.689-2.710			1.053-19.269
	Overall Average		1.284			5.555

Note: Comparative ratio is the modeled concentration at SS2 divided by the actual concentration at SS2.

simulation of the actual 1993 sampling events provided excellent correlation between the model and observed conditions.

As discussed previously, the model did not attempt to account for losses of contaminants via mechanisms such as volatilization, photo-oxidation, or biological uptake/degradation. These mechanisms are likely part of the reason why the model computed contaminant concentrations that were an average of 1.284 times higher than the concentrations detected in observed sampling events.

20.6 Predicted Improvement in Creek Quality

Following the determination of all input parameters, and the calibration of the model, assimilation calculations were completed for the following conditions:

- UA_1 WSS groundwater uncontained, (i.e. current conditions); and
- UA_1 WSS groundwater contained.

Table 20.3 presents the results of the model for UA_1 WSS uncontained (current conditions).

The model computed that iron, toluene, aniline, and bis(2-ethylhexyl)phthalate would exceed their respective PWQO or guideline concentration at SS2. The predicted exceedences of iron and toluene are consistent with the parameters at SS2 which historically exceed their PWQO in sampling at SS2.

While computed aniline concentrations exceed the proposed PWQO, the actual mean aniline concentration at SS2 is less than the PWQO. The assimilation model overestimates aniline concentrations by a factor of 73. It is not known why the model significantly overestimates aniline concentrations, whereas the model accurately computes similar compounds such as benzothiazole and carboxin.

Likewise, while computed bis(2-ethylhexyl)phthalate concentrations exceed the PWQO, the observed mean concentration at SS2 is less than the PWQO. Bis(2-ethylhexyl)phthalate is a common analytical laboratory contaminant, and false detections are often reported by laboratories. The computed exceedence of bis(2-ethylhexyl)phthalate is due to the elevated level assumed in the calculations for the background Creek concentration, which has resulted from suspected false detections.

With the exception of aniline, the modeled SS2 values were very good predictors of observed Creek concentrations.

Table 20.4 presents the results of the assimilation calculations for UA_1 WSS contained. The assimilation calculations estimated that toluene and aniline would be reduced below their PWQO/G and would be reduced to non-detect concentration levels.

Table 20.3 Assimilation calculations. UA1 WSS uncontained.

	C Back	Modeled SS+855	PWQO [1]	Ratio Model / PWQO
General Chemistry (mg/L)				
Ammonia as N [2]	- -	- -		- -
Chloride	28.6	44.2		N/A
Cyanide	ND(0.005)	ND(0.005)	5	- -
Formaldehyde	0.126	- -	0.8	- -
Sulfate	34.6	45.1		N/A
Metals (mg/L)				
Cadmium	ND(0.003)	ND(0.003)	0.0002	[3]
Iron	0.704	0.663	0.3	2.21
Lead	ND(0.025)	ND(0.025)	0.005	[3]
Sodium	13.7	26.2		N/A
Strontium	0.235	0.303		N/A
Volatile Organic Compounds (µg/L)				
Benzene	ND(0.5)	ND(0.5)	15	- -
Chlorobenzene	ND(0.5)	2	8	0.13
Ethylbenzene	ND(0.5)	ND(0.5)	2	- -
m,p-Xylenes	ND(0.5)	ND(0.5)	40	- -
o-Xylenes	ND(0.5)	ND(0.5)	0.8	- -
Toluene	ND(0.5)	9.8	20	12.24
Trichloroethylene	ND(1)	ND(1)		- -
Base/Neutral/Acid Extractables (µg/L)				
2,3,4- Trichlorophenol	ND(2)	ND(2)	18	- -
2,4,5- Trichlorophenol	ND(2)	ND(2)	18	- -
2,4,6- Trichlorophenol	ND(1.2)	ND(1.2)	18	- -
2,4- Dichlorophenol	ND(1.2)	ND(1.2)	0.2	[3]
2,6- Dichlorophenol	ND(1.2)	ND(1.2)	180.2	[3]
2- Chlorophenol	ND(2)	ND(2)		- -
2- Mercaptobenzothiazole	ND(10)	73.7		N/A
Aniline	ND(2)	40.9		20.43
Benzothiazole	ND(2)	4.1		N/A
Bis(2-Ethylhexyl) phthalate	2.8	2.7	0.6	4.53
Carboxin	ND(2)	15.1		N/A
m,p- Cresols	ND(2)	ND(2)	1	[3]
NDBA	ND(10)	ND(10)		N/A
NDEA	0.017	0.103		N/A
NDMA	ND(0.01)	0.115		N/A
NDPA/DPA	ND(5)	- -	7	N/A
NMOR	ND(10)	ND(10)		N/A
o-Cresol	ND(2)	ND(2)	1	[3]
p-Chloro-m-Cresol	ND(2)	ND(2)		N/A
Phenol	ND(1.1)	ND(1.1)	5	- -
Herbicides/Pesticides (µg/L)				
2,4- D	0.067	0.106	4	0.03
Lindane	ND(0.003)	0.004	0.01	0.43
Dioxins/Furans (ηg/L)				
2,3,7,8- Tetra CDD	0.0003	0.0003 [4]		N/A
2,3,7,8- Tetra CDF	0.0002	0.0002 [4]		N/A
Total TEQ	0.0013	0.0012 [4]		N/A

Notes:

[1] PWQO - Provincial Water Quality Objective or Interim Provincial Water Quality Objective

[2] See Table 20.5 for ammonia assimilation calculations.

[3] The PWQO for this compound is less than the detection limits.

[4] No standard detection limits available for dioxins and furans. Detection limits not stable due to the very low levels of detection involved.

- - Data not available

N/A No applicable PWQO

ND Concentration deemed non-detect because it is lower than the standard detection limits shown in parenthesis.

[box] Indicates concentration exceeds the PWQO.

Table 20.4 Assimilation calculations. UA$_1$WSS contained.

	C Back	Modeled SS+855	PWQO [1]	Ratio Model/PWQO
General Chemistry (mg/L)				
Ammonia as N [2]	- -	- -		- -
Chloride	28.6	47.0		N/A
Cyanide	ND(0.005)	ND(0.005)	5	- -
Formaldehyde	0.126	- -	0.8	- -
Sulfate	34.6	47.0		N/A
Metals (mg/L)				
Cadmium	ND(0.003)	ND(0.003)	0.0002	[3]
Iron	0.704	0.654	0.3	2.21
Lead	ND(0.025)	ND(0.025)	0.005	[3]
Sodium	13.7	29.0		N/A
Strontium	0.235	0.308		N/A
Volatile Organic Compounds (μg/L)				
Benzene	ND(0.5)	ND(0.5)	15	- -
Chlorobenzene	ND(0.5)	ND(0.5)	8	0.13
Ethylbenzene	ND(0.5)	ND(0.5)	2	- -
m,p-Xylenes	ND(0.5)	ND(0.5)	40	- -
o-Xylenes	ND(0.5)	ND(0.5)	0.8	- -
Toluene	ND(0.5)	ND(0.5)	20	- -
Trichloroethylene	ND(1)	ND(1)		- -
Base/Neutral/Acid Extractables (μg/L)				
2,3,4- Trichlorophenol	ND(2)	ND(2)	18	- -
2,4,5- Trichlorophenol	ND(2)	ND(2)	18	- -
2,4,6- Trichlorophenol	ND(1.2)	ND(1.2)	18	- -
2,4- Dichlorophenol	ND(1.2)	ND(1.2)	0.2	[3]
2,6- Dichlorophenol	ND(1.2)	ND(1.2)	0.2	[3]
2- Chlorophenol	ND(2)	ND(2)		- -
2- Mercaptobenzothiazole	ND(10)	ND(10)		N/A
Aniline	ND(2)	ND(2)		20.43
Benzothiazole	ND(2)	ND(2)		N/A
Bis(2-Ethylhexyl) phthalate	2.8	2.7	0.6	4.53
Carboxin	ND(2)	15.1		N/A
m,p- Cresols	ND(2)	ND(2)	1	[3]
NDBA	ND(10)	ND(10)		N/A
NDEA	0.017	0.100		N/A
NDMA	ND(0.01)	0.112		N/A
NDPA/DPA	ND(5)	- -	7	N/A
NMOR	ND(10)	ND(10)		N/A
o-Cresol	ND(2)	ND(2)	1	[3]
p-Chloro-m-Cresol	ND(2)	ND(2)		N/A
Phenol	ND(1.1)	ND(1.1)	5	- -
Herbicides/Pesticides (μg/L)				
2,4- D	0.067	0.096	4	0.03
Lindane	ND(0.003)	0.004	0.01	0.43
Dioxins/Furans (ηg/L)				
2,3,7,8- Tetra CDD	0.0003	0.0003 [4]		N/A
2,3,7,8- Tetra CDF	0.0002	0.0002 [4]		N/A
Total TEQ	0.0013	0.0012 [4]		N/A

Notes:

[1] PWQO - Provincial Water Quality Objective or Interim Provincial Water Quality Objective
[2] See Table 20.5 for ammonia assimilation calculations.
[3] The PWQO for this compound is less than the detection limits.
[4] No standard detection limits available for dioxins and furans.
Detection limits not stable due to the very low levels of detection involved.
- - Data not available
N/A No applicable PWQO
ND Concentration deemed non-detect because it is lower than the standard detection limits shown in parenthesis.
[box] Indicates concentration exceeds the PWQO. (Boxed values: Iron 0.704 and 0.654; Bis(2-Ethylhexyl) phthalate 2.8 and 2.7.)

Iron concentrations (a natural constituent of Canagagigue Creek surface water) would be relatively unaltered and were computed to still exceed the PWQO. Bis(2-ethylhexyl)phthalate concentrations were computed to also be unaltered and also exceed its PWQO. However, the calculated exceedence of bis(2-ethylhexyl)phthalate is solely a function of the elevated background concentration. These calculations are significantly biased by the detection limits available for this compound and by the preponderance of evidence which has established that this compound is a common laboratory contaminant and should be regarded as a sampling artifact (see the USEPA Contract Laboratory Program National Functional Guidelines for Organic Data Review, (USEPA, 1993)).

Some parameters such as cadmium, lead, 2,4- and 2,6-dichlorophenol, and m,p- and o-cresols were modeled to be non-detect, however, current detection limits for these compounds are above their respective PWQOs.

Computed ammonia concentrations are calculated differently than other parameters because:

1. ammonia is typically reported as total ammonia (ammonia plus ammonium); and
2. the PWQO for ammonia relates only to the un-ionized fraction of the total ammonia.

The percentage of un-ionized ammonia in an aqueous ammonia solution can be calculated if the pH and temperature of the liquid is known. Therefore, when determining in-stream compliance with the PWQO, the pH and temperature of the receiving stream (in this case Canagagigue Creek) must be known.

The MOEE provided CRA with total ammonia, pH, and temperature records for sampling Station CC1-A (located in Canagagigue Creek upstream of Uniroyal). The data record for this station extends back to 1975, however, field pH measurements were not routinely obtained until 1987. Since the pH can vary significantly between field values and laboratory values, only the data from 1987 to date were considered in the assimilation assessment.

The ammonia data were summarized by month and these data were used as C_{BACK} for ammonia. It is noteworthy that exceedences of the PWQO have occurred at least once at CC-1A in all months except January and December.

Ammonia assimilation calculations performed for the UA_1 WSS contained are presented in Table 20.5

The ammonia assimilation calculations performed for the UACS showed that with the UACS operating and low flow conditions in the Creek, un-ionized ammonia was predicted to exceed the PWQO in the warmer months of the year, specifically June, July, and September. The calculations showed that the PWQO would be exceeded by a maximum factor of 1.40 at SS2; however, during the month of June, the mean upstream Creek water quality at CC-1A already exceeds the PWQO due to natural background water quality or upstream human-induced conditions.

Table 20.5 Ammonia assimilation calculations. UA_1 WSS contained.

Ammonia Concentrations (mg/L)	C Back Total Ammonia (1)	Average Conversion Factor (1)	C Back Un-ionized Ammonia	Modelled SS2 Concentration (Un-ionized Ammonia)	Actual SS2 (2) (Un-ionized Ammonia)	Ratio Model Actual	Ratio Model PWQO
January	0.416	1.09 %	0.005	0.006		--	0.286
February	0.592	1.85 %	0.011	0.012		--	0.623
March	0.891	1.02 %	0.009	0.009	0.001	9.42	0.471
April	0.475	3.47 %	0.016	0.020		--	1.00
May	0.173	5.17 %	0.009	0.017	0.011	1.51	0.832
June	0.384	5.43 %	0.021	0.027	0.012	2.26	1.35
July	0.277	6.87 %	0.019	0.028	0.102	0.28	1.40
August	0.187	4.42 %	0.008	0.015	0.047	0.31	0.737
September	0.227	7.55 %	0.017	0.028	0.010	2.77	1.39
October	0.194	5.14 %	0.010	0.017	0.002	8.72	0.872
November	0.216	2.47 %	0.005	0.009	0.010	0.88	0.442
December	0.107	2.20 %	0.002	0.006	0.001	5.86	0.293

Note:

(1) Based on MOEE data from Canagagigue Creek Sampling Station CC-1A.

(2) Actual SS2 concentrations based on quarterly Creek sampling conducted by Uniroyal.

☐ Indicated concentration exceeds the PWQO of 0.02 mg/L.

The C of A for the UACS provides Creek water quality compliance requirements that are closely tied to the assimilation model results. For most parameters considered in the assimilation model, the in-stream compliance concentration at SS2 is set well below the PWQO. Should the predicted improvements of Creek water quality not occur after operation of the UACS, Uniroyal could potentially be in violation of a condition of the C of A.

Assimilation assessment calculations were also performed downstream of the point of compliance for Uniroyal. Other downstream sources of flow to the Creek were considered for these analyses, such as sewage treatment plant discharges and flows from tributaries to the Creek. These assimilation calculations were also presented in the C of A support document for the UACS.

20.7 Conclusions

The Uniroyal Creek model formed the basis for CRA's conclusion that the proposed UACS would improve water quality in the Creek to provincial water quality standards under low flow conditions, except where background water quality would not permit sufficient improvements. The model then gave rise to a series of recommended contingency measures which would be implemented if the water quality standards were not met. These contingency measures included:

- a hydraulic containment contingency plan; and
- a surface water quality contingency plan.

The MOEE concurred that contingency plans were required and a considerable effort was expended by both Uniroyal and the MOEE to ultimately agree on the scope of the contingency plans.

However, the MOEE and CRA were not able to reach agreement on the predictions made by the Creek model. The MOEE's criticisms centered on three central issues, as follows:

- the assumed value of the Creek Low Flow (7Q20);
- the handling of non-detects; and
- the calculations of representative parameter concentrations using geometric means from the various sources of input to the Creek.

The basis for CRA's determination of the 7Q20 for the Creek and the development of input parameters was described above. The MOEE believed that the appropriate value of the 7Q20 was 30 to 70 L/s, versus the 170 L/s evaluated by CRA. The MOEE also believed that arithmetic means of contaminant concentrations should be used in the assimilation calculations, despite the excellent calibration results obtained by using geometric means as discussed above.

In the end, the MOEE chose to disagree with our modeling effort in general, but to approve the implementation of the UACS on the basis that water quality would be improved and further groundwater containment measures would be implemented if they were shown to be necessary. The criteria selected to determine whether or not downstream surface water was significantly impacted by the uncontained discharges from the UA were based principally upon the output of the model. Other factors were taken into account, such as analytical quantitation limits and analytical variability. However, the model was utilized in the new C of A as the principal measuring stick for the performance of the UACS.

Construction of the UACS treatment system is expected to begin in April 1996, pending weather conditions and receipt of a final C of A for one component of the treatment system. The extraction wells have already (February 1996) been installed. Operation of the UACS is scheduled to begin in the fall of 1996.

The reliability of the model will be tested by in-stream measurements of surface water quality shortly after start-up and over the life of the operation of the system. By the time we can fully validate the model, we will know more about the relationship between the UA and the Creek than this model could have ever predicted.

References

CRA, (Conestoga Rovers Assocs) 1994a, Upper Aquifer Feasibility Study, Uniroyal Chemical Ltd., Elmira, Ontario, April 1994, c550pp incl. appendices.

CRA, 1994b, Support Document, Amendment of Certificate of Approval (No. 4-0166-90-938), Upper Aquifer Groundwater Containment and Treatment, Uniroyal Chemical Ltd., Elmira, Ontario, November 1994, c300pp incl. appendices.

Inland Waters Directorate, 1992, Historical Stream Flow Summary, Ontario, Water Resources Branch, Water Survey of Canada, Environment Canada, c250pp.

USEPA 1993, USEPA Contract Laboratory Program National Functional Guidelines for Organic Data Review, United States Environmental Protection Agency, Publication 9240.1-05, February 1993, p82.

Chapter 21

Setting of Total Maximum Daily Loads for an Urban Water Body

David Crawford and Mary Abrahms

The State of Oregon has established water quality standards for meeting the goals and requirements of the United States Clean Water Act. The normal model for standards is followed in Oregon where the standard is composed of designated beneficial uses and water quality criteria or action levels which are set to maintain these beneficial uses. Section 303 of the Clean Water Act requires that water bodies that fail to meet water quality, usually through non attainment of beneficial uses and measured through chronic violation of water quality criteria such as temperature, chlorophyll a (as an indicator for excessive algae growth) and dissolved oxygen, be identified as "water quality limited." When this water quality limited designation is in place for a water body, the state must establish a total maximum daily load (TMDL). The Columbia Slough in Portland, Oregon, has been designated as water quality limited.

The Columbia Slough is an urban water body (see Figure 21.1) that is significant to the metropolitan Portland community, because of large undeveloped reaches and potential as a valuable fish and wildlife habitat within an urbanized area. The area also contains significant commercial and industrial facilities such as the Portland International Airport. Like most water bodies, the Columbia Slough is usually thought of as a single entity and usually studied with

© *Advances in Modeling the Management of Stormwater Impacts - Vol. 5* W. James, Ed.
Pub. by CHI, Guelph, Canada 1997. ISBN 0-9697422-7-4. Fax: +519 767-2770

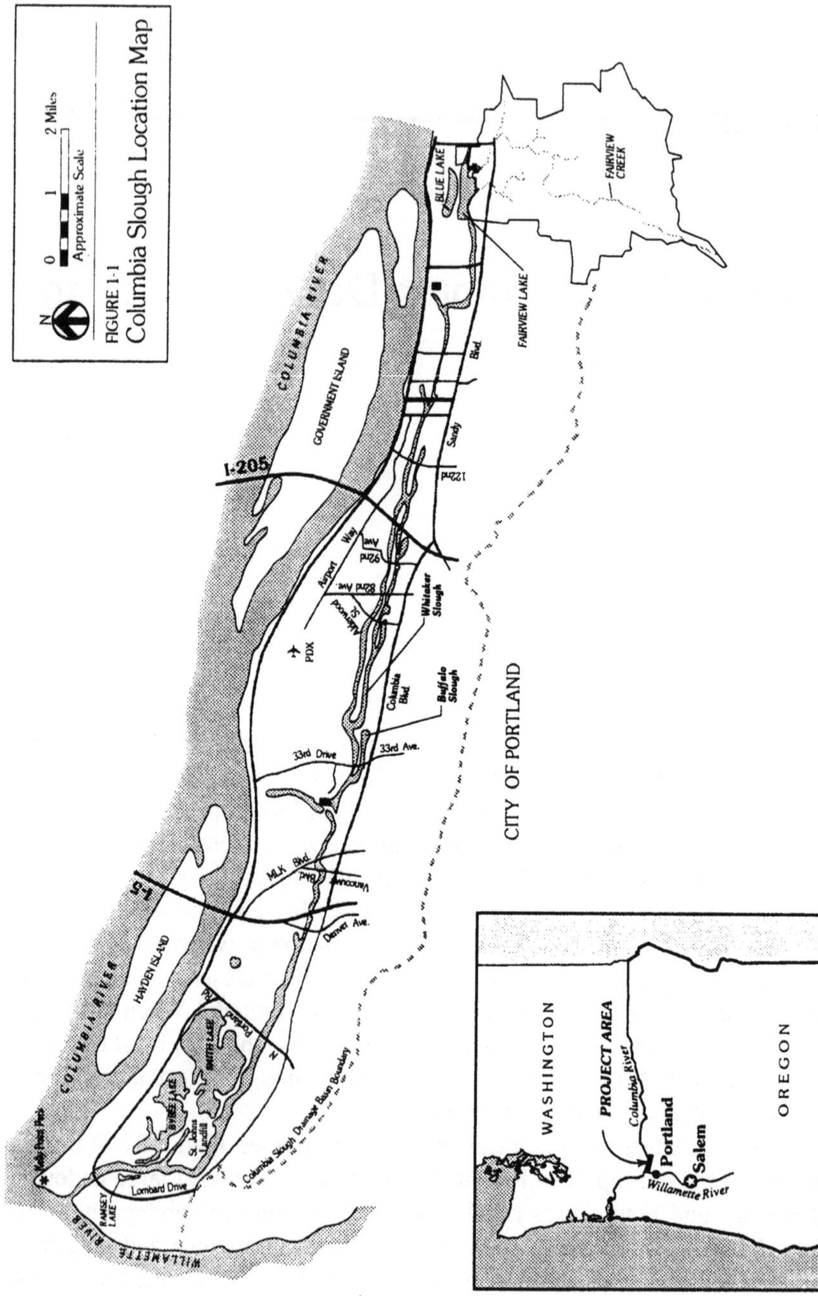

Figure 21.1 Columbia Slough location map.

catch-all solutions for the whole water body. The Slough, however, is more appropriately classified into several distinct reaches with unique hydraulic regimes, pollutant sources, and water quality conditions. This classification of the Slough into reaches for TMDL-setting facilitated the analysis of water quality problems and solutions. The classification into reaches did not, however, deter the consideration of watershed effects and in fact improved the chances of implementing more comprehensive watershed-wide plans.

The City of Portland and other affected municipalities and the Oregon Department of Environmental Quality have worked cooperatively to determine appropriate TMDLs for a number of pollutants that are degrading water quality. This group formed the "Core Team" that effectively came together and produced a rational plan to improve water quality of the Columbia Slough within the real fiscal and regulatory constraints imposed.

The many processes followed by the Core Team to determine appropriate TMDLs are discussed below. This includes the development of the overall TMDL process and designation and acceptance of the possible forms of TMDLs. The studies (CH2M HILL, 1995) performed included development of databases for water quality and other data management and analysis, estimation of stormwater and other pollutant loadings using SWMM and other techniques, and instream water quality modeling using CE-QUAL-W2 (Wells et al., 1995).

21.1 The TMDL Process

The process for developing a TMDL for the Columbia Slough followed the procedures developed by EPA through various guidelines and agency workshops. The EPA procedures (shown generally in Figure 21.2) proved too general and linear for application on the Slough and so an overall process diagram was developed by the Core Team to map the various paths and possible outcomes of the study and overall TMDL setting. The development of the process diagram also helped in focusing the team on the key problems and the interaction of various groups with responsibility for portions of the Slough. The process diagram developed is shown in Figure 21.3. [*Editor's Note: see pages 361-362*].

The key components of the process are the phased approach suggested and the recognition that TMDLs can be either directing responsible parties to collect additional data, can be descriptive remedies, or can be numerical limitations on specific pollutants. The process also reflected the need for verification of water quality improvements due to implementation of controls and other review mechanisms.

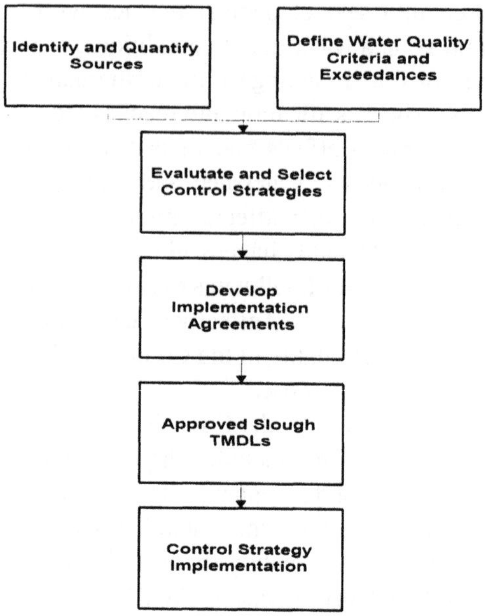

Figure 21.2 Simplified TMDL process.

21.2 Beneficial Uses and Water Quality Limitations

In general, the water quality problems of the Columbia Slough are associated with low dissolved oxygen, high temperatures, and algae blooms. The Columbia Slough receives combined sewer overflows and stormwater runoff from urban and agricultural areas. At times airport operations discharge high biological oxygen demand (BOD) loads which under appropriate hydraulic condition cause depressed dissolved oxygen levels. Groundwater discharges, which dominate the summer flow regime, carry a high nutrient load caused by former use of septic fields for sewage disposal.

The assessment of water quality and effects upon beneficial uses started with a qualitative assessment of beneficial uses within the Slough and a detailed review and analysis of available water quality data. The beneficial uses of the Slough are determined from requirements set for the Willamette River and in some cases do not coincide with existing or possible uses of the Slough. No detailed survey of uses was conducted. The phased approach adopted accepts the possibility of detailed beneficial use surveys coupled with a use attainability study to provide compliance with the intent of the TMDL process and possible revision of water quality standards.

Water quality data have been collected on the Slough for over 20 years for a variety of objectives with different data quality requirements. A database using Microsoft Access was developed to analyze over 40,000 data points. Figure 21.4 shows the database structure followed and the long-term data management and analysis program under development for the Slough.

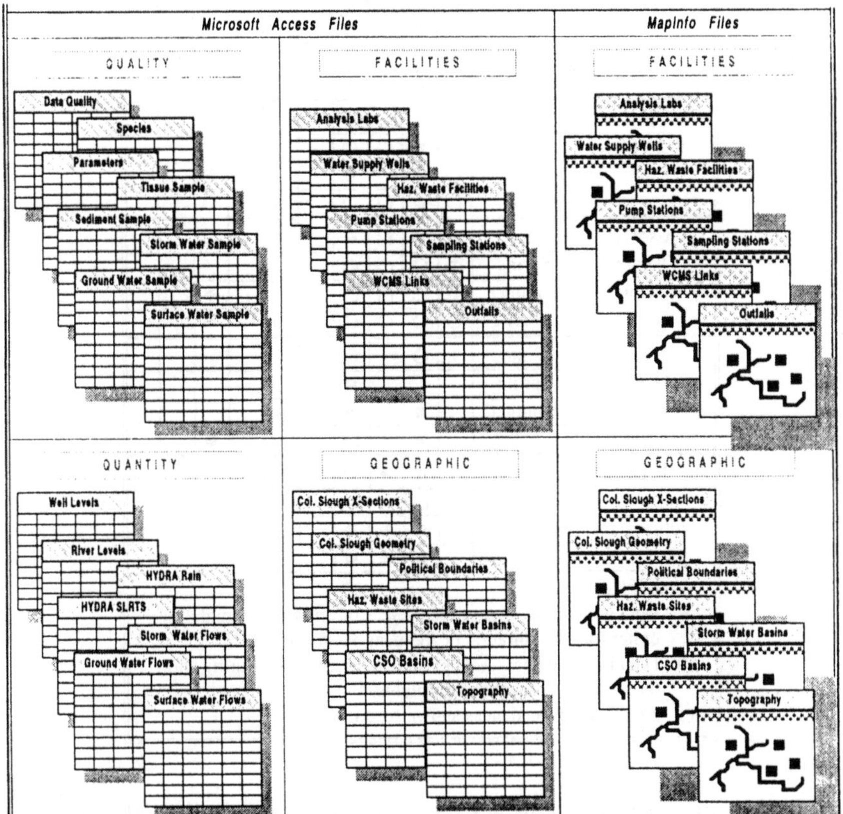

Figure 21.4 Data management structure.

Some of the deficiencies found in the data revolved around quality control and assurance procedures and adequate selection of monitoring sites. The lack of appropriate or consistent QA/QC protocols resulted in discounting the significance of certain data during the process of identifying and determining the extent of water quality problems or lack of problems. Transient water quality effects from combined sewer discharges, tidal influences, and stormwater runoff from industrial and commercial areas also proved problematic.

Table 21.1 shows a portion of a table that summarizes the water quality assessments for those locations with sufficient and reliable data. Extrapolation of

Table 21.1 Summary of *known* surface water quality concerns by reach and season.

Exceedences of Criteria											
Parameter	Season or Constituent	Reach									
		1A	1B	1C	2A	2B	2C	2D	3	4	5
Temperature (proposed: ≤ 17.8°C)	Winter	○	○	○	○				○		
	Spring	●	●	●	●					●	
	Summer	●	●	●	●				●	●	
	Fall	●	●	●	●		□		●	●	
pH (≥6.5 &≤8.5)	Winter	○	○	○	○				□		
	Spring	○	●	○	○						
	Summer	○	●		●				○		
	Fall	○	●	○	○		○		○		
Dissolved Oxygen (≥6.0 mg/L)	Winter	●	●	□	●				○		
	Spring	●	○	○	○						
	Summer	●	□	□	□		○		●		
	Fall	○	○	○	○		○		●		
Algae (<15 μg/L)	Winter		○		○						
	Spring				●		○				
	Summer		●		●		●				
	Fall		●		□		●		●		
Fecal Coliforms (≤400/100 mL)	Winter	●	●	●	□						
	Spring		□		●						
	Summer		●	●	●		□		□		
	Fall		●	●	●		□		○		

Symbol Key
- ● Data indicate frequent exceedences of criterion or guideline either for long duration or frequent occurrence.
- □ Data indicate occasional exceedences primarily during short-term events or localized problems.
- ○ No or rare exceedences shown by data.

data and qualitative assessments whenever data are scarce is illustrated in Table 21.2. These tables were used extensively by the Core Team to focus efforts on the main problem areas and parameters. Major problems are denoted by solid circles and boxes and limited problems by open circles and boxes when data are sufficient to make such a judgment. The boxes or fields in Table 21.1 without symbols indicated insufficient data to make a judgment which precipitated the evaluation given by Table 21.2.

A typical water quality data analysis is shown in Figure 21.5 which shows an example of the box and whisker plots produced for most water quality parameters. The analysis followed the reach definitions and predefined seasons. Both the segregation of the Slough into reaches and seasons helped define the water quality problems more succinctly and aided in the targeting of pollutant sources and effective control strategies.

Table 21.2 Summary of *potential* surface water quality concerns by reach and season.

Exceedences of Criteria											
Parameter	Season or Constituent	Reach									
		1A	1B	1C	2A	2B	2C	2D	3	4	5
Temperature (proposed: ≤ 17.8°C)	Winter					■	■	■		■	■
	Spring					"	"	■	○	○	"
	Summer					"	"	■		○	○
	Fall					"		■		■	"
pH (≥6.5 &≤8.5)	Winter					■	■	■		■	■
	Spring					"	○	■	■		■
	Summer			○		■	■	■			○
	Fall					■		■			■
Dissolved Oxygen (≥6.0 mg/L)	Winter					"	"	"		○	○
	Spring					■	■	■	"	"	■
	Summer					■		■		○	○
	Fall					■		■		○	○
Algae (<15 µg/L)	Winter	■		■		■	■	■	"	■	■
	Spring	○	○	○		○		■	○	○	"
	Summer	○		○		○		■	○	○	"
	Fall	○		○		○		■		○	"
Fecal Coliforms (≤400/100 mL)	Winter					"	"	■	"	"	"
	Spring	"		○		"	"	■	■	■	■
	Summer	■				"		■		■	■
	Fall	■				"		■		■	■

Symbol Key
- ○ Potential for frequent or long-duration exceedences.
- " Potential for occasional or localized exceedences.
- ■ Some available water quality data or other information (e.g. upstream data or absence of sources) suggest no exceedences.

21.3 Water Quality Modeling

The setting of appropriate TMDLs relies on interpretation of field data and application of computer and other models. The models attempt to simulate the responses of the system to changing physical and weather conditions and to highly variable pollutant loadings. The water quality model is driven by the loadings from the volume and quality of water from the known and various unknown but suspected sources. The results of quantity monitoring and models are a primary input to the water quality modeling tasks. The quantity model is based upon a calibrated rainfall-runoff model using SWMM (XPSWMM ,1995). The SWMM models were calibrated for specific events from data collected as

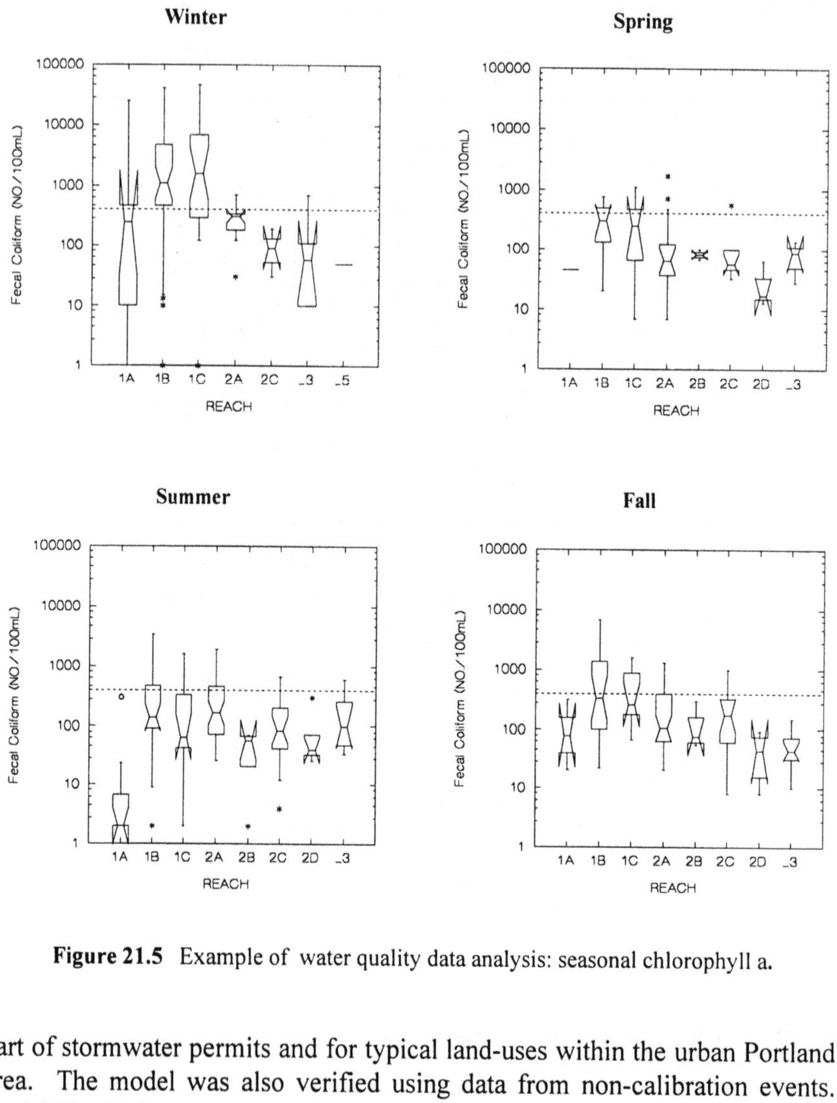

Figure 21.5 Example of water quality data analysis: seasonal chlorophyll a.

part of stormwater permits and for typical land-uses within the urban Portland area. The model was also verified using data from non-calibration events. Variability in the stormwater data was large, which resulted in the use of event mean concentrations for the development of loading quantities for the range of land-uses found in the Columbia Slough watershed. Annual loadings have been produced to show the loading balances between the Columbia Slough reaches. Loading estimates for typical, dry, and wet summers and for other unique periods have also been produced. Volume estimates and flow rates have been checked against hydraulics of the system during events and simulations in the water quality model. Additional details of the SWMM modeling can be found in the Water Body Assessment Report (CH2M HILL, 1995).

Water quality modeling estimates the effects that the various loadings have on the instream water quality and simulates the processes within the Columbia Slough. The Columbia Slough has been modeled in two segments: the Lower Slough Model consisting of Reach 1 (1a, 1b, and 1c) which receives output from the Upper Slough Model of Reach 2 (2a, 2b, and 2c) and Reach 3. No water quality modeling has been performed on Reach 2d or Reaches 4 and 5.

The model has been successfully calibrated and verified to monitoring data collected over the past several years for a variety of weather and flow conditions (Wells,1994,1995). The quantity and quality models have been used in conjunction to estimate the water quality effects of control alternatives. Parameter specific alternatives have been developed but often parameter specific control alternative will impact other water quality parameters. For example, control of combined sewer overflows impacts coliform and nutrient loadings. The TMDL process developed includes evaluation of alternatives for the most significant water quality problems found. An example of the progressive process is given in Figure 21.6.

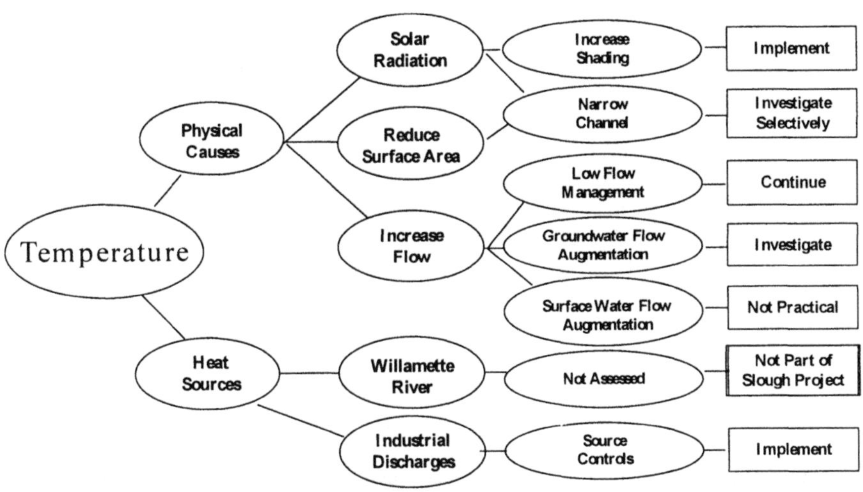

Figure 21.6 Example of evaluation process for parameter specific controls.

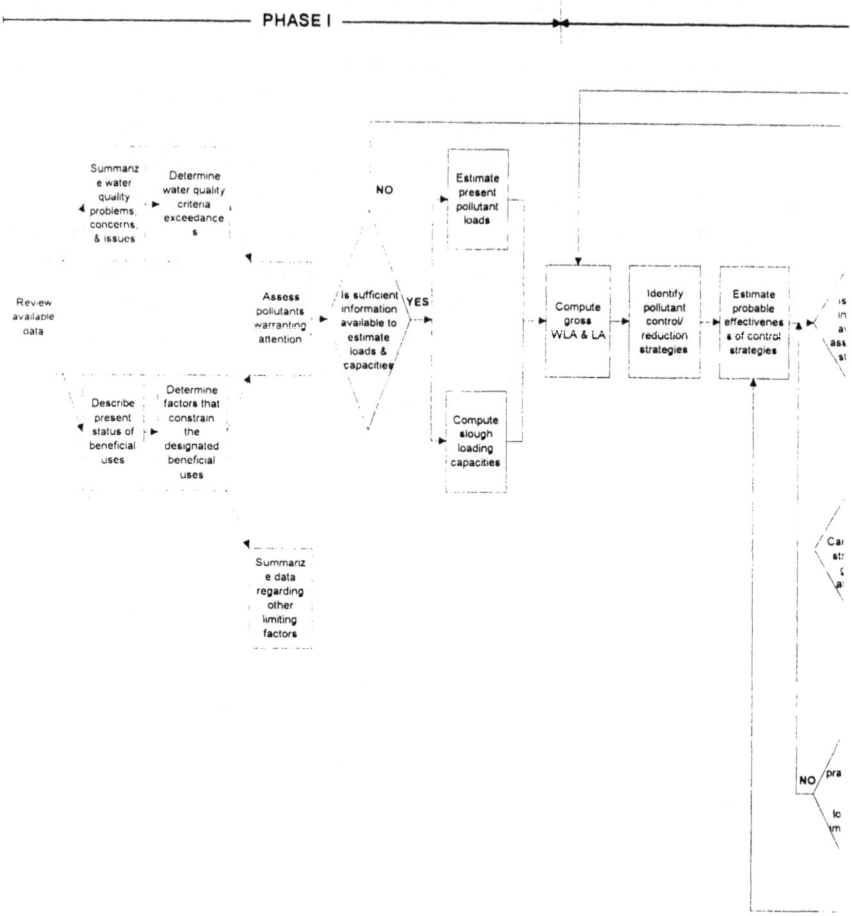

Figure 21.3 Phased approach and process developed for the Columbia Slough.

21.4 Conclusions

The identification of the water quality problems and the adoption of a seasonal and reach approach have facilitated the development of alternatives that can be implemented. The solutions are believed to be effective in providing improvements to water quality and hence protect the actual beneficial uses of the Columbia Slough. Examples of the alternatives proposed are given in Table 21.3.

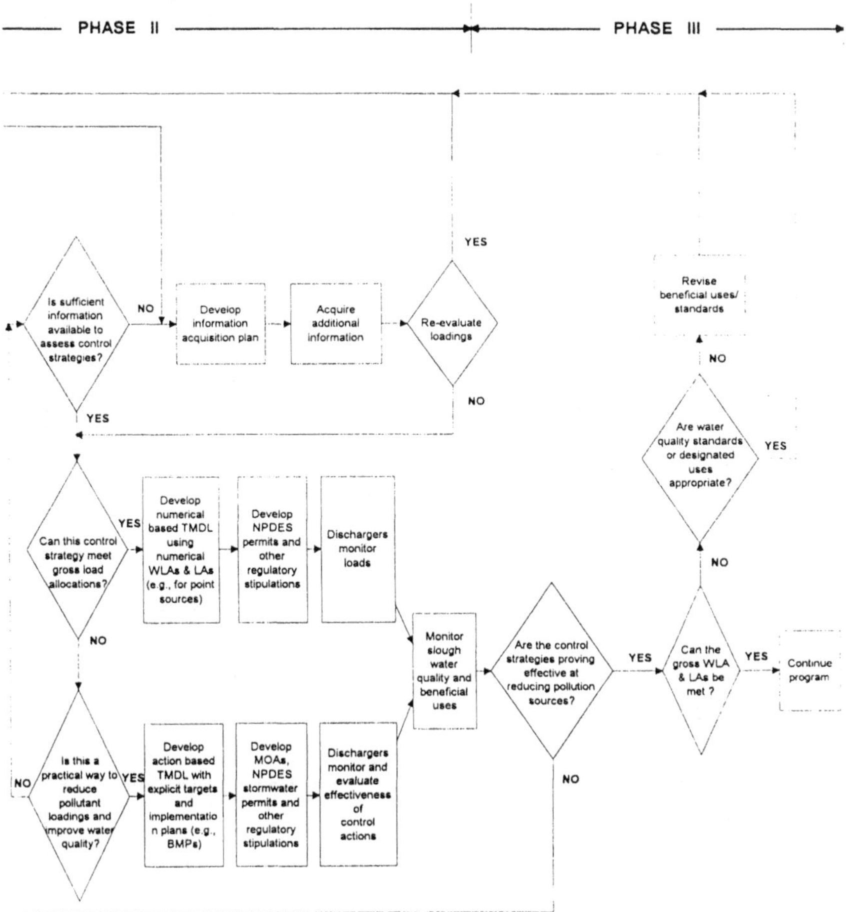

Figure 21.3 continued Phased approach and process developed for the Columbia Slough.

Although the TMDLs are currently under development by the DEQ and will be finalized after EPA and public review, the controlling agencies, such as the City of Portland, have begun implementing selected controls. The effectiveness of these controls will be evaluated over the next few years. Pursuing them now before definitive TMDLs are set concurs with the phased and iterative approach suggested by the TMDL setting process developed for the Columbia Slough.

Table 21.3 Columbia Slough water quality improvement programs; existing and planned control strategies.

Program	Responsible Agency	Affected Reaches	Pollutants Addressed	Pollutant Source	Control Technology or Management Practice	Status
NPDES Municipal Stormwater Program Co-application with City of Portland, Multnomah Co., Port of Portland, ODOT, Multnomah Drainage District #1, and Peninsula Drainage Districts #1 and #2, NPDES Permit Application (May 1993) 1. Develop operations and maintenance plans for all public storm facilities with goal of reducing pollutant discharges.	City of Portland-BES Multnomah County	1, 2, 3, 4, 5	Sediments Nutrients Metals Organics Oil and grease	Stormwater	Develop written procedures for routine inspection and maintenance of stormwater inlets and sumps.	Completed (BES) Current and ongoing (Multnomah)
2. Evaluate stormwater maintenance practices that affect water quality at existing stormwater quality facilities.	City of Portland-BES	1, 2, 3	Sediments Nutrients Organics Metals Oil and grease	Stormwater	Review and revise operation and maintenance manual for existing facilities to improve treatment capability.	To be completed in 1995
3. Evaluate stormwater maintenance practices that affect water quality at existing stormwater quality facilities.	Port of Portland	2A	Sediments Nutrients Organics Metals Oil and grease	Stormwater	Review and revise operations and maintenance plans for 4 operating areas: marine terminals, Portland International Airport, commercial real estate, and Portland Ship Repair Yard.	Current and proposed

References

CH2M HILL, 1995. Water Body Assessment: Columbia Slough TMDL Development. Report to City of Portland, 88pp.

Wells, S.A. and Berger, C. 1993. *Hydraulic and Water Quality Modeling of the Upper and Lower Columbia Slough: Model Calibration, Verification, and Management Alternatives Report.* Technical Report submitted to HDR Engineering and City of Portland, 202pp.

Wells, S.A. and Berger, C. 1993. *Hydraulic and Water Quality Modeling of the Upper and Lower Columbia Slough: Model Description, Geometry, and Forcing Data.* Technical Report submitted to HDR Engineering and City of Portland, 119pp.

Wells, S.A. and Berger, C. 1995. *Hydraulic and Water Quality Modeling of the Upper and Lower Columbia Slough: Model Calibration, Verification, and Management Alternatives Report 1992-1994.* Draft Technical Report EWR-2-95, Department of Civil Engineering, Portland State university, Portland, Oregon Prepared for the City of Portland, Bureau of Environmental Services. March 1995, 258pp.

XP Software. 1995. XP-SWMM Storm Water management Model with XP Graphical Interface, User's Manual, Version 2, Volumes 1 and 2. Tampa, Florida.

Chapter 22

Wastewater Information Management System: Flow Modeling and Sewer Connection Permit Applications

David Crawford and Felix Limtiaco

The City and County of Honolulu, Department of Wastewater Management (DWWM) is charged with the wastewater service of approximately 900,000 persons living on the Island of Oahu. Effective management of the wastewater system by DWWM is being aided through construction of an island-wide wastewater information management system (WIMS).

Modules within WIMS automate to a large extent the standard procedures for managing system inventory, estimating flows, and evaluating system hydraulics and capacity. The use of the various modules within WIMS results in speedier evaluation of basin flows and impact of new sewer connections, improved project evaluations. It also provides a means to account for system-wide impacts that it was not practical to account for in the past. This chapter outlines two new modules developed for WIMS: the Sewer Flow Analysis System (SFAS) and the Sewer Connection Application System (SCAS). Figure 22.1 shows the interconnection between SFAS, SCAS and other modules within WIMS.

SFAS and SCAS use the ArcInfo geographic information system (GIS), commercial packages for data management and hydraulic modeling, and customized computer programs to complement other components of WIMS. The two new WIMS modules integrate several different information management technologies:

© *Advances in Modeling the Management of Stormwater Impacts - Vol. 5* W. James, Ed. Pub. by CHI, Guelph, Canada 1997. ISBN 0-9697422-7-4. Fax: +519 767-2770

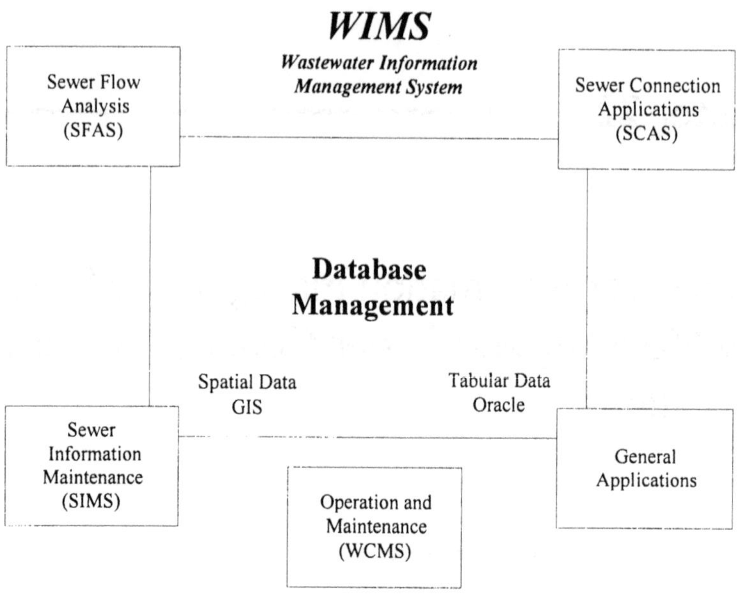

Figure 22.1 Wastewater information management system: City and County of Honolulu.

- Database management is used to record, evaluate, and manage new sewer system connections.
- Database management and spreadsheets are used to analyze extensive flow monitoring data and to calibrate the flow estimating and routing models.
- GIS is used to manage sewer system connectivity and inventory, and identify all the wastewater facilities impacted by new sewer connection permits.
- A custom GIS tool-set is used to define sewer system collection system basins and the basin characteristics based on population estimates, land-use, and traffic analysis zone data.
- The GIS layers are used in combination or separately to estimate current and future wastewater flow for defined basins or the system contributing to a wastewater pump station, flow monitor area, or treatment plant.

- Dynamic and kinematic hydraulic models are constructed from the GIS generated data and analyzed for historical or design storm events.
- Model results are analyzed in the model package or through links to the WIMS infrastructure database for generation of a sewer capacity table for use in SCAS and for display and analysis of results. SFAS results are used for evaluation of new sewer connection applications and for management of the sewerage system.

22.1 System Complexity

The creation and integration of SFAS and SCAS had to be completed between July 1, 1995, and a deadline of December 31, 1995 set by Consent Decree between the City and County of Honolulu and the U.S. Environmental Protection Agency (EPA). CH2M HILL's team worked collaboratively with diverse groups at the DWWM in developing the SFAS and SCAS modules of WIMS. Time was spent in determining and refining a system that meets the needs of the City and County staff who are charged with management of the various Department programs. It was also important to become familiar with the needs of DWWM, the various business processes within the DWWM, and the WIMS system already in place. CH2M HILL is the third consultant working on WIMS and will be continuing development of the modules in cooperation with DWWM staff. The first two consultants set up the WIMS framework prototype modules for SFAS and SCAS and other WIMS modules.

The sewerage system contains over 1,300 miles (2000 km) of main sewers, 79 pump stations, eight treatment plants, and over 250 flow splits and other hydraulically challenging features. A large proportion of the sewer system is below sea level. Approximately 9,500 individual sewers, consisting of sewers 10 inch (254 mm) and larger, are analyzed in the flow routing model. About 500 individual inflow basins contributing flows are created using the GIS and database tools developed. The largest single hydraulic model constructed consisted of over 4,800 pipes. Figure 22.2 illustrates the complexity of the system, showing the Kailua-Kaneohe area of Oahu and the tool tracing sewers within the basin for later processing for flow estimation and export of sewer data for the hydraulic model.

Challenges met by the City/Consultant team included:

- Identification of data errors, and data gaps, and the correction of these data deficiencies. This represented approximately a correction of about 15% of the physical systems built and the completion of the representation of the entire sewer system connectivity for adequate construction of sewer basins for flow estimation.

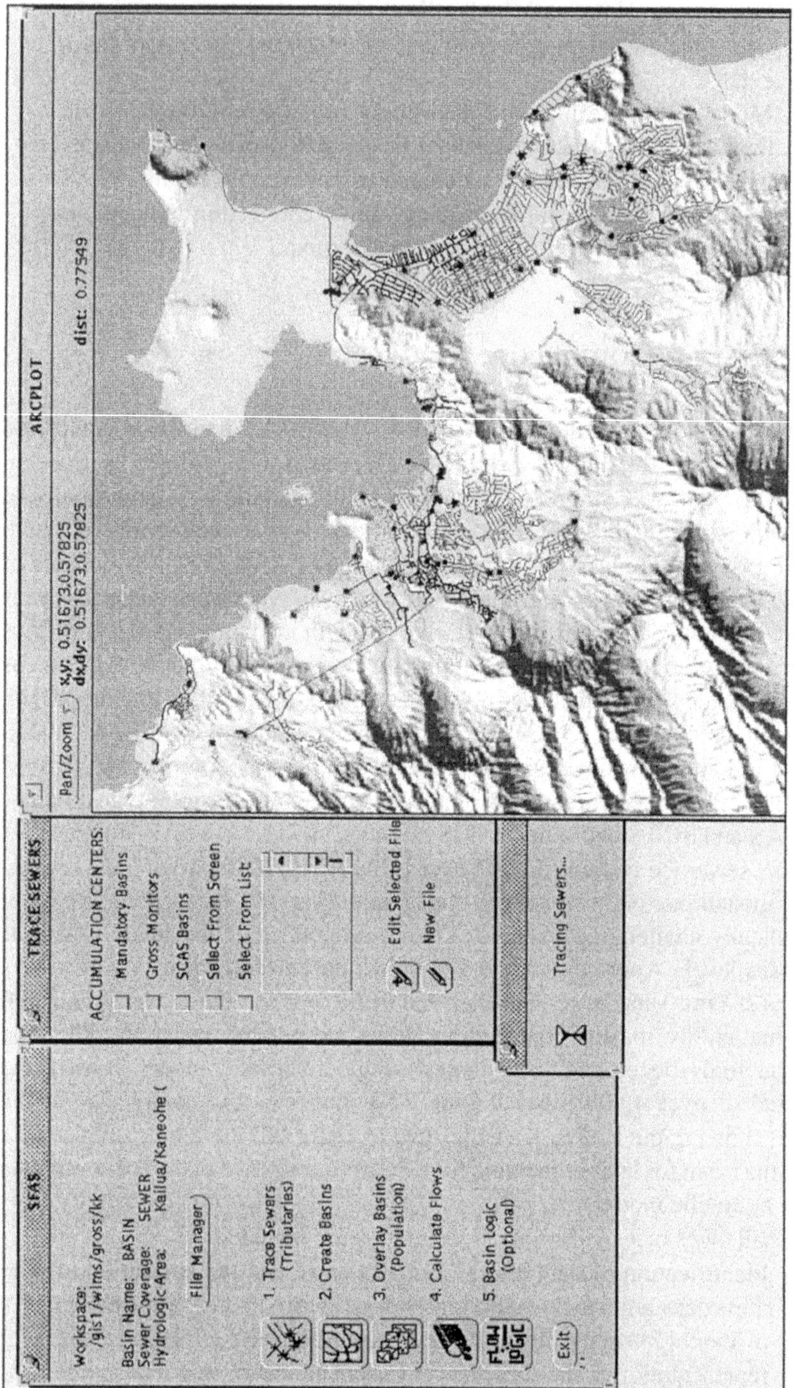

Figure 22.2 Example of Sewer Traces and Model Creation.

- Generation and calibration of representative flow estimates from a large amount of flow data for a range of hydrologic conditions.
- Dynamic hydraulic modeling of a complex system with flow splits, backwater effects, and adverse sloped sewers.

22.2 Sewer Flow Analysis System (SFAS)

For most of this century civil engineers have used the same basic procedures for estimating wastewater and storm induced inflow and infiltration and linking the cumulative flows to hydraulic and flow routing models. Impacts from new developments and system retrofitting and repair are a common theme for evaluation by most communities. Before SFAS, the City estimated flows and calculated the capacity of its sewer lines using several manual methods, spreadsheets, and a static wastewater flow accumulation model. Not all of these methods provided consistent results, which caused uncertainty in estimating capital improvement projects. With SFAS, sewer system subbasins are created automatically from the sewer network maps in the WIMS and the subbasins are overlaid with population and other demographic data to estimate flow at any point in the wastewater system. The City uses the flow basins to determine realistic flow rates generated from a variety of sources, accounting for differences in the residential populations, number of hotel rooms, employment in the commercial areas, and major point discharges in each subbasin. Figure 22.3 shows a typical screen for the calculation of sewer flows.

Hydraulic modeling, using XP-SWMM, is linked to the GIS to route both wet and dry weather flow through the selected sewerage system. The SFAS module and hydraulic model will be used for planning new projects, evaluation of existing system hydraulics and capacity, and to enhance infiltration and inflow reduction analysis.

The creation of the model pipe data, flow files, and production of model results and transfer back to the GIS environment results in the production of numerous files. These files are managed and interpreted though the use of Microsoft Access database tables and queries and several custom-constructed programs, such as PSSCRAT which converts comma separated files to binary files suitable for reading by XP-SWMM.

Microsoft Access was used extensively to query data files, particularly in identifying inconsistent physical data such as pipe offsets or large slope changes. The databases were used to generate the physical data input files for XP-SWMM and the flow interface file. Figure 22.4 shows the Microsoft Access tables for manhole and conduit data and Figure 22.5 shows examples of the query to produce model input files.

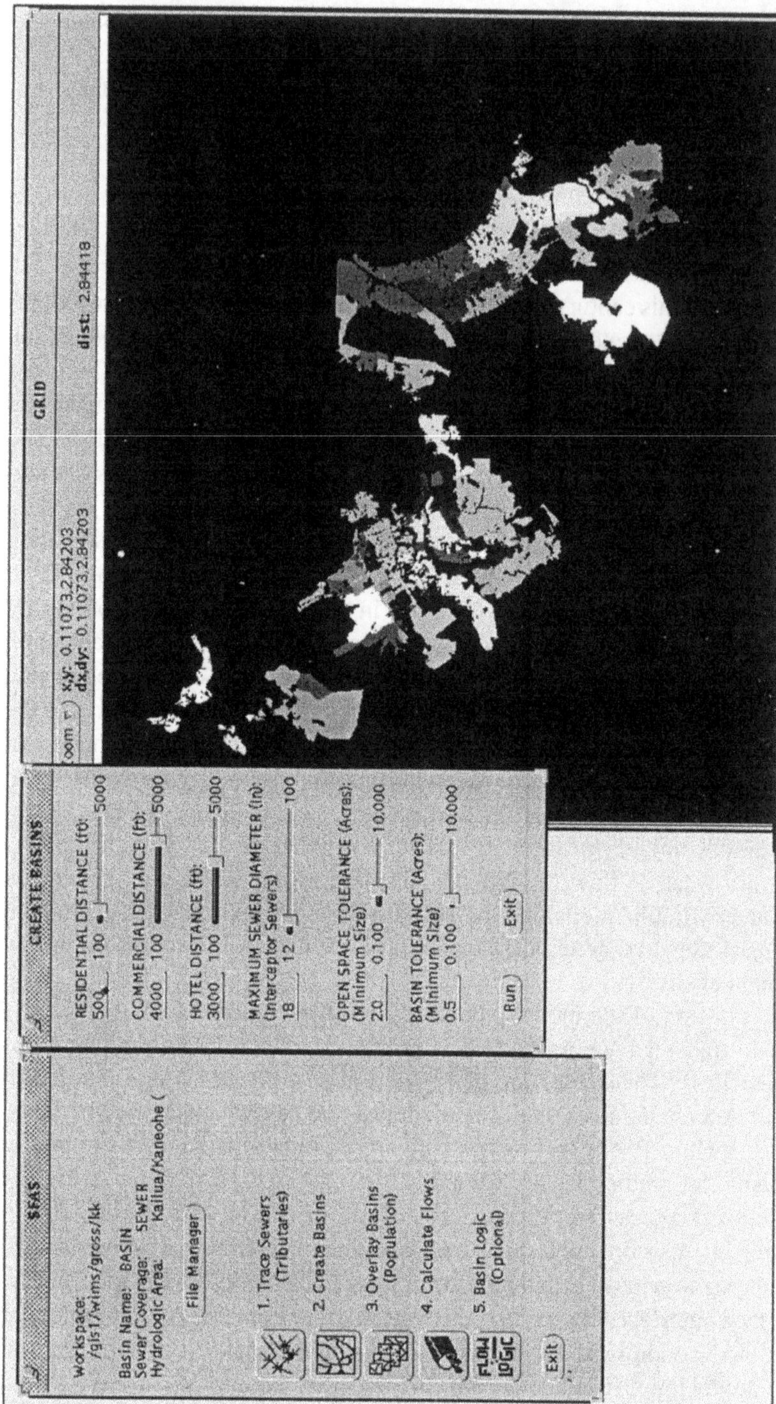

Figure 22.3 Creation of sewer basins and flow estimates.

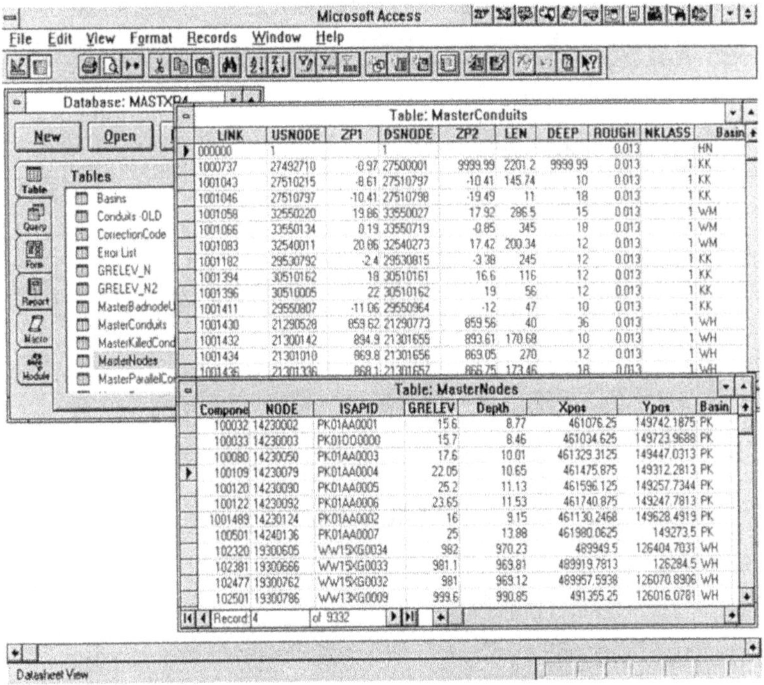

Figure 22.4 Example of manhole and conduit tables.

22.3 Sewer Connection Application System (SCAS)

In the SCAS system, applications for a new sewer permit (for example), are processed on line to facilitate locating the connection point to the sewer system and, depending upon the sewer basin, to facilitate the estimation of sewer flow generated from the new development.

SCAS will allow the City to process sewer connection application forms, store and track the sewer application data, and effectively manage the system by producing warnings whenever sewer capacity is reaching critical levels. The SCAS tool-set then checks for sufficient flow capacity in all downstream wastewater facilities (Figure 22.6), and issues sewer connection permits and payment receipts if sufficient downstream capacity is available. Figure 22.7 shows the interconnection between SFAS and SCAS.

The capacity status of every sewer reach and pumping plant is stored in a capacity allocation table. As new permits are approved, the SCAS allocation table is updated to reflect the erosion in available capacity. When capacity reaches pre-set limits, notification of these triggering flows is automatically issued and

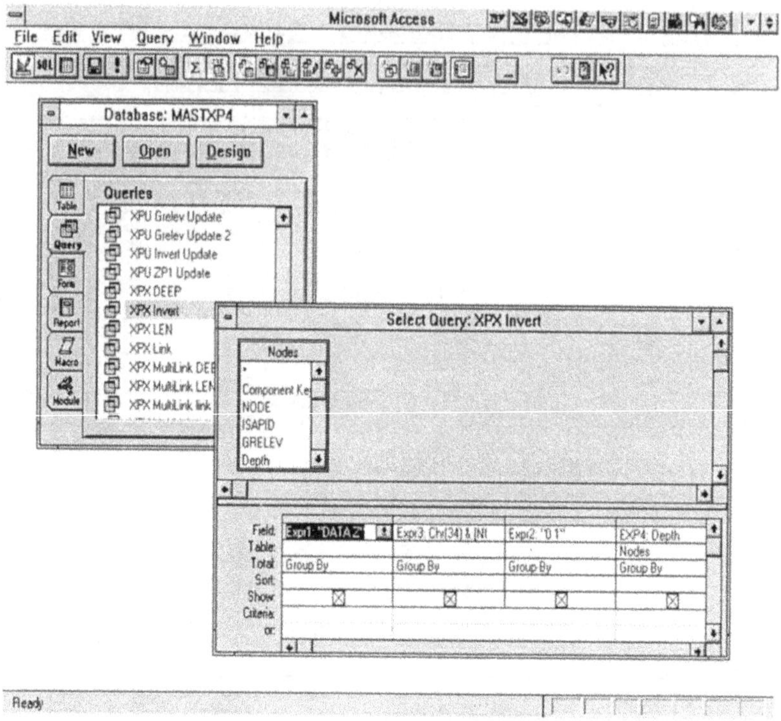

Figure 22.5 Example of query to generate model input.

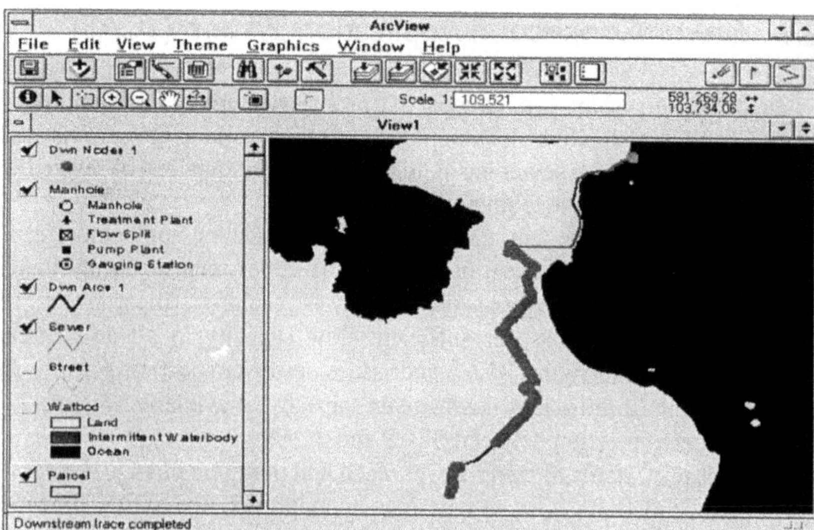

Figure 22.6 On-line sewer trace to check for capacity.

planning functions for correction of the capacity deficiencies can be started. Use of the capacity allocation table improves WIMS' response and eliminates the need to run the dynamic model for each sewer connection.

Wastewater flows from new connections are estimated by using a set of land use and equivalent population values calibrated from a detailed analysis of flow monitors located throughout the system. With SCAS, the City will effectively manage growth and better understand the cumulative effects of new connections on the sewerage system.

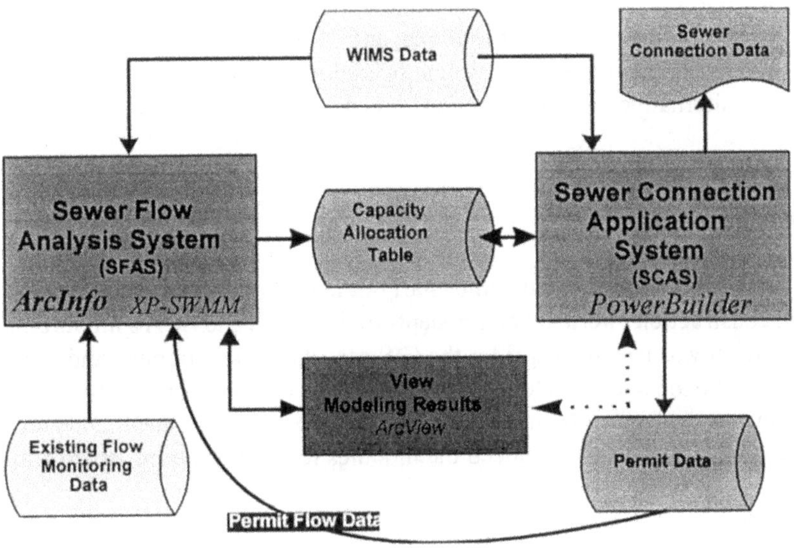

Figure 22.7 Relationships between SFAS and SCAS.

22.4 Conclusions

The sewer flow analysis and sewer connection application system modules provide the City's engineers with a standardized methodology to estimate flows and the capacity of their sewerage system, to make intelligent decisions about whether or not to approve a sewer connection permit, and to trigger facility plan updates.

These modules offer several advantages for wastewater management:

- the adoption of standard and proven civil engineering practices with modern computer information management systems to provide timely analyses and to allow consideration of system-wide impacts;
- rapid review of sewer connection applications that allows new connections when unused capacity is available, and identification of those parts of the system that prevent the acceptance of new flows when unused capacity is not available;
- better management of the collection system by using up-to-date flow monitoring data, and current land use and demographic data;
- more effective sewer infrastructure improvements by the consideration of system-wide impact and the use of consistent flow estimating and evaluation procedures; and
- optimization of sewer system operation by developing and testing alternative system operations procedures and scenarios.

Acknowledgement

Tina Ono, Ed Pier, and other staff of the DWWM provided valuable input and construction of critical components of the SFAS and SCAS modules. Dave Bramwell was task manager for the GIS aspects of the programs and produced most of the tools for developing basins and sewer networks in ArcInfo. Westley Chun was the Project Manager and Jim McKibben the Technical Director. Peat Marwick KPMG staff produced the modules for on-line permit applications.

Chapter 23

Application of WASP5E to Model Phosphorus Removal Dynamics in a Stormwater Wetland

Karina Lopez, William James, Isobel Heathcote and John Fitzgibbon

Water pollution abatement has received considerable attention from the research community over the recent years. Treatment of urban stormwater runoff has likewise been the focus of sizeable amounts of research. Phosphorous is a key indicator parameter for eutrophication problems and was therefore the parameter of focus in this research. Urban activities create sources of phosphate such as fertilizers, animal waste, detergents, etc. Phosphate loads in urban runoff vary greatly due to variations in rainfall characteristics, watershed features and urban activities (EPA, 1993). Reported ranges of total phosphate loads vary from 0.2 to 2.0 kg/ha for residential areas, and from 0.9 to 6.0 kg/ha for industrial areas (Novotny and Olem, 1994). Constructed wetlands have been widely promoted for urban stormwater management because of their inherent capacity for water storage and water quality improvement.

The Ontario Land Use Planning Act (1990) defines "wetlands" as: "lands that are seasonally or permanently covered by shallow water, as well as lands where the water table is close to or at the surface. In either case the presence of abundant water has caused the formation of hydric soils and has favoured the dominance of either hydrophytic plants or water tolerant plants"

This research, conducted at the School of Engineering, University of Guelph, studied the feasibility of modeling phosphorus assimilative capacity by stormwater wetlands in cold climates. The over-riding aim of this research was

© *Advances in Modeling the Management of Stormwater Impacts - Vol.5* W. James, Ed.
Pub. by CHI, Guelph, Canada 1997. ISBN 0-9697422-7-4. Fax: +519 767-2770

the application of computer simulation in evaluating water quality improvement efficiency of constructed wetlands in an effort to enhance their proficiency and promote their utilization.

For this research, a vast amount of background data relating to urban stormwater, wetlands and modeling was collected and examined. Previous studies and field data were reviewed to find useful and adequate data sets for modeling purposes. Due to the lack of data from a constructed wetland system, a data set for the Hidden Valley wetland, a natural stormwater wetland (*Typha* marsh), was chosen and adopted to test the selected water quality models.

A number of water quality models were scrutinized to establish their adequacy for modeling phosphorus dynamics in wetland systems. The US EPA's WASP5 simulation program and the vegetation growth/phosphorus uptake model ECOL1 were selected. The linked phosphorus cycle kinetics of WASP5 to ECOL1 developed here is called WASP5E. The two primary objectives of the study were to assemble into a computer simulation process the main components of phosphorous dynamics in stormwater wetlands and to test the utility of the process in simulating those phosphorous dynamics.

Phosphorus dynamics in wetlands systems have been widely described and analyzed in literature. Phosphorus wetland processes are schematically illustrated in Figure 23.1. The dynamics of phosphorus removal in wetlands is an interaction of mechanisms such as sedimentation, chemical precipitation and incorporation into biomass, plants and algae. For a description of P cycle in wetland systems, refer to Tchobanoglous and Schroeder (1987), Vymazal (1995), Kadlec and Knight (1996), Strecker *et al.* (1992), Shaver (1995), Kadlec and Kadlec (1978), Bayley (1985), Kadlec (1987), Reddy and DeBusk (1987), Good and Patrick (1987), Hossner and Baker (1988), and Davis *et al.* (1978).

23.1 Models Description

This section presents a description of the two computer models used in this research, WASP5 and ECOL1. The main focus of this study is on the model algorithms related to the phosphorus cycle dynamics. Algorithms of each model affecting the phosphorus dynamics will also be addressed.

23.1.1 WASP5

WASP5 is a dynamic water quality simulation program. Conventional pollutants in a body of water such as nutrients, are simulated in this program by the sub-model EUTRO5. The model network represented in WASP5, is a set of segments that together represent the physical configuration of the water body. The network water column segments can be of two types: (i) surface water

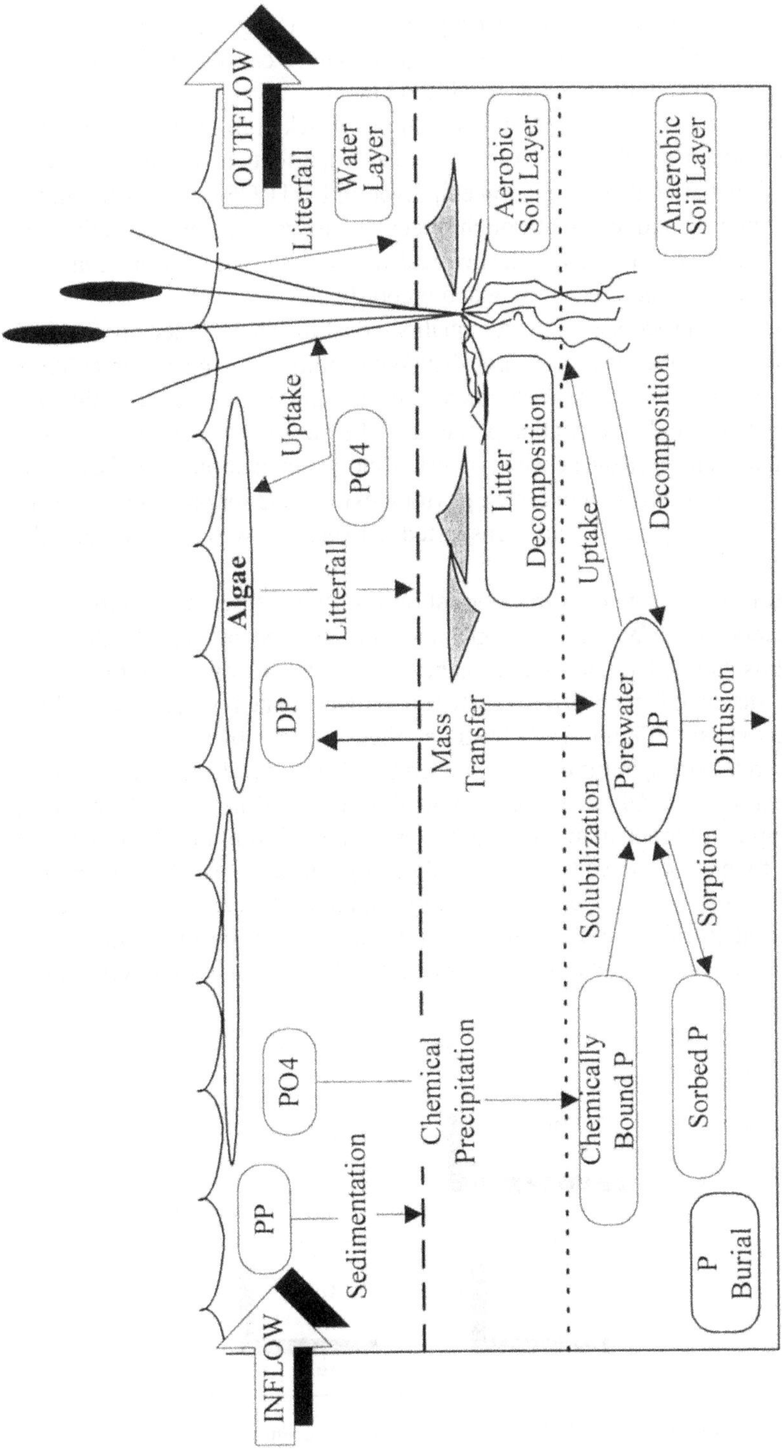

Figure 23.1 The wetland phosphorus cycle.

(epilimnion); and (ii) subsurface (hypolimnion). Along with water column segments, benthic segments can be included. These can represent: (i) upper benthic segments; and (ii) lower benthic segments.

All information presented in this section was taken from the WASP5 Part A: Model Documentation (Ambrose et al. (1993a); refer to this manual for more details). The mass balance equation employed in EUTRO5 accounts for all the material entering and leaving through direct and diffuse loading, advective and dispersive transport, and physical, chemical, and biological transformation. The mass balance incorporates the x, y and z coordinates.

Transport processes of water quality constituents included in WASP5 simulation are advective flow and dispersive mixing in the water column, movement of pore water in the sediment bed, transport of particulate pollutants, and evaporation and precipitation processes. The transport of particulate pollutants includes settling, resuspension, scouring and sedimentation of solids. By this transport in EUTRO5, inorganic, phytoplankton, and organic phosphorus, sorbed onto solid particles, are transported between the water column and the sediment bed.

Within the phosphorus transformation processes, three phosphorus variables are modeled in WASP5: phytoplankton, organic and inorganic (orthophosphate) phosphorus. Organic and inorganic phosphorus are divided into particulate and dissolved concentrations, based on designated variable dissolved fractions for each. Figure 23.2 illustrates the state variables and interactions involved in the phosphorus cycle simulated by EUTRO5, where PO_4 represents orthophosphate and OP relates to organic phosphorus. During simulation, PO_4 is taken up by phytoplankton for growth, and returned from the phytoplankton biomass to both dissolved and particulate organic and dissolved inorganic phosphorus through endogenous respiration and mortality. A portion of the OP is converted to PO_4 through mineralization or bacterial decomposition. The phosphorus cycle rate equations are presented below with a brief description of the variable.

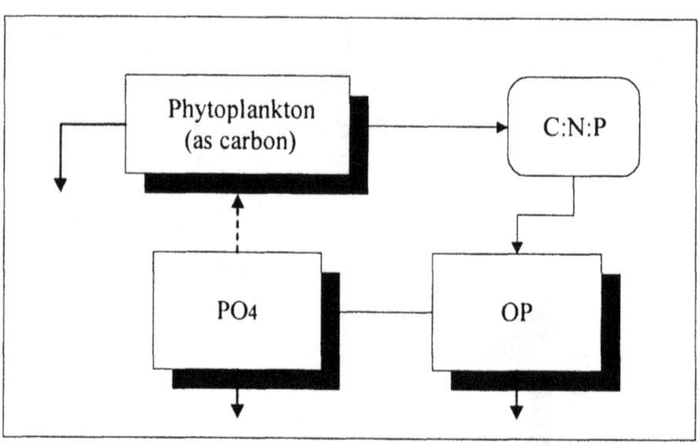

Figure 23.2 EUTRO5 phosphorus variables interactions.

Phytoplankton Phosphorus Rate Equation: The change of phosphorus as phytoplankton biomass with time is represented in WASP5 (Ambrose and Martin, 1993a):

$$\frac{\partial(C_4 a_{pc})}{\partial t} = \underset{\text{growth}}{\underline{G_{p1} a_{pc} C_4}} - \underset{\text{death}}{\underline{D_{p1} a_{pc} C_4}} - \underset{\text{settling}}{\underline{\frac{V_{S4}}{D} a_{pc} C_4}} \tag{23.1}$$

where:

C_4 = phytoplankton concentration, (mg L^{-1});
G_{pl} = specific growth rate, (day^{-1});
D_{pl} = biomass reduction rate, (day^{-1});
V_{S4} = phytoplankton settling velocity, (m day^{-1});
a_{pc} = phosphorus to carbon ratio, (mg P mg^{-1} C); and
D = depth of the waste column or model segment, (m).

Phytoplankton biomass is represented by an aggregated variable, chlorophyll *a*, which is characteristic of all phytoplankton. For internal computational purposes, EUTRO5 uses phytoplankton carbon as a measure of algal biomass using a carbon to chlorophyll *a* ratio. The growth rate (G_{p1}) is specified by a fixed maximum value which is a function of temperature, light limitation and phosphorus limitation. The light limitation factor takes into account the seasonal depth and turbidity light attenuation and photo-inhibition effects on phytoplankton population growth. WASP5 offers a choice of two similar light modeling formulations. For this study the formulation developed by DiToro given in the WASP manual (Ambrose and Martin, 1993a) was chosen; it averages conditions over a given depth and over a fixed interval of time. WASP models the phytoplankton reduction term (D_{p1}) as a function of respiration, death from parasitization and herbivorous zooplankton grazing. If the respiration rate of the phytoplankton as a whole is greater than the growth rate, there is a net loss of phytoplankton biomass.

Organic Phosphorus Kinetics. The kinetic rate of change of organic phosphorus in the water column and the benthic-water column exchange in the system are given by Equations 23.2a and 23.2b, respectively:

$$\frac{\partial C_8}{\partial t} = \underset{\text{death}}{\underline{D_{p1} a_{pc} f_{op} C_4}} - \underset{\text{mineralization}}{\underline{k_{83} \theta_{83}^{T-20} \left(\frac{C_4}{K_{mpc} + C_4} \right) C_8}} - \underset{\text{settling}}{\underline{\frac{V_{S3}(1-f_{D8})}{D} C_8}} \tag{23.2a}$$

$$\frac{\partial C_8}{\partial t} = \underbrace{k_{pzd}\theta_{pzd}^{T-20} a_{pc} f_{op} C_4}_{\textit{algal decomposition}} - \underbrace{k_{opd}\theta_{opd}^{T-20} f_{D8} C_8}_{\textit{mineralization}} \tag{23.2b}$$

where:

C_8 = organic phosphorus concentration, (mg L^{-1});
f_{op} = fraction of dead and respired phytoplankton recycled to the organic P pool;
K_{83} = dissolved organic phosphorus mineralization at 20° C; (day^{-1});
θ_{83} = temperature coefficient;
K_{mPc} = half saturation constant for phytoplankton limitation of P cycle, (mg C L^{-1});
f_{D8} = fraction of dissolved organic P;
V_{S3} = organic matter settling velocity, (m day^{-1});
k_{pzd} = anaerobic algal decomposition rate, (day^{-1});
θ_{pzd} = temperature coefficient;
k_{opd} = organic P decomposition rate, (day^{-1});
θ_{opd} = temperature coefficient; and

other variables as already defined.

Inorganic Phosphorus Kinetics. Finally, the inorganic phosphorus kinetic rate of change in the water column and the benthic-water column exchange in the system are given by Equations 23.3a and 23.3b, respectively:

$$\frac{\partial C_3}{\partial t} = \underbrace{D_{p1}a_{pc}\left(1-f_{op}\right)C_4}_{\textit{death}} + \underbrace{k_{83}\theta_{83}^{T-20}\left(\frac{C_4}{K_{mpc}+C_4}\right)C_8}_{\textit{mineralization}} - \underbrace{G_{p1}a_{pc}C_4}_{\textit{growth}} - \underbrace{\frac{V_{S5}\left(1-f_{D3}\right)}{D}C_3}_{\text{settling}} \tag{23.3a}$$

$$\frac{\partial C_3}{\partial t} = \underbrace{k_{pzd}\theta_{pzd}^{T-20} a_{pc}(1-f_{op}) C_4}_{\textit{algal decomposition}} + \underbrace{k_{opd}\theta_{opd}^{T-20} f_{D8} C_8}_{\textit{mineralization}} \tag{23.3b}$$

where:

C_3 = inorganic phosphate concentration, (mg L^{-1});
V_{S5} = inorganic sediment settling velocity, (m day^{-1});
f_{D3} = fraction of dissolved inorganic P in the water column;

and other variables as defined above.

23.1.2 ECOL1

ECOL1 is a dynamic aquatic plant growth/nutrient uptake simulation model. It incorporates biomass yields and water quality concentration in the water body. This model contains two subroutines: ECOL that contains the primary algorithms to determine growth, respiration and death/washout for the plants; and ASSIM, which simulates the short term exchange rates of oxygen, phosphorus and nitrogen between the biomass compartment and the water column phase. All information was taken from the Aquatic Plant model -Derivation and Application manual (Walker *et al.*, 1982; refer to this manual for ECOL1 details).

Net production of biomass simulated by the ECOL subroutine is determined by subtracting biomass respiration and washout from biomass productivity. Productivity is the amount of new biomass produced, determined by an adjusted growth rate. The optimal vegetation growth rate is adjusted for temperature, solar radiation and nutrient present in the system. A light-temperature limited growth rate (THP) is determined for vegetation from a light-limited growth factor (RADC) and a temperature-limited growth rate (THCP). The amount of P available for plant uptake is calculated from the incoming P concentration (PSUPLY). The total demand of P (TOTP) in the system is calculated in ECOL by:

$$TOTP = THP * PASS * BIOMASS * AREA * ReqFac \qquad (23.4)$$

where:

THP = light-temperature limited growth rate; (h^{-1});
$PASS$ = phosphorus assimilation ratio by vegetation, (g P g^{-1} biomass);
$BIOMASS$ = density of vegetation, (g m^{-2});
$AREA$ = surface area of the segment, (m^2); and
$ReqFac$ = efficiency factor for nutrient utilization for vegetation.

By this method, the instantaneous phosphorus demand at each time step is calculated and is subjected to an efficiency factor. This factor, based on vegetation physical or biological conditions, and environmental condition (i.e. temperature), is greater than one and is adjusted during model calibration. PASS is a stoichiometric ratio of P uptake related to vegetation synthesized and submitted in the input file. Upon determination of phosphorus system demand, a P availability factor (PFAC) is determined by dividing the PSUPLY by the TOTP by vegetation. The estimated PFAC represents the nutrient limitation on growth and determines whether growth of new biomass can proceed. Biomass produced each time step is calculated by ECOL1 by:

$$PRODC = THP * BIOMASS * PFAC * NFAC * TS * INHIBT \quad (23.5)$$

where:

$PFAC$ = phosphorus availability;
$NFAC$ = nitrogen availability;
TS = time step, (hr); and
$INHIBT$ = plant growth inhibition coefficient.

ECOL algorithms account for P luxury storage by the vegetation pool. If demand of P exceeds supply, the vegetation is able to draw upon stored supplies of P (STFAC) to meet the demand. The vegetation respiration rate (RESP) is based on the unit respiration rate at 20° C, corrected by the temperature correction factor (TFR). This rate is dependent upon the uptake rate of dissolved oxygen and is given by:

$$RESP = GR_{20} * BIOMASS * TFR * TS * \frac{O_2}{(O_2 + 1.5)} \quad (23.6)$$

where:

GR_{20} = unit respiration rate at 20° C, (g g^{-1} h^{-1});
TFR = temperature correction factor;
O_2 = oxygen concentration (mg L^{-1}) ; and
other variables as above defined.

Vegetation death is calculated as vegetation productivity (PRODC) minus respiration (RESP). The last section of the ECOL subroutine calculates the total amount of phosphorus uptaken and released by vegetation (PUP and PREL, respectively) at each time step. The PUP is a function of the biomass present, the P availability (PFAC) and the vegetation P assimilation ratio (PASS). Phosphorus is released to the water as vegetation respires and dies. The PREL calculated in ECOL is a function of the vegetation respiration rate (RESP) and the PASS by vegetation. PUP and PREL are given as mass per area (g m^{-2}) unit.

The ASSIM subroutine calculates concentration of P in the system upon vegetation uptake and release at each time step. The program performs a mass balance procedure to account for the amount of P coming into the system (reach), the P concentration already in the system, and the amount of P taken up and released by vegetation. The computed output concentration is given as mg L^{-1}.

23.2 Description of the Study Area

The Hidden Valley wetland is an 18.4 ha natural cattail (*Typha glauca*) marsh located west of the Grand River in Kitchener, Ontario, adjacent to the south side of Highway 8 (see Figure 23.3) (Gehrels, 1988; and Limnoterra, 1988). The wetland occupies the central portion of a 100 ha drainage basin and has two major points of inflow and one of outflow. The main source of surface inflow is located at the western limit of the wetland area, draining 50 hectare of industrial/ commercial parking areas developed west of Hidden Valley. The second point of inflow is a small natural watercourse located along the northern perimeter of the study, draining an 80 ha developed area containing unpaved land, roadways, and some housing. The outflow of the study area drains directly into the Grand River and is located at the southeastern corner of the wetland (Gehrels, 1988; Ecologistics, 1979; and Limnoterra, 1988).

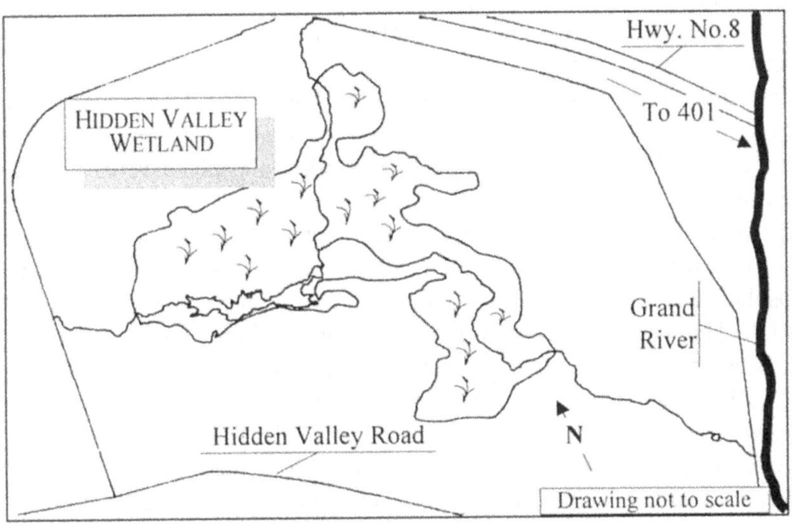

Figure 23.3 Hidden Valley location.

The Hidden Valley is located in an area with a mean annual length of growing season between 200-210 days, 'growing season' being defined as that period in an average year during which the mean daily temperature is equal or higher than 5.5° C (plant growth stops at lower temperatures) (Brown *et al.*, 1980). The climatological seasons for this area are: *Winter* (December - February) is cold (mean temperature of -6.1 °C), *Spring* (March - May) presents a noticeable warming trend (mean temperature of 5.5 °C), *Summer* (June - August) is warm (mean temperature of 18.7 °C), and *Fall* (September - November) is characterized by a marked decrease in temperature (mean temperature of 8.7 °C). Monthly mean precipitation during these seasons is not very different, with winter showing the lowest mean precipitation (60 mm) and summer the highest (81 mm) (Environment Canada, and Gehrels, 1988).

Hydrology and water quality data available for the Hidden Valley wetland used for calibration and validation (later discussed) were obtained from a study performed by Gehrels and Mulamoottil at the University of Waterloo, Waterloo, Ontario. These data were collected from June 1986 until July 1987 (Gehrels, 1988). Inputs to the hydrologic cycle wetland budget measured by Gehrels are precipitation, stream inflow, and groundwater inflow. The outputs are evapotranspiration, surface outflow and groundwater outflow. Reported flowrates of surface water entering the wetland through the main western drain range from 0.01 $m^3\ s^{-1}$ to 2.16 $m^3\ s^{-1}$, with an average non-storm flow of 0.04 $m^3\ s^{-1}$; the northern inflow presented flowrates ranging from 0.01 $m^3\ s^{-1}$ to 0.67 $m^3\ s^{-1}$, with and average non-storm flow of 0.015 $m^3\ s^{-1}$. Reported flowrates of surface water discharging from the wetland ranged from 0.00 to 0.18 $m^3\ s^{-1}$, and an average non-storm flow of 0.004 $m^3\ s^{-1}$.

Water quality data available for the Hidden Valley wetland include total and reactive phosphate (ortho phosphate), pH and chlorides. For simulation purposes, adequate organic phosphorus load estimates were determined by subtracting measured ortho-phosphates from measured total phosphorus. During summer, 10.0 kg of total phosphate and 2.0 kg of ortho-phosphate were calculated to enter the wetland through the northern drainage, and 12 kg of total phosphorus and 2 kg of ortho-phosphate were reported to enter through the western drainage. Reported surface outflow from the wetland showed ortho-phosphate and total phosphate exports increasing from the spring to summer and reaching a peak in the fall. The discharge of ortho-phosphate doubled in the fall.

For simulation purposes, estimates of water quality data and biological and physical features of the Hidden Valley wetland that are directly or indirectly involved in phosphorus dynamics were established. These parameters included nitrogen-to-phosphorus ratio, phytoplankton (represented as chlorophyll-*a* concentration) and vegetation biomass, and temperature in the water column.

23.3 Modeling Methodology

23.3.1 Model Integration

Model integration comprised linkage of vegetation phosphorus kinetics routines in ECOL1 and phosphorus cycle kinetics in WASP5. The procedure is illustrated in Figure 23.4, and called WASP5E herein.

WASP5E assigns the fraction of incoming ortho-phosphate available for vegetation and phytoplankton (represented by x in Figure 23.4) assimilation. This fraction is calculated as a function of PO_4 assimilation capacity and maximum growth rate for each component, and the ratio of vegetation/phytoplankton biomass present in the wetland (Equation 23.7). Environmental conditions represented by season may influence these factors, therefore, x was calculated on a seasonal basis.

$$x = \frac{\text{PASS}}{a_{pc}} \times \frac{Gp_{\text{veg}}}{Gp_{\text{phy}}} \times \frac{\text{Biomass}_{\text{veg}}}{\text{Biomass}_{\text{phy}}} \tag{23.7}$$

where:

Gp_{veg} and Gp_{phy} = vegetation and phytoplankton growth rate, (day^{-1});
$Biomass_{veg}$ and $Biomass_{phy}$
= vegetation and phytoplankton biomass, (g m^{-2});
a_{pc} = phosphorus to carbon ratio (mg/mg); and
other variables as defined above.

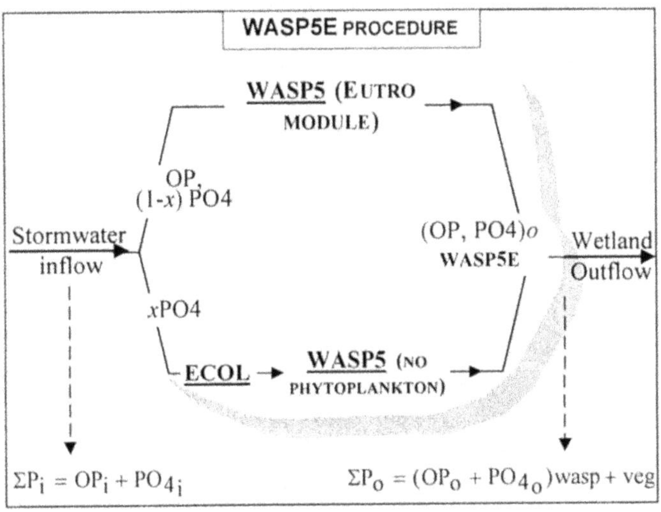

Figure 23.4 Representation of the computer modeling WASP5E procedure; x = fraction of ortho-phosphate assigned to vegetation, OP = organic phosphorus and PO_4 = ortho-phosphate.

In the WASP5E modeling, phosphorus cycle dynamics are simulated by WASP5 which involves organic phosphorus, ortho-phosphate and phytoplankton phosphorus cycle dynamic simulations on a half hour time step. On the other hand, ECOL1 simulates biomass yields and P concentration in the water body on a 2 hour time step. Phosphorus released by vegetation is given in ECOL1 algorithms as total phosphate. Therefore, reported values of the fraction of vegetation phosphorus recycled through respiration and mortality to the organic phosphorus pool and to the ortho-phosphate pool in wetland systems were applied to obtain computed ortho-phosphate and organic phosphorus concentration released by vegetation to the water column. The fraction of vegetation phosphorus recycled to the organic phosphorus has been reported to vary between 0.25 and 0.75 (Ambrose and Martin, 1993a; Mendelsohn and Rines, 1985; Vymazal, 1995). Dead vegetation matter and ortho-phosphate remaining in the system after net uptake by vegetation is loaded into WASP5. The linkage of ECOL1 followed by WASP5 with the omission of phytoplankton algorithms in the system performs this process, allowing benthic P dynamic simulation.

Mass of PO_4 and organic phosphorus (OP) resulting from each model procedure are combined within the WASP5E system to give the total computed outflow concentration (OP_o and PO_{4o}) after a given period of time (e.g. 2 hours). The general mass balance equation derived for WASP5E system is given by Equation 23.8. This mass balance is performed separately for organic phosphorus and inorganic phosphorus. Residential mass change within the system is represented by ΔMass/Δt in Equation 23.8. The mass inputs ($Mass_{in}$) and outputs ($Mass_{out}$) to and from the wetland, and the mass generated ($Mass_{gen}$) in the system are presented in Table 23.1 for organic phosphorus and ortho-phosphate respectively. The WASP5E water balance is given by Equation 23.9.

$$\frac{\Delta Mass}{\Delta t} = Mass_{in} - Mass_{out} \pm Mass_{gen} \quad (23.8)$$

$$\frac{\Delta V}{\Delta t} = P + SW_i + GW_i - E - SW_o - GW_o \quad (23.9)$$

where:

P = precipitation (L^3/T);
E = evapotranspiration (L^3/T);
$SW_{i,o}$ = surface water flow (input, output) (L^3/T);
$GW_{i,o}$ = groundwater flow (input, output) (L^3/T); and
ΔV = change in volume storage (L^3).

As part of each computer modeling phase described above, the following sequence of tasks was carried out: study area discretization, parameter selection, sensitivity analysis and model calibration and validation.

23.3.2 Study Area Discretization

The study area was considered as a single "block" system. Thus the system is assumed to be completely mixed, where the concentration of a constituent within the wetland and that exiting the wetland are equal. This assumption has been discussed as applicable for wetland systems with short hydraulic residence time and small length to width ratios (Dortch and Gerald, not published). Our

Table 23.1 WASP5E mass balance components.

Notation	Description	*G
	Ortho - phosphate mass balance terms	
MPO4in		
·MPO_4sw	stormwater discharge (north and west drainage)	
·MPO_4np	surface water non-point source runoff	
MPO4out		
·MPO_4wout	surface water wetland outflow	
·MPO_4gw	groundwater discharge	
MPO4gen		
·MPO_4phy-up	phytoplankton uptake	-
·MPO_4phy-rel	phytoplankton release	+
·MPO_4op-min	organic phosphorus mineralization	+
·MPO_4sed	ortho-phosphate settling	-
·MPO_4veg-up	vegetation uptake	-
·MPO_4veg-rel	vegetation release	+
	Organic phosphorus mass balance terms	
MOPin		
·MOPsw	stormwater discharge (north and west drainage)	
·MOPnp	surface water non-point source runoff	
MOPout		
·MOPwout	surface water wetland outflow	
·MOPgw	groundwater discharge	
MOPgen		
·MOPphy-rel	phytoplankton release	+
·MOPop-min	organic phosphorus mineralization	-
·MOPsed	ortho-phosphate settling	-
·MOPveg-rel	vegetation release	+

* Generation: + = source of mass; - = sink of mass.

wetland has L:W $\approx$ 1.5. Other considerations must also be taken into account on this assumption, such as, wind stress, and water column depth. Within the system, three layers were distinguished, together representing the physical configuration of the system. The top layer, the "surface water layer", represents the shallow water column in the wetland. The other two layers represent the sediment within the wetland. The "upper benthic layer" is made up of the organic matter (peat) layer within the sediment, and the "lower benthic layer" consisted of silt sandy soils underlying the peat layer in the sediment. The initial volume of the wetland layers was calculated for the beginning of the simulated year. Initial volumes were estimated by average depth, length and width of the system layers (Novotny and Olem, 1994). The water depth was calculated, as suggested by Duever (1988) and Tchobanoglous (1987), by correlating major vegetation types with water depths, and depth of the sediment layers was obtained from a soils map of the area. Typical cattail systems water depth reported in literature range from 0.3 to 0.6 m (Tchobanoglous, 1987).

23.3.3 Parameter Selection

WASP5 Parametrization: The parameters involved in WASP5 simulation were divided into: 1. transport coefficients; and 2. phosphorous cycle kinetics constants.

Exchange and transport of constituents between the system segments and outside the system were estimated for simulation purposes. Advective exchange occurs due to the bulk motions of the flowing water, while nonadvective exchange occurs mainly due to mechanical mixing during fluid advection along with molecular diffusion. Available data for surface and groundwater flow, as measured by Gehrels, were used to calculate advective flows and dispersive coefficients within the study area. Seasonal average net groundwater flow varied from 0.005 to 0.011, where negative values indicate a net recharge flow to the groundwater. Dispersion values obtained in this analysis are within the range reported in literature, ranging between 10^{-8} to 10^{-10} m^2 s^{-1} for silty sand soils (diameter of 0.004 to 0.05 mm) (Bouwer, 1978; and Dortch and Gerald, not published). This range was considered for sensitivity analysis.

The transport of phosphorus sorbed to suspended solids, which undergoes sedimentation, was also addressed. Settling velocities were set within the range of Stokes' velocities corresponding to the suspended particles size distribution. Stokes' velocities for organic, inorganic and phytoplankton matter given by Mendelsohn and Rines (1985) were utilized in this study. Mendelsohn and Rines (1985) modeled the phosphorus cycle in a lake system applying the phosphorus kinetics equations from WASP program. Stokes' velocities for these three parameters given in their study range from 0.05 to 0.5 m d^{-1}.

The selection of phosphorus cycle kinetic constants was divided into three components of the phosphorus cycle as represented in WASP5: organic phosphorus, inorganic (orthophosphate) phosphorus and phytoplankton phosphorus kinetics. The expected values for the coefficients and parameters involved in phosphorus cycle kinetics are tabulated in Table 23.2, with a brief description of the variables. These values were mostly obtained from literature (otherwise indicated). It must be kept in mind that the model is capable of simulating more

Table 23.2 Phosphorus cycle kinetics: expected value of parameters.

Notation	Description	Expected value	Units
Phytoplankton Net Growth Terms			
Exogenous Variable			
f	Fraction of day that is daylight [1, 2]	0.5	--
Z	Zooplankton population	0	$mg\ C\ L^{-1}$
Rate Constants			
k_{1c}	Phytoplankton max. growth rate [1, 2, 3]	2.0	day^{-1}
θ_{1c}	Temperature coefficient [1, 2]	1.068	--
K_c	Phytoplankton self-light attenuation [1, 2]	0.017	$m^2\ mg^{-1}Chla$
θ_c	Carbon - Chlorophyll *a* ratio [1, 2]	35	--
I_s	Saturation light intensity [1, 2]	300	$ly\ day^{-1}$
K_{mP}	Half - saturation cnt for phosphorus [1,2]	0.001	$mg\ P\ L^{-1}$
k_{1R}	Endogenous respiration at 20° C [1, 2]	0.125	day^{-1}
θ_{1R}	Temperature coefficient [1, 2]	1.045	--
V_{S4}	Phytoplankton settling velocity [1, 2]	0.1	$m\ day^{-1}$
k_{1d}	Phytoplankton death rate [1, 2]	0.02	day^{-1}
Phosphorus Reaction Terms			
a_{pc}	Phosphorus to carbon ratio [1, 2, 3, 4]	0.025	$mgPmg^{-1}C$
k_{83}	Dissolved organic P mineralization at 20° C [1, 2]	0.22	day^{-1}
θ_{83}	Temperature coefficient [1]	1.08	--
k_{mPC}	Half saturation constant for phytoplankton limitation of P recycle	0.001	$mg\ C\ L^{-1}$
f_{op}	Fraction of dead phytoplankton recycled to the organic P pool [1, 2]	0.5	--
f_{d3}	Fraction dissolved inorganic P in the water column [1, 2]	0.8	--
V_{S3}	Organic matter settling velocity [2]	0.32	$m\ day^{-1}$
V_{S5}	Inorganic matter settling velocity [2]	0.32	$m\ day^{-1}$
Benthic Phosphorus Reaction Coefficients			
k_{pzd}	Anaerobic algal decomposition rate [1]	0.02	day^{-1}
k_{opd}	Organic P decomposition rate [1]	0.0004	day^{-1}
θ_{opd}	Temperature coefficient [1]	1.08	--
f_{D3j}	Fraction inorganic P dissolved in benthic layer [1]	0.045	--

[1] Ambrose and Martin, 1993a; [2] Mendelsohn and Rines, 1985; [3] Vymazal, 1995; and [4] Tchobanoglous, 1987.

variables than there are data available to support. During parameter selection, lack of data necessary to support some variables required the use of common literature values. With more time, these could be more fully characterized.

ECOL1 Parametrization: The parameter values involved in phosphorus dynamics within the ECOL1 vegetation pool were compiled from literature. These parameters represent vegetation growth and death kinetics in the system. Parameters that involved dynamics of *Typha spp.* (cattails) in freshwater systems are present in Table 23.3.

Table 23.3 Vegetation kinetics ECOL1 parameters.

Notation	Term	Base	Range	Units
Cgr20	Unit respiration rate at 20 °C [1, 2]	0.0027	0.0015-0.0045	$g\ O_2\ g^{-1}\ h^{-1}$
ReqFac	Efficiency factor for nutrient utilization [2]	2.25	1.2-3.5	---
K_{max}	Optimum growth rate for cattails [1, 2, 3, 4]	0.10	0.06-0.24	$g\ g^{-1}\ h^{-1}$
Ic	Light model constant [1, 2]	0.75	0.50-1.0	$ly\ min^{-1}$
Pass	Assimilation ratio for P for cattails [1, 3, 4]	0.0015	0.001-0.0025	$g\ g^{-1}$
Nass	Assimilation ratio for N for cattails [1, 2, 3]	0.02	0.015-0.04	$g\ g^{-1}$
$P_{in}P$	Initial P conc. in plants [1, 3]	0.0004	0.0004-0.004	$g\ g^{-1}$

[1] Vymazal, 1995; [2] Walker *et al.*, 1982; [3] Reddy and DeBusk, 1987; and [4] Brix, 1994.

23.3.4 Sensitivity Analysis

Sensitivity analysis was performed for both computer models, focusing intensively on the summer months (June, July and August), since they present the highest fluctuations in phosphorous concentration and flowrates in the wetland's inflow and outflow. After a series of runs, it was possible to gain a sense of the model sensitivity for some parameters. No attempt is made here to present the comprehensive results for each parameter analysed; parameters that yield highest sensitivity for each model simulation are presented.

During the analysis, the parameters were changed one-by-one using the lowest and highest value from each parameter value range, and the percent change in computed output was recorded. On the graphical representation (below), the middle point represents simulation of the system using expected values. The % change in computed output was related to the change in input parameter, where the computed output obtained using expected value parameters correspond to 0.0 % change in computed output.

Parameters analyzed for sensitivity for WASP5 simulation were the advective flows, nonadvective flows, sediment transport parameters, and the phosphorus cycle kinetics constants (see Tables 23.2 and 23.4). These parameters were altered one by one to determine their effect on both orthophosphate and organic phosphorus computed average concentration in the water column over summer.

Table 23.4 Phosphorus cycle kinetics: range of rate parameters tested for sensitivity.

Notation	Description	Expected value	Range	Units
Phytoplankton Net Growth Terms				
Rate Constants				
k_{1c}	Phytop. max growth rate [1, 2, 3]	2.0	1.0 - 4.0	day^{-1}
θ_c	Carbon - Chlorophyll *a* ratio [1, 2]	35	25 - 35	- -
k_{1R}	Endogenous resp. at 20°C[1, 2]	0.125	0.05-0.13	day^{-1}
V_{S4}	Phytoplankton settling vel. [1, 2]	0.1	0.05 - 0.5	$m\ day^{-1}$
k_{1d}	Phytoplankton death rate [1, 2]	0.02	0.02 - 0.6	day^{-1}
Phosphorus Reaction Terms				
k_{83}	Dissolved organic phosphorus mineralization at 20° C [1, 2]	0.22	0.22-0.44	day^{-1}
f_{op}	Fraction of dead phytoplankton recycled to the organic P pool [1, 2]	0.5	0.25-0.75	- -
f_{d3}	Fraction dissolved inorganic P in the water column [1, 2]	0.8	0.7-0.85	- -
V_{S3}	Organic matter settling vel. [2]	0.32	0.25-0.5	$m\ day^{-1}$
V_{S5}	Inorganic matter settling vel. [2]	0.32	0.25-0.5	$m\ day^{-1}$
Benthic Phosphorus Reaction Coefficients				
f_{D3j}	Fraction inorganic P dissolved in benthic layer [1]	0.045	0.001-0.045	- -

[1] Ambrose and Martin, 1993a; [2] Mendelsohn and Rines, 1985; and [3] Vymazal, 1995.

Figure 23.5 presents relative sensitivity for computed average PO_4 concentration for summer months. From this analysis, computed orthophosphate outflow concentration was observed to be most sensitive to organic and inorganic matter settling velocity (V_{S3} and V_{S5}), to dissolved organic phosphorus mineralization (K_{83}) and to the fraction of dissolved inorganic phosphorus in the water column (Fd_3). Similarly, this analysis revealed that the computed organic phosphorus concentration is most sensitive to settling velocity (V_{S3}) and to the dissolved organic phosphorus mineralization parameter (K_{83}).

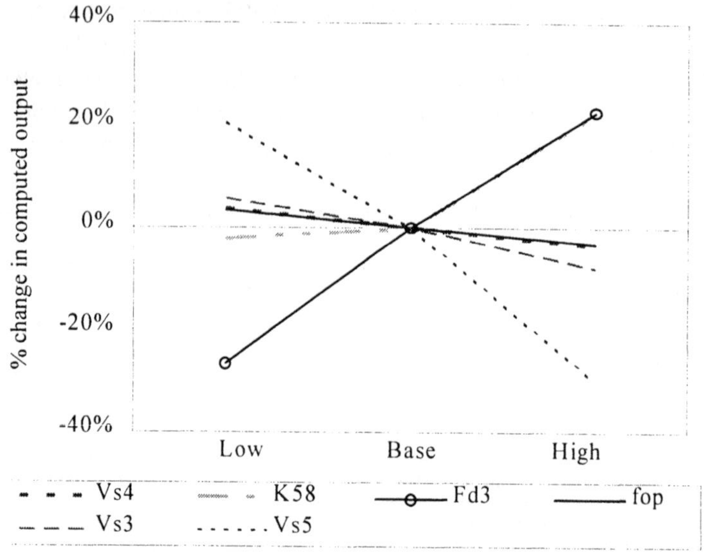

Figure 23.5 Relative sensitivity of computed output for WASP5 for ortho-phosphate input parameters.

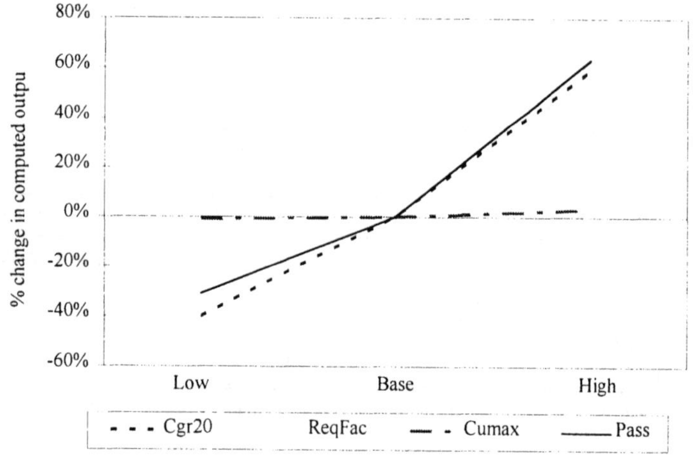

Figure 23.6 Relative sensitivity of computed output for ECOLI for phosphate uptake by vegetation input parameters.

Sensitivity analysis for ECOL1 was performed to determine the effect of vegetation kinetics parameters on the computed average phosphate release and uptake by vegetation. In addition, the analysis was performed for the computed average phosphorus concentration in the water column. Figure 23.6 illustrates the relative sensitivity for the computed average phosphorus uptaken by vegetation. These analyses revealed a high degree of sensitivity of the computed output for kinetics parameters such as the vegetation unit respiration rate (Cgr20), the factor for nutrient utilization (ReqFac), and the cattails phosphorus assimilation ratio (PASS). The phosphate uptaken by vegetation was also found to be sensitive to the optimum growth rate of cattails (K_{max}).

23.4 WASP5E Application to the Hidden Valley Wetland

The summer and fall hydrologic and water quality data set was used to calibrate the various coefficients and parameters described above for phosphorus cycle kinetics, as it represents the time of the year with the highest hydrologic and phosphorus loads. The key calibration parameters and coefficients were the ones identified previously from sensitivity analysis that yield the highest impact on the model response. The range of values for parameters and coefficients used for WASP5E calibration were identical to those used for sensitivity analysis. The values for parameters and coefficients adjusted by WASP5E calibration are presented in Table 23.5. Other parameters and coefficients were kept at the expected value as presented in Tables 23.2 and 23.3. Ortho-phosphate and organic phosphorus are the only water quality variables available for the Hidden Valley wetland to which the model can be calibrated. Hence, this limits the reliability of the calibration for any particular model parameter.

Table 23.5 Phosphorus cycle kinetics: parameter values adjusted from calibration.

Not.	Description	Adjusted values	Units
WASP5 Parameters			
V_{S4}	Phytoplankton settling velocity	0.13	m day^{-1}
k_{83}	Dissolved organic phosphorus mineralization at 20° C	0.35	day^{-1}
f_{op}	Fraction of dead and respired phytoplankton recycled to the organic phosphorus pool	0.40	- -
f_{d3}	Fraction dissolved inorganic P in the water column	0.70	- -
V_{S3}	Organic matter settling velocity	0.25	m day^{-1}
V_{S5}	Inorganic matter settling velocity	0.25	m day^{-1}
Vegetation kinetics ECOL1 parameters			
Cgr20	Unit respiration rate at 20 °C	0.0015	$gO_2\ g^{-1}\ h^{-1}$
ReqFac	Efficiency factor for nutrient utilization	1.2 - 3.5	---
K_{max}	Optimum growth rate for cattails	0.24	$g\ g^{-1}\ h^{-1}$
Ic	Light model constant	0.50	ly min^{-1}
Pass	Assimilation ratio for P for cattails	0.001-0.0025	$g\ g^{-1}$

In ECOL1, the phosphorus assimilation capacity ratio (PASS), reflects the stoichiometry of phosphorus composition on the vegetation population which determines the relation between phosphorus uptake/mass of vegetation synthesized. The value for this parameter varies due to the varying cellular content of phosphorus which is, in turn, a function of the external nutrient concentrations and the past history of the vegetation population. The use of a constant ratio in the simulation by ECOL1, is therefore questionable. For this reason, we considered it appropriate for the Hidden Valley simulation to calibrate this ratio by season within the given range (0.001 to 0.0025 g g^{-1}).

Comparison of observed and simulated ortho-phosphate and organic phosphorus concentrations for the calibration period is presented in Figure 23.7. Adjusted calibrated parameters were used to simulate organic and ortho-phosphate concentration during winter and spring (1986/1987). Comparison of observed and simulated PO_4 and OP concentrations for this period is presented in Figure 23.8. In Figures 23.7 and 23.8, the observed instantaneous concentration data are presented with their corresponding estimated errors, 30% error for outflow organic phosphorus and 25% error for outflow ortho-phosphate. A plot of the continuous results based on a 2 hour time step and the instantaneous observed concentrations for the entire year is presented in Figure 23.9. Correspondent parameters were adjusted seasonally.

Several approaches to statistically evaluate the goodness of fit between computed and observed data have been widely discussed (James, 1994). For this research, the standard error of estimate (SEE) and the relative error (RE) were calculated for WASP5E seasonal and entire year performance. As a representative statistic to measure the accuracy of fit between WASP5E computed data and the observed data, the SEE was calculated by Equation 23.10 (James, 1994):

$$SEE = \sqrt{\frac{\sum_{i=1}^{n}\left(C_M - C_D\right)^2}{n-2}} \qquad (23.10)$$

where:

C_M = the model computed value;
C_D = the measured value (observed data); and
n = the number of points compared.

The relative error statistic provides a measurement of model performance that is comparable among different variables because they are normalized to the value of each variable. The mean relative error was calculated using Equation 23.11 (Mendelsohn and Rines, 1985):

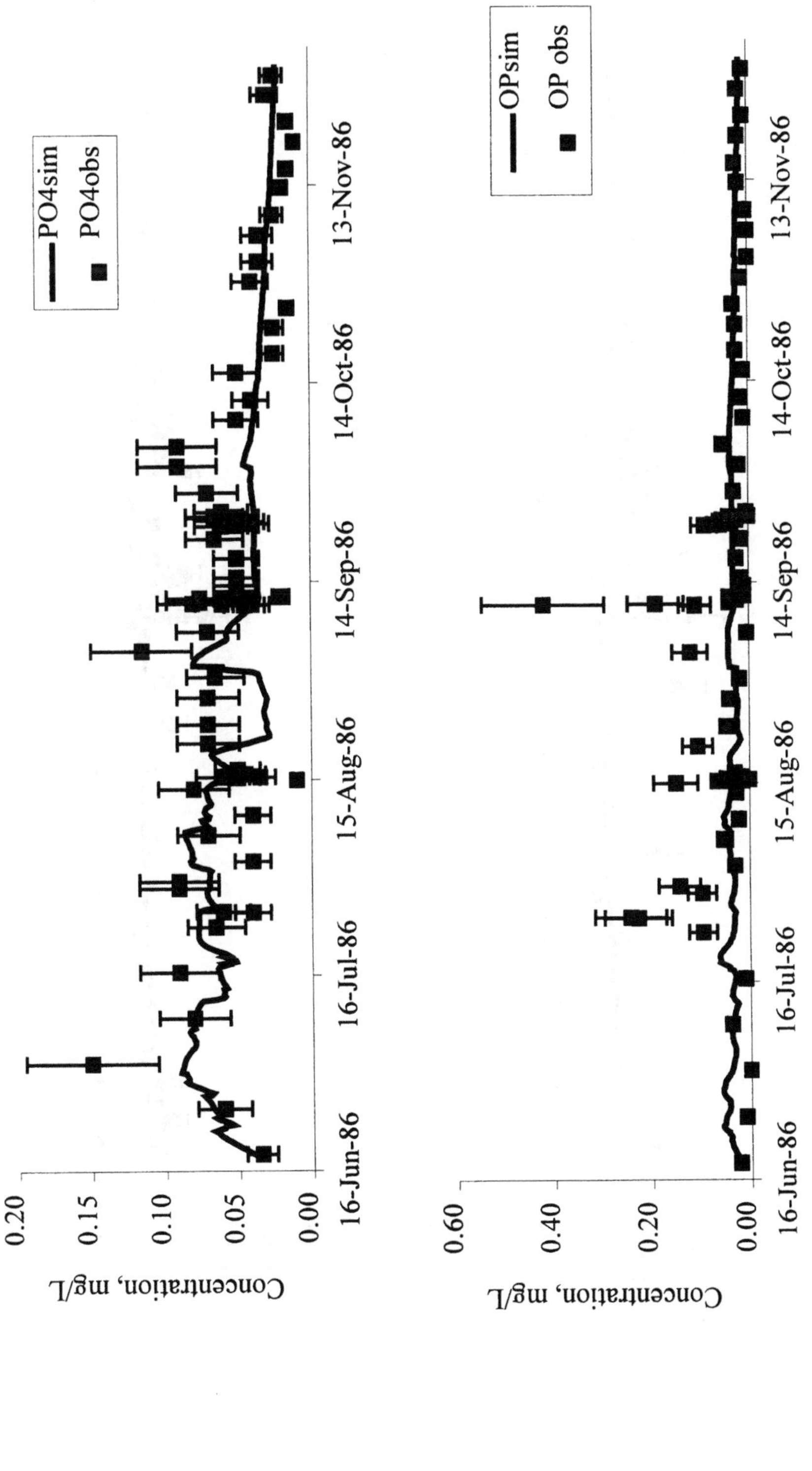

Figure 23.7 Observed and simulated ortho-phosphate (above) and organic phosphorus (below) concentration for summer and fall 1986.

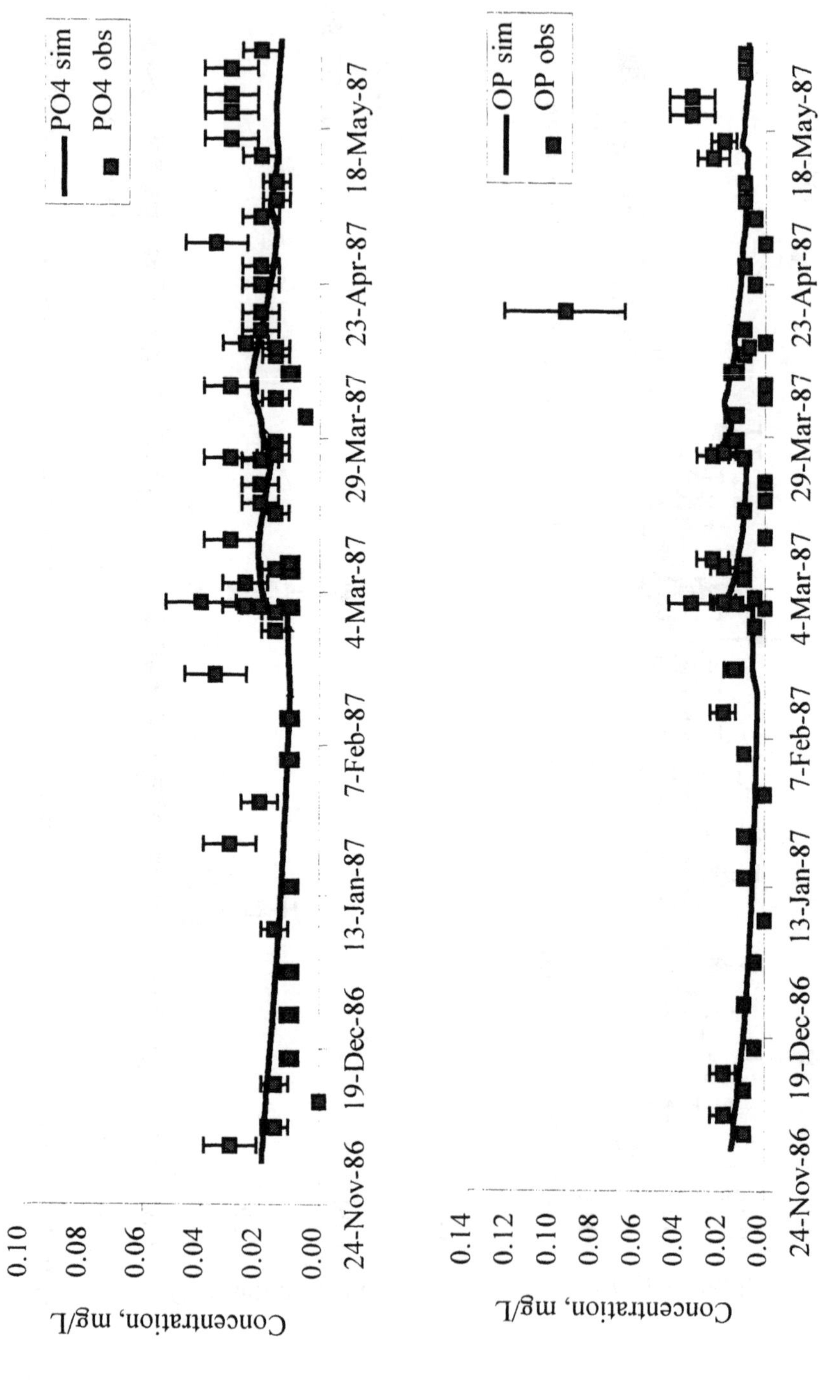

Figure 23.8 Observed and simulated ortho-phosphate (above) and organic phosphorus (below) concentration for winter and spring (1986-1987).

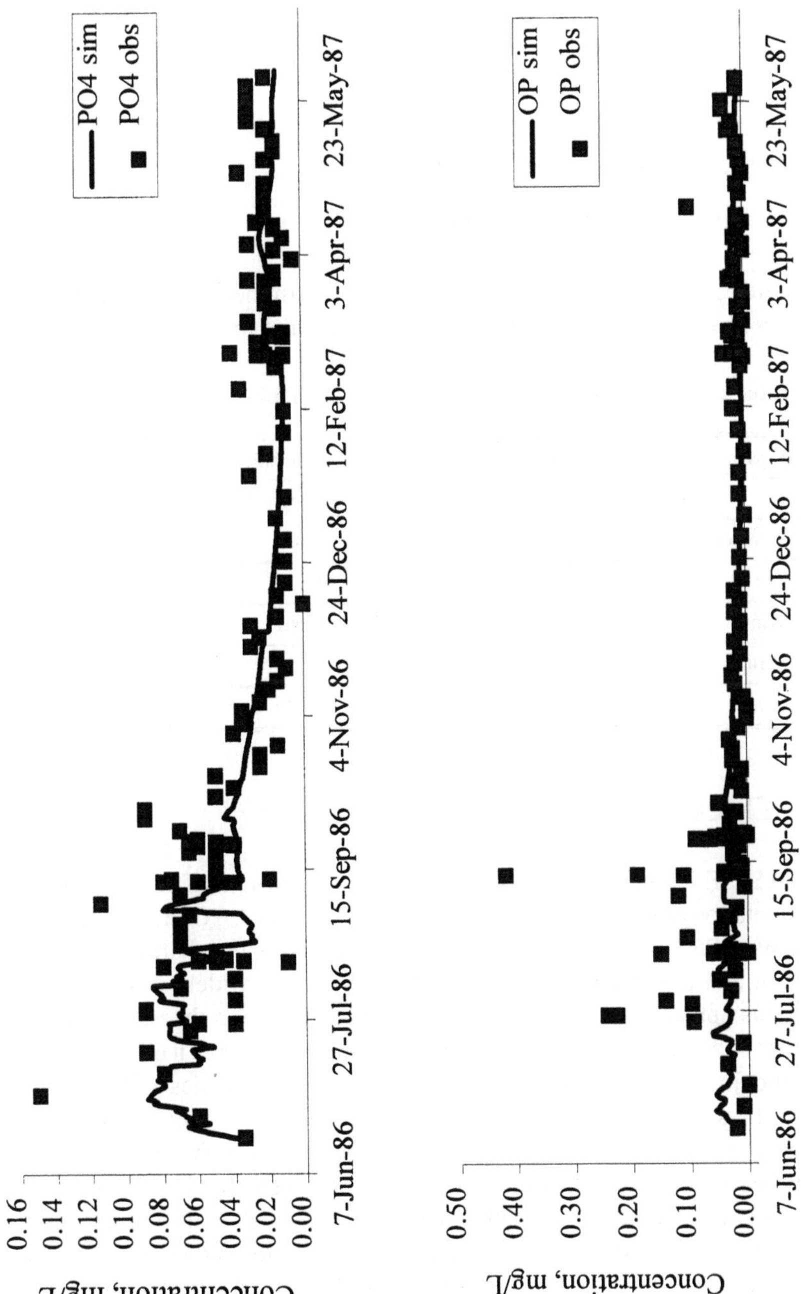

Figure 23.9 Continuous ortho-phosphate (above) and organic phosphorus (below) WASPE5 simulation from June 1986 to June 1987.

$$RE = \frac{\left|\overline{C_D} - \overline{C_M}\right|}{\overline{C_D}} \qquad (23.11)$$

where:

$\overline{C_D}$ = the mean measured value; and

$\overline{C_M}$ = the mean computed value.

Table 23.6 presents a statistical evaluation determined from the WASP5E simulation, based on a 2 hour time step, by season and for the entire year.

Table 23.6 Summary of statistics analysis for WASP5E simulation results.

Season	Ortho-phosphate		Organic phosphorus	
	SEE (mg L^{-1})	RE %	SEE (mg L^{-1})	RE %
Summer	0.026	4.1	0.071	45.6
Fall	0.021	26.9	0.037	15.5
Winter	0.011	11.7	0.007	14.8
Spring	0.010	10.9	0.017	10.9
Year	0.018	15.9	0.040	27.2

23.5 Discussion

WASP5E links the phosphorus dynamics processes simulated by WASP5 to the vegetation growth/phosphorus uptake module called ECOL1. Through this linkage WASP5E assigns the fraction of incoming ortho-phosphate available for vegetation and phytoplankton assimilation. It has been widely discussed that phosphorus uptaken by vegetation occurs primarily through the roots from the sediments. The ECOL1 model overlooks this phenomena since it only simulates P dynamics in the water column but excludes simulation dynamics of P in the pore water in the sediment. The model performs the simulation of ortho-phosphate taken up by vegetation by assigning an efficiency factor for phosphorus utilization by vegetation. This factor is specified in the input file and although its interpretation is not clearly defined in the ECOL1 manual, in the current study, adjustment of this factor was the mechanism employed to account for the P that, if simulated, would go to sedimentation and could be present in the sediment layer in a form available for vegetation uptake. It was recognized that this approach may be somewhat coarse but it was considered that due to restrictions such as lack of phosphorus concentration data for sediments, other feasible approaches may not have been any better.

In addition, algorithms of phosphorus dynamics in ECOL1 alone exclude full consideration of vegetation phosphorus sedimentation and organic matter decomposition in the sediment. The linkage of ECOL1 followed by WASP5 with the omission of phytoplankton algorithms in the system (see Figure 23.4), partially addresses this problem by loading dead organic matter from ECOL1 vegetation to the WASP5 benthic segment, allowing P dynamics in sediment and organic matter decomposition to be simulated.

The results from WASP5E simulation were verified against standard material balance equations (for water and PO_4 and OP). This mass balance verification was within 5% for organic phosphorus and ortho-phosphates and within 9% for water. This affirms that the WASP5 procedure computes the P balance with 5 to 9% error for this dataset.

An understanding and measurement of the rate of processes involved in phosphorus dynamics simulation in a wetland system is a key factor to select an appropriate simulation time step. The ECOL1 model used a 2 hour time step for phosphorus vegetation simulation. This time step was chosen to capture the short term variation of key environmental conditions for this process such as solar radiation and temperature profiles through a typical day. For the WASP5 simulation, a 0.5 hour time step was applied by the model itself, based on the rate of processes involved in phosphorus dynamics accounted for in WASP5. The rate of these processes tends to vary within the same system, where they are affected by different environmental or physical factors, such as surface water velocity and hydraulic retention time, thus complicating the task of estimating an appropriate time step. Time steps are constrained within a specific range to maintain stability and minimize numerical dispersion, or solution inaccuracies. The 0.5 hr. results were then averaged to yield a 2 hr step consistent with ECOL1. Given the relatively slow response of the wetland to changing flow, radiation and chemical conditions, it is believed that the 2 hour time step is adequate to capture P dynamics in wetlands of this type.

The relative error calculated for seasonal simulation (Table 23.6) revealed that WASP5E accurately simulates organic phosphorus and ortho-phosphate outflow during winter and spring, at a relative error of 15% or less. During summer, the model simulation yielded a significantly higher relative error for organic phosphorus simulation, whereas for inorganic phosphorus WASP5E yielded a much lower relative error (4%). During fall, the organic phosphorus simulation resulted in a low relative error while for ortho-phosphate a higher relative error was observed.

Although the effects are not clearly defined, it has been generally reported (Vymazal, 1995) that exceedingly high or low temperatures affect the normal performance of wetland processes. Vegetation and biological processes can be highly restricted under these conditions. The WASP5E simulation during summer and fall (when average temperature are higher than 10°C) may have been restricted by the way in which WASP5E integrated models handle fluctuating temperatures. The kinetic rate parameters, such as unit respiration rate for vegetation (Cgr20 of ECOL1), are typically set for standard conditions (e.g. 20°C). The temperature correction relation employed in ECOL1 may oversimplify the actual relationship.

The Hidden Valley wetland was characterized during summer mainly by wet periods, high temperatures, low flow rate at the outlet of the wetland with sporadic peaks and low concentrations of organic phosphorus and ortho-phosphate entering the wetland through the surface water. This combination of hydrologic factors and physical factors may have created critical conditions for removal mechanisms of phosphorus, such as quick release of phosphorus by vegetation under high temperature and low flows, or concentration of phosphorus under high evapotranspiration periods and subsequent washout by continuous rainfall events. These processes were reflected in sudden high levels of organic phosphorus and ortho-phosphate observed at the outflow of the wetland.

During the fall, a new set of environmental characteristics were observed. Conditions that may have affected the efficiency of wetland vegetation for ortho-phosphate uptake, potential death and washout of vegetation over high outflows, as well as possible release of phosphorus from sediments, are: high precipitation, decrease in temperature, increase in outflow rates and higher concentration of phosphorus entering the wetland via surface water.

During the winter and spring, the Hidden Valley wetland exhibits hydrodynamic and physical characteristics that are quite different. In winter the temperature and precipitation is low and the inflow concentrations and outflow rates are significantly reduced. The wetland vegetation effects are essentially absent during this season. These conditions may cause relatively low process rates in the wetland, reflected by low constant outflow concentrations discharging from the wetland. Spring presented a relative warming trend, consistent with increasing precipitation and outflow rates. Concentrations at the outflow begin to vary, best described by relatively consistent low organic phosphorus and ortho-phosphate concentrations.

The ortho-phosphate simulation performed by WASP5E yields a slight underestimation at high observed concentrations and an overestimation at low observed concentrations. For organic phosphorus the computed underestimation occurred at higher observed concentrations, which occur in summer and fall.

23.6 Sources of Error

The main modeling phases are listed in Table 23.7

Natural variability in the real ecosystem. The variability of the processes begins with the variability and frequency of stormwater introducing phosphorus into the wetland.

Table 23.7 WASP5E simulation performance evaluation.

Natural variability in the real system error.
Observation and sampling error of observed inputs and outputs.
Algorithms structure of the model system and model aggregation.
Area discretization - start-up error.
Input datafile / parameter selection error.
Parameter optimization.

Observation and sampling error of observed input and output. Errors in data observation are mainly related to field instrumentation, whereas the error related to sampling may be associated with the timing and location of the field equipment (James, 1994 and Gehrels, 1988). In estimating the error, the limited frequency of sampling and the lack of samples duplicates were a critical factor, and sometimes its effect is underestimated. The reported estimated errors of the observed data could represent a conservative error estimation, and higher relative error could be expected.

Algorithm structure of the model system and model aggregation. This error addresses algorithms absent from the model structure. This will require future investigation, but includes:

1. WASP5 does not modify the benthic P mineralization rates or the P dissolved fraction as a function of DO or other chemistry. Anaerobic /aerobic conditions in the water-sediment interface may promote exchange of phosphorus between this interface. It has been suggested (Gehrels, 1988; and Sloey *et al.*, 1978), that if the dissolved oxygen drops below 2 mg L^{-1}, phosphorus can be released from the sediment to the interstitial water and subsequently to the water column. The omission of this algorithm within the model could cause underestimation, especially in summer and fall, of organic phosphorus and ortho-phosphate outflow levels.
2. Linkage of a hydrodynamic model which properly simulates flow routing in the system to WASP5 may result in a substantial enhancement of the simulation. Factors to consider include cross-section, profile and overbank geometry. The integration of a hydrodynamic model could potentially provide information on wetted perimeter which will allow exposed sediment areas to be calculated. WASP5 assumes that its benthic segments are always under water, which excludes the simulation of areas that are temporarily exposed and therefore subjected to changes in redox conditions affecting movement of phosphorus accumulated in the

sediment. Under such conditions, decomposition rate as a function of wet and dry state would have to be defined.

3. In addition, tighter linkage between ECOL1 and EUTRO5 algorithms would have improved simulation of plant material depositing to benthic segments and undergoing mineralization.

Area discretization-"start-up error". Spatial discretization of the study area was restricted mainly by the limitation of the spatial distribution of the observed data (e.g. at the two inflows and the outflow of the wetland). The potential "start-up error" could be mainly attributed to calculation of initial volume of the system layers as well as initial concentration of organic and inorganic phosphate in each layer.

Input datafile-parameter selection error. Some of the most common errors in model simulation result from incorrect data specification in the input file and selection of input parameters.

Parameter optimization error. Parameter optimization plays a fundamental role in water quality simulation. In this process, the key is to calibrate the model appropriately, keeping the parameter values within their reasonable ranges and pay particular attention to parameters that were revealed as more crucial during the sensitivity analysis. As previously mentioned, the only observed variables to which the model can be calibrated were organic phosphorus and ortho-phosphorus. Lack of observed data on vegetation and phytoplankton biomass may have greatly limited the WASP5E calibration process.

23.7 Conclusions

The following conclusions were drawn:

1. WASP5E was successful in assembling into a computer procedure some of the main components of phosphorus dynamics that take place in a wetland system.
2. WASP5E seasonal simulation revealed simulation to a relative error of 15% or less of organic phosphorus and ortho-phosphate outflow during winter and spring.
3. WASP5E simulation for organic phosphorus and ortho-phosphate slightly underestimated outflow concentrations at high observed concentrations and a slight overestimation at low observed concentrations. This was clearly illustrated for summer and fall.
4. The temperature correction employed in ECOL1 may have oversimplified the actual relationship, thus disturbing WASP5E simulation performance particularly in summer simulations.
5. Benthal release of phosphorus from sediment caused by change in redox condition in the water-sediment interface and the litter

decomposition in the Hidden Valley wetland may have occurred during summer and fall 1986. These processes are not accounted for in the WASP5E algorithms, thus underestimating organic phosphorus and ortho-phosphate outflow levels.

6. Other wetland processes, such as filtration of particulates, is not included in WASP5E algorithms.
7. The wetland level of discretization applied in this study was adequate for calibration purposes and coarse management decisions. More detailed discretization could be desired for specific management practices, which in turn will require an extensive collection of chemical and hydrological data and thus increase the overall costs.
8. The simulation time step applied adequately captured phosphorus dynamics occurring in the wetlands.
9. WASP5E calibration processes were greatly dependent on observed data pertaining to vegetation and phytoplankton biomass for the Hidden Valley wetland, which were limited.
10. The integration of a hydrodynamic model could potentially provide more elaborate information on wetted perimeter which will permit the calculation of exposed sediment areas. In addition, the simulation of short circuiting flows, which scour the sediment and flush the vegetation, should also be considered in a hydrodynamics model.

Notation

a_{PC}	phosphorus to carbon ratio, (mg P mg^{-1} C);
AREA	surface area of the segment, (m^2);
BIOMASS	density of vegetation, (g m^{-2});
BIOW	washout factor;
C_3	inorganic phosphate concentration, (mg L^{-1});
C_4	phytoplankton concentration, (mg L^{-1});
C_8	organic phosphorus concentration, (mg L^{-1});
D	depth of the waste column or model segment, (m).
D_{pl}	biomass reduction rate, (day^{-1});
DV	change in volume storage (L^3 / T).
E	evapotranspiration (L^3 / T);
f_{D3}	fraction of dissolved inorganic P in the water column;
f_{D8}	fraction of dissolved organic P;
f_{op}	fraction of dead and respired phytoplankton recycled to the organic P pool;
G_{pl}	specific growth rate, (day^{-1});
GR_{20}	unit respiration rate at 20° C, (g g^{-1} h^{-1});
$GW_{i,o}$	groundwater flow (input, output) (L^3 / T); and

INHIBT	plant growth inhibition coefficient.
K_{83}	dissolved organic phosphorus mineralization at 20° C; (day^{-1});
K_{mPc}	half saturation constant for phytoplankton limitation of P cycle, (mg C L^{-1});
k_{opd}	organic P decomposition rate, (day^{-1});
k_{pzd}	anaerobic algal decomposition rate, (day^{-1});
NFAC	nitrogen availability;
O_2	oxygen concentration (mg L^{-1}) ;
P	precipitation (L^3 / T);
P_{ASS}	phosphorus assimilation ratio by vegetation, (g P g^{-1} biomass)
PFAC	phosphorus availability;
q_{pzd}	temperature coefficient;
q_{83}	temperature coefficient;
q_{opd}	temperature coefficient;
ReqFac	efficiency factor for nutrient utilization for vegetation.
STFAC	storage factor for P in plants;
$SW_{i,o}$	surface water flow (input, output) (L^3 / T);
TFR	temperature correction factor;
THP	temperature-adjusted growth rate; (h^{-1});
TS	time step, (hr); and
V_{S3}	organic matter settling velocity, (m day^{-1});
V_{S4}	phytoplankton settling velocity, (m day^{-1});
V_{S5}	inorganic sediment settling velocity, (m day^{-1});
WASH	total amount of biomass washout, (g biomass m^{-2} hr^{-1});

References

Ambrose, Robert B. Martin, James L., and Wool, T.A. 1993a. *The water quality analysis simulation program, WASP5. Part A: model documentation.* Environmental Research Laboratory. USEPA. Athens, Georgia. 209 p.

Bayley, S. E 1985. *Water quality functions of wetlands: natural and managed systems.* The Ecological Considerations in Wetlands Treatment of Municipal Wastewaters. Ed. Godfrey, *et al.* Van Nostrand Reinholf Co. Inc. New York, U.S.A. Pp: 180-189.

Bouwer, H. 1978. *Groundwater Hydrology.* McGraw-Hill Inc. U.S.A. ISBN: 0-07-006715-5. 480 p.

Bowland, H. 1993. *SWMM Version 4.2 (SWMM4) Workshop.* XP Software Inc. Tampa, FL. 50 p.

Brix, Hans. 1994. *Functions of macrophytes in constructed wetlands.* Water Science and Technology. ISSN: 0273-1223. **29**(4): 71-78.

Brown, D. M., McKay, G. A., and Chapman, L. J. 1980. The Climate of Southern Ontario. Climatological studies number 5. Environment Canada. Atmospheric Environment Service. Toronto, Ontario. 67 p.

Davis *et al.*, 1978. *Natural fresh water wetlands as nitrogen and phosphorus traps for land runoff.* Wetland Functions and Values : The State of Our Understanding. American Water Resources Association. Minneapolis, Minnesota. U.S.A. Pp: 457-465.

Dortch, M. S. 1992. *Literature analysis addresses the functional ability of wetlands to improve water quality.* The Wetlands Research Program Bulletin. US Army Corps of Engineers. December 1992. **2**(4): 1-4.

Dortch, M. S., and Gerald, J. A., not published. *Screening-Level model for estimating pollutant removal by wetlands.*

Duever, M. J. 1988. *Hydrologic Processes for Models of Freshwater Wetlands.* Wetland Modelling. Elsevier. U.S.A. Pp: 9 - 39.

Eastlick, B. K. 1993. *Wetland wastewater treatment.* Wetland Design Group. Calgary, Alberta. Canada. 71 p.

Ecologistics limited. 1979. Hidden Valley Inventory of Environmental Features & Functions. Prepared for Major Holdings & Developments Limited. October 1979.

Environment Canada. Monthly Meteorological Summary. Waterloo - Wellington 'A', Ontario. Atmospheric Environment Service.

EPA, 1993. Natural Wetlands and Urban Stormwater: Potential Impacts and Management. United States Environmental Protection Agency. February 1993. 843-R-001. Washington, DC. 76 p.

Gehrels, J. 1988. *The hydrology of Hidden Valley Wetland. Water Balance and phosphate budgets.* University of Waterloo. Master of Arts in Geography. 211p.

Gehrels, J. and Mulamoottil, G. 1990. *Hydrologic processes in a southern Ontario wetland.* Hydrobiologia. **208**: 221-234.

Good, B. J., and Patrick, W. H. Jr. 1987. *Root-water-sediment interface processes.* Aquatic Plants for Water Treatment and Resource Recovery. Edited by Reddy, K. R. and Smith, W. H. Magnolia Publishing Inc. Orlando, Florida. U.S.A. ISBN: 0-941463-00-1. Pp:359-371.

Hossner, L. R., and Baker, W. H. 1988. *Phosphorus transformations in flooded soils.* The Ecology and Management of Wetlands. Edited by Hook, D. D. Timber Press. Portland, Oregon. U.S.A. Pp: 293-306.

Hotchkiss, Neil. 1972. *Common Marsh, Underwater & Floating-leaved Plants of the United States and Canada.* Dover Publications Inc. New York, N.Y. ISBN: 0-486-22810-X. 124p.

James, W. 1992. Stormwater Management Modelling, Conceptual Workbook. Computational Hydraulic International, Inc. Guelph, Ontario, Canada. 200 pp.

James, W. 1994. Rules for Responsible Modelling. Computational Hydraulic International, Inc. Guelph, Ontario, Canada. 144 pp.

Kadlec, R. H. 1987. *The hydrodynamics of wetland water treatment systems.* Aquatic Plants for Water Treatment and Resource Recovery. Edited by Reddy, K. R. and Smith, W. H. Magnolia Publishing Inc. Orlando, Florida. U.S.A. ISBN: 0-941463-00-1. Pp:373-392.

Kadlec, R. H. and Knight, R. L. 1996. Treatment Wetlands. CRC Press, Inc. Boca Raton, Florida. ISBN: 0-87371-9304.

Kadlec, R. H., and Kadlec, J. A. 1978. *Wetlands and water quality.* Wetland Functions and Values: The State of Our Understanding. Edited by Greeson *et al.* American Water Resources Association. Minneapolis, Minnesota. U.S.A. Pp: 436-456.

Limnoterra Limited. 1988. Environmental Impact Statement. Hidden Valley Residential

Subdivision City of Kitchener. Prepared for Major Holdings & Developments Limited. October 1988.

Mendelsohn, D. L. and Rines, H. 1985. *Development and application of a full phosphorus cycle water quality model to Lake Champlain.* Modern Methods for Modelling the Management of Stormwater Impacts. Edited by James, William. ISBN 0-9697422-4-X. Pp: 231-258.

Novotny, V. and Olem, H. 1994. *Water Quality. Prevention, Identification, and Management of Diffuser Pollution.* Van Nostrand Reinhold. New York. U.S.A. ISBN: 0-442-00559-8. 1054 p.

Reddy, D. R. and DeBusk, W.F. 1987. *Nutrient storage capacity of aquatic and wetland plants.* Aquatic Plants for Water Treatment and Resource Recovery. Edited by Reddy, K. R. and Smith, W. H. Magnolia Publishing Inc. Orlando, Florida. U.S.A. ISBN: 0-941463-00-1. Pp: 337-357.

Reddy, K. R., DeLaune, R. D., DeBusk, W.F. and Koch, M. S. 1993. *Long-term nutrient accumulation rates in the everglades.* Soil Science Society of America Journal. **57**:1147-1155.

Shaver, Earl, personal communication. 1995. Environmental Engineer, Delaware Department of Natural Resources and Environmental Control. Division of Soil and Water Conservation, Sediment and Stormwater Program. Dover, DE. USA

Sloey, W., Spangler, F., and Fetter, C. W. 1978. *Management of freshwater wetlands for nutrient assimilation.* Freshwater Wetlands: Ecological Processes and Management Potential. Edited by Good R. E. *et al.* Academia press, New York. Pp: 321-340.

Strecker, E. W., Kersnar, J. M., Driscoll, E. D. and Horner, R. R. 1992. *The use of wetlands for controlling stormwater pollution.* The Terrene Institute. Washington, D.C. 66 p.

Tchobanoglous, C. 1987. *Aquatic plant systems for wastewater treatment: engineering considerations.* Aquatic Plants for Water Treatment and Resource Recovery. Edited by Reddy, K. R. and Smith, W. H. Magnolia Publishing Inc. Orlando, Florida. ISBN: 0-941463-00-1. Pp: 27-48.

Tchobanoglous, G. 1993. *Constructed wetlands and aquatic plant systems: research, design, operational, and monitoring issues.* Constructed Wetlands for Water Quality Improvement. Edited by Moshiri, Gerald A. Lewis Publishers. ISBN: 0-87371-550-0. Pp: 23-34.

Tchobanoglous, G. and Schroeder, E. D. 1987. *Water Quality: Characteristics, Modelling and Modification.* Addison-Wesley Publishing Co. U.S.A. ISBN: 0-201-05433-7. 768 p.

Vymazal, J. 1995. *Algae and Element Cycling in Wetlands.* Lewis Publishers. U.S.A. ISBN: 0-87371-899-2. 689 p.

Walker, R., Weatherbe, D. G., and Willson, K. 1982. Aquatic Plant Model-Derivation and Application. Grand River Implementation Committee. Quality Protection Section. Water Resuurces Branch. Ontario Ministry of The Environment. Toronto, Ontario, Canada. 95 pp.

Chapter 24

Modeling Retrofitted Extended-Detention Wet Ponds and Wetland Pockets

Alan S. Lam and R. Mark Palmer

The restoration of stressed watercourses in urbanized watersheds can best be achieved by taking advantage of opportunities such as the retrofitting or improvement of existing best management practices (BMPs). Within the Sixteen Mile Creek and Emery Creek Watersheds, two existing on-line stormwater management (SWM) ponds were recently retrofitted for water quality enhancement: the Ninth Line SWM facility (Mississauga) and the Pine Valley SWM pond (Vaughan). These ponds were originally constructed in the 1980's for flood control, but without the pollutant removal capabilities of today's extended-detention wet pond and wetland/wet pocket systems.

Each component of these retrofit BMPs was designed to employ a variety of pollutant removal pathways to increase the efficiency of sediment removal and improve the quality of water discharging downstream. Typical pollutant removal pathways, in order of dominance, are sedimentation, adsorption by vegetation, physical filtration, and nutrient uptake by the wet pond and wetland/wet pocket compartments plants and algae. The original flood control and riparian storage functions were preserved in each retrofit design.

In order to quantify the sediment removal capabilities of each retrofit, the QUALHYMO and STORM computer programs were used with water budget models to design the forebay, wet pond, wet pocket, and extended-detention storage features. Annual sediment loadings and evaporation characteristics, during periods of minimal precipitation, were then used to develop a maintenance schedule for the new retrofit features.

© *Advances in Modeling the Management of Stormwater Impacts - Vol. 5* W. James, Ed.
Pub. by CHI, Guelph, Canada 1997. ISBN 0-9697422-7-4. Fax: +519 767-2770

A dynamic-wave routing analysis was also necessary for the Ninth Line SWM facility. This analysis used the OTTHYMO, QUALHYMO and BOSS-DAMBRK computer programs and confirmed that the relatively flat gradient of the receiving watercourse and two undersized culverts downstream will not cause a backwater effect significant enough to submerge the new outlet structure during the water quality design storm. This analysis also determined the "realtime" drawdown operation of the on-line facility, for both continuous and less frequent event storm simulations, as well as the submergence problems affecting the original flood control outlet structure.

24.1 Introduction

Over the last two decades, substantial evidence has been collected about the pervasive impacts of urbanization on stream hydrology, geomorphology, water quality, habitat and ecology. In response, growing municipalities such as the municipalities of Mississauga and Vaughan have implemented stream restoration and environmental protection strategies for the urban Sixteen Mile Creek and Humber River (which includes Emery Creek) watersheds, respectively, in partnership with the local Conservation Authorities, Government Ministries, community stakeholders, and land development industries.

The restoration of stressed or degraded stream systems is perhaps the most challenging watershed management objective for achieving sustainable "targets" such as water quality enhancement. The restoration of degraded watercourses within urbanized watersheds, to any meaningful degree, can be economically achieved by taking advantage of "opportunities" such as the retrofitting or improvement of existing BMPs in the drainage network. Opportunities for urban retrofitting are limited in developed watersheds, but they can be revealed through detailed on-site evaluations during the preparation of sub-watershed plans. Typically, the best sites for urban retrofits are found at the terminus of a storm drainage system, across or within artificial open channels, adjacent to a natural or engineered open channel, or within an older BMP system, such as a dry stormwater detention pond (Anacostia Restoration Team, 1992).

The range of possible retrofit techniques includes the following:

- source control;
- open channel;
- natural channel;
- off-line storm sewer;
- on-line storm sewer; and
- "end-of-pipe" surface storage.

Each technique differs with respect to where the retrofit would be located within an existing storm drainage system. This chapter presents two retrofit projects which utilized the open channel and end-of-pipe surface storage techniques.

Open channel retrofits are installed in an engineered channel, immediately below a single or network of storm sewer outfalls, and preferably where a road embankment or other structure crosses the watercourse. On-line options typically consist of "extended-detention - shallow wetland" or "wet pocket - wet pond" systems. Alternatively, off-line open channel retrofits employ a flow splitter to divert the target water quality runoff volume to a similar bio-diverse system. However, the retrofit features are located in or adjacent to the floodplain. The first case study presented in this chapter concerns an open channel retrofit in the City of Mississauga - the largest of its kind in the Greater Toronto Area.

Retrofits for surface storage facilities include the modification of older, end-of-pipe dry stormwater detention or flood control ponds to improve their runoff and storage treatment capacity. The new storage created, either by excavation, adding to the height of the embankment, and/or constrictions from low flow orifices, is used to provide:

1. permanent pool(s), such as forebay and wet pond features,
2. extended-detention storage,
3. shallow wetland or wet pocket compartments, or
4. combinations of the above categories.

For example, pond retrofitting has been the primary focus of restoration efforts in the Washington Metropolitan Area (Herson, 1989) and has typically involved converting older, dry stormwater ponds into extended-detention, wet pond and wetland systems. The second case study is located in the City of Vaughan and involves a dry stormwater management pond, which was originally constructed for flood control in 1981.

24.2 Description of Projects

24.2.1 Ninth Line Stormwater Management Facility

In the City of Mississauga, the Lisgar Region Water Quality Management Plan (Winter, 1993) was approved in 1994. This sub-watershed plan set out the best achievable targets for water resources and environmental management for the 26.8 km^2, warm-water East Tributary of Sixteen Mile Creek. It also identified constraints and opportunities in the sub-watershed to achieve the desired targets. The management targets included:

1. reducing erosion and flooding susceptibilities,
2. maintaining or enhancing base flows to emulate natural conditions,
3. maintaining existing wetlands, and
4. enhancing aquatic/terrestrial habitats and water quality to comply with the Provincial Water Quality Objectives (MOEE, 1994b).

The plan was developed to address stormwater management and drainage requirements for existing activities within the East Tributary sub-watershed and future land use changes.

The relatively flat gradient of the tributary was identified as a constraint during the 1993 study since the watercourse is prone to flooding during more frequent storms (i.e. every summer storm). This affects the on-line SWM open channel facility, upstream of Ninth Line, which often fills to near capacity during rainstorms. Together, the downstream channel slope, two undersized culvert crossings downstream of the SWM facility, and the temporal distribution of rainfall events affect the drawdown operation of this flood control facility.

One of the key recommendations from the Lisgar Region Water Quality Management Plan was to retrofit the 38.9 ha-m flood control/open channel facility, referred to as the Ninth Line SWM facility, for water quality enhancement of runoff from a 13 mm rainfall event. An environmental study to examine various retrofitting options and to design the preferred alternative was completed, so that a number of community developments discharging to the facility could be constructed (Fred Schaeffer Associates (FSA), 1994).

This retrofit project also investigated the steady-state hydraulics of the East Tributary, and submergence problems at the Ninth Line SWM facility's outlet structure caused by the two downstream undersized culverts. It was confirmed that the new retrofitted water quality outlet control structure would not be submerged during the 13 mm design storm event. However, the outlet would be submerged by storms greater than a two year event (FSA, 1994).

A culvert replacement study was conducted by the City for the two downstream structures that were causing problems at the Ninth Line SWM facility (C.C. Tatham Associates (CCTA), 1995). A hydraulic or dynamic wave routing analysis was undertaken for the East Tributary using output data from event storm and continuous storm hydrologic models of the Lisgar Region sub-watershed. While this additional analysis recommended replacement structures for the two undersized crossings, in order to restore the original flood control design outflows and storage water levels at the Ninth Line SWM facility, it also confirmed that the SWM facility's new retrofitted water quality outlet would operate properly during the 13 mm design storm event.

A monitoring program to study the performance of the water quality facility was undertaken. The program will run over five years and results will be published at selected intervals.

24.2.2 Pine Valley Stormwater Management Pond

Recently, the Metropolitan Toronto and Region Conservation Authority (MTRCA) initiated the *Humber River Watershed Management Strategy*, which includes the Emery Creek Watershed within the City of Vaughan. However, prior to this initiative, the MTRCA and Ministry of Natural Resources (MNR) identified in 1993 preliminary water resources and environmental management targets for the 3.2 km^2 headwater development area of the warm-water Emery Creek watershed. Stormwater peak flow control and quality enhancement for future developments were two of the target objectives. These conditions were included in the plan of subdivision approval requirements for a number of Woodbridge residential and commercial development applications, north of Highway 7, within the watershed.

Due to existing constraints within the receiving storm sewer system and previously approved functional drainage requirements (ABA, 1981 and 1983) for the future headwater development area, the only option available for peak flow control was on-site within each development block. However, two surface storage options for stormwater quality enhancement were possible, namely: at the source within each development, or by taking advantage of an off-site retrofit end-of-pipe opportunity within an older 16.9 ha-m dry stormwater pond known as the Pine Valley SWM pond. This pond forms part of the Emery Creek drainage network and is located at the end of the storm sewer that services the 3.2 km^2 headwater area. Through consultation with the City of Vaughan, MTRCA, and MNR, it was agreed that site plan approvals would be granted to all affected developments, provided that the developers would construct the water quality retrofitting improvements to the Pine Valley SWM pond (FSA, 1993).

24.3 Project Goals and Objectives

The primary goals of each retrofit project, which was constructed in 1994, were as follows:

1. retain the original flood control and riparian storage capabilities of the existing facility; and
2. rehabilitate or enhance downstream water quality and associated resources by way of the most appropriate retrofitting works.

The retrofit designs were selected from a variety of options, using best management practice matrix evaluations.

Both selection processes were undertaken in accordance with the ecosystem approach, regulatory policies and guidelines at the time, and the Class Environmental Assessment for Municipal and Wastewater Projects (1993) document. In addition, the following objectives were adhered to in the selection of the designs:

1. retain, as a minimum, the original peak flow shaving and riparian storage design relationships;
2. minimize any impacts to local base flows, relative to existing conditions;
3. optimize suspended solids and heavy metal removal efficiencies;

4. optimize nutrient uptake potential and diversity of plants to provide terrestrial and aquatic habitats;
5. maximize aesthetic features in keeping with the overall master landscape or park use plan for the area;
6. consider the limitations imposed by urban planning issues related to the preferred retrofit option in the overall context of the watershed;
7. consider operation and maintenance costs (i.e. requirements and frequency) as part of the selection process;
8. acceptance by public and regulatory agency;
9. consider construction costs;
10. consider other watercourse structures downstream that may affect the hydraulics of the new retrofit features (e.g. recommendations were provided on the hydraulic capacities of two culvert crossings downstream of the Ninth Line SWM facility, as well as replacement structures, after investigating the submergence effects on the actual versus original design flood control relationships of the facility);
11. implement, after the construction of the preferred design, a post-construction monitoring program at the inlet(s) and outlet to measure water quality - pursuant to the retrofit compliance requirements of the Ministry of Environment and Energy's (MOEE) Certificate of Approval.

At the time of each project, the Province's manual on stormwater management practices (MOEE, 1994a) was not available; however, each project incorporated the design principles specified in that manual.

24.4 Description of Preferred Retrofit Options

24.4.1 Ninth Line Stormwater Management Facility

General

A long list of eleven practical design options was developed from the following open channel retrofit categories:

- an extended-detention basin with no excavational storage improvements (i.e. only 7 mm storm runoff enhancement possible) to the existing facility (*note*: this option had the lowest capital cost but also had many performance disadvantages, relative to the other alternatives, and, in addition, the stormwater enhancement volume was much less than the target storage specified in the Lisgar Region Water Quality Management Plan);
- an extended-detention basin with additional excavation to treat runoff from a 13 mm storm;

- an extended-detention basin (13 mm storm) and creation of wetland/wet pockets;
- a retention basin (i.e. wet pond covering either one-third, two-thirds, or the entire surface of the facility) and extended-detention storage (13 mm storm); and
- combination of the above open channel retrofit categories.

Each retrofit option included:

1. a control berm and water quality outlet, immediately upstream of the existing flood control outlet structure,
2. a sediment forebay with a multi-dispersed pipe outlet, bio-engineered berm and spillway,
3. a landscape and open space enhancement strategy,
4. a maintenance and operation plan, and
5. a post-construction water quality monitoring program.

As a result of the environmental-assessment public-consultation process, the forebay features were revised to address nuisance concerns (e.g. shape and location of the forebay, appearance and materials to be used). In addition to all of the above features, the preferred retrofit option from the general "combination" category included:

1. additional excavation for extended-detention storage,
2. a wet pond covering about one-third of the basin floor, next to the new control berm and outlet structure,
3. creation of four large wet pockets upstream of the wet pond, and
4. removal of the existing concrete-lined low flow channel through the facility.

Functional Design Parameters

The functional design parameters of the retrofitted Ninth Line SWM facility included:

1. the capture of the storm runoff volume from the first 13 mm rainfall depth (i.e. total volume of 43,000 m^3);
2. the detention of the storm runoff from the first 13 mm of rainfall for between 24 and 48 hours;
3. optimized sediment removal associated with the 40 μm criteria, as recommended in the Lisgar Region Water Quality Management Plan;
4. as a minimum, the matching of the quality enhancement criteria (i.e. in terms of first order pollutants and total sediment concentrations), as determined from the approved continuous modeling for the Lisgar Region Water Quality Management Plan; and
5. the retention of the existing flood control capabilities of the facility.

During the hydrologic investigations of the environmental assessment study (FSA, 1994), it was determined that the water level associated with a 2-year storage volume would be about 188.7 m, based on the "ultimate" design storage relationship of the facility. The FSA study also recommended that the quality enhancement storage volume of 43,000 m^3 be obtained at a maximum water level of 188.5 m. Rainfall runoff volumes (i.e. 2-year through Regional Storm events) in excess of the retrofitting design storage would then overtop the proposed quality control berm of the retrofitted facility and be controlled by the existing concrete flood control weir. In other words, the approved discharge-storage function of the facility for flood control and riparian storage could be retained with the new retrofit features:

1. the retention of the existing inlet and outlet locations at opposite ends of the facility to prevent short circuiting and minimize potential re-suspension due to the on-line configuration of the facility;
2. the incorporation into the design of a means to provide downstream runoff volume maintenance, if required later on;
3. the provision of a sediment forebay to pre-treat the stormwater runoff by trapping large debris and coarse suspended solids, and reduce the inlet velocity prior to flows entering the SWM facility;
4. the provision of shallow wet pockets to promote a variety of pollutant removal pathways such as sedimentation, adsorption, and nutrient uptake; and,
5. the provision of a permanent pool depth of 1.6 m within the wet pond compartment to further enhance the sediment removal capabilities of the facility.

Table 24.1 summarizes the design volumes and water levels associated with each component of the preferred retrofit option.

Detailed Modeling Procedures Leading to the Design

Each component of the retrofitted SWM facility was designed to employ a variety of pollutant removal pathways to increase the efficiency of sediment removal and improve the quality of the water discharging downstream. Typical pollutant removal pathways, in order of dominance, are: sedimentation, adsorption by vegetation, physical filtration, and nutrient uptake by wet pond and wet pocket plants and algae.

With respect to quantifying pollutant removal by way of sedimentation, the computer models QUALHYMO and STORM were used in the detail design. Since sedimentation is the dominant pollutant removal pathway, it was used as an indicator of water quality enhancement.

Table 24.1 Summary of design volumes and water levels for the Ninth Line SWM facility.

Retrofit Feature	Design Water Level (m)	Maximum Design Depth (m)	Minimum Required Volume (m^3)	Proposed Design Volume (m^3)
Sediment Forebay (1)	188.5	2.0	4,300 [A.]	8,900 (4,800 active)
Wet Pond (1)	187.2	1.6	--	27,000
Wet Pocket (4)	187.2 to 187.4	0.7 to 0.9	--	6,900
Extended-detention Storage	188.5	1.3	43,000	61,900 [B.]

A. Typically, the active forebay volume should comprise at least ten percent of the entire treatment volume (i.e. 10% of 43,000 m^3) within the SWM facility.

B. Additional volume provided as a result of: 1) regrading constraints to provide 7:1 sideslopes within the extended-detention storage compartment, 2) compensating for a future 25 years of ±12,000 m^3 of sediments within the extended-detention compartment, as part of the maintenance program, and 3) providing the balance of the approved design storage below 188.5 m (ie. ±4,300 m^3) which was not present within the existing facility.

QUALHYMO Modeling

The continuous hydrologic model QUALHYMO (Version 2.11) was used to model the retrofitting design and the water quality results were compared with the approved modeling enhancement criteria from the Lisgar Region Water Quality Management Plan. A "pre-retrofit" model of the watershed (i.e. existing Ninth Line SWM facility without the preferred retrofit features) was used as the baseline model and modified (i.e. "post-retrofit" condition) to include the extended-detention storage, wet pond, and wet pocket compartments of the design. This was done to assess whether the preferred design, at a minimum, could match the approved enhancement criteria from the Lisgar Region Water Quality Management Plan. Model changes were in the form of revised stage-storage and stage-discharge curves, while the balance of the original model in terms of land use, rainfall, etc. was not modified.

For the purpose of the analysis, the permanent pool of the forebay and wet pocket volumes were added to the wet pond volume in the "post-retrofit" modeling analysis. Given the low permeability of the soils within the pockets and lined forebay bottom, it was assumed that the forebay and wet pockets would function similarly to the wet pond feature with respect to sediment removal. This assumption was appropriate for the wet pockets given the observed standing water in an existing wet meadow/cattail depression, which had formed along the northern perimeter of the existing SWM facility.

The precipitation records used for the QUALHYMO analysis were from the Lester B. Pearson International Airport for the period May to November, 1972 inclusive. This series of data was used as an "average year" for the QUALHYMO water quality simulations that were used to develop the Lisgar Region Water Quality Management Plan.

STORM Modeling

The continuous hydrologic model STORM was also used in the detail design to determine the overflow volumes and sub-watershed runoff amounts under two series of rainfall data. The first model used the same series of precipitation records from the "post-retrofit" QUALHYMO modeling analysis discussed above. The second STORM analysis used a longer period of rainfall data (i.e. February, 1977 to January, 1982 inclusive) from the Airport, as this series of rainfall data has been previously used by FSA in the design of many stormwater quality ponds throughout the Greater Toronto Area.

The results from both STORM models were used in conjunction with the Chen (1975) methodology, which uses the Camp formula (1946) to predict sediment removal efficiency:

$$F = 1 - \left[1 + \frac{1}{n} \frac{V_S}{Qr / A} \right]^{n} \qquad (24.1)$$

where:

F = sediment removal efficiency
Q_r = average constant release rate from the SWM facility
A = active storage surface area of the SWM facility
V_s = settling velocity
n = turbulence or short circuiting parameter

Fair and Geyer suggest an empirical relationship between performance and the value of "n", which is: n=1(very poor); n=3 (good); n>5 (very good). When n= ∞, the equation reduces to the Chen formula: $F=1-e^{-Vx/Qr/A}$. In this study, a conservative value of 1 was assumed for "n".

The distribution of suspended solid fractions were chosen based on the US National Urban Runoff Pollution Program and were considered representative for the sub-watershed. The distribution is listed in Table 24.2. The fractional distributions and settling velocities of suspended solids used in the STORM analysis were the same distributions used in the QUALHYMO model for the Lisgar Region Water Quality Management Plan.

It was felt that in the event of an overflow, the trap or sediment removal efficiency would be reduced due to the increase in velocity in the Ninth Line SWM facility. The results from the STORM model were used to determine the average

Table 24.2 Distribution of fractions of suspended solids used in the QUALHYMO and STORM modeling.

Class	Total %	Avg. Diameter (mm)	V_s Average mm/s
i	20	4.00	5.5
ii	20	0.40	0.59
iii	20	0.13	0.13
iv	20	0.06	0.025
v	20	0.02	0.0027

annual runoff *(R)* and overflow volume *(R_o)*. Using a value of 0.5 for trap efficiency *(E_t)* for events causing overflows, the average annual trap efficiency, E, was calculated as:

$$E = \frac{R_O \times E_t \times (R - R_O) \times F}{R} \tag{24.2}$$

Water Budget Analysis

A water budget analysis was also undertaken to determine the water level fluctuations in the "full" permanent pool storage volumes of the forebay and wet pond compartments of the SWM facility during periods of minimal precipitation. In addition, another simulation was completed assuming both the forebay and wet pond were essentially empty prior to the first rainfall event in each of the years examined. The record from May to September, 1986-1990 (inclusive) was analyzed for both the full and empty condition, because of the unusually dry and hot summer weather experienced in southern Ontario during that time. Daily precipitation data used in the analysis were obtained from the Lester B. Pearson International Airport records. It was assumed that these weather conditions would serve as a good indicator for ensuring that both the wet pond and forebay would not evaporate and therefore detract from the aesthetic and operational characteristics of the retrofitted facility. Also, water level fluctuations were considered to be important for determining what varieties of plants would be best suited to survive under these conditions.

The water budget spreadsheet analysis included the following:

- The basic equation for the water budget:

$$\Delta S = P + R - E - I \tag{24.3}$$

where:

S = change in storage,
P = precipitation directly on surface areas of forebay and wet pond,
R = runoff from the sub-watershed,
E = evaporation from the forebay and wet pond, and
I = infiltration through bottom of forebay and wet pond.

- Evaporation rates were calculated from the findings of an earlier study in the Town of Richmond Hill by FSA.
- Daily runoff to the SWM facility was calculated using the volumetric runoff coefficient from the OTTHYMO model for ultimate sub-watershed development conditions from the Lisgar Region Water Quality Management Plan and daily precipitation volumes.
- Infiltration was calculated using a constant permeability coefficient of 8.6 mm/day (i.e. typical of silty clay soils in the bottom of the wet pond compartment). This approach was also conservatively applied in the forebay water budget analysis, even though a hardened bottom was recommended.
- Any overtopping constituted the exceedence of the dead storage volume, which meant that water would rise into the active volume, where it discharges through an outlet pipe.

BOSS-DAMBRK Modeling

The construction of the original Ninth Line SWM facility in 1988 for flood control included a 42.0 m wide stepped concrete weir with a lateral contraction. This flood control weir and the facility's ultimate design storage (which has not been fully constructed yet) was designed in 1983 to provide only peak flow reduction for the 2- to 100-year flood events. The facility was not designed for incremental riparian storage. However, additional storage was included in the original design in order to:

1. maintain the peak time lags and shapes of the 1983 or "pre-development" hydrographs for post-development conditions, during the 2- to 100-year events, and
2. maintain the total floodplain storage for the Regional Storm.

In 1993, the Lisgar Region Water Quality Study identified the relatively flat gradient of the Sixteen Mile Creek East Tributary as a constraint since the watercourse is prone to flooding during frequent storm events. The study concluded that the "pre-retrofit" Ninth Line SWM facility often filled to near-capacity during rain storms and the cause appeared to be two undersized culverts downstream. In addition, backwater from the relatively flat channel gradient was submerging these two culverts and flood control weir of the Ninth Line SWM facility.

During the 1994 retrofit design by FSA, a preliminary steady-state hydraulic analysis of the tributary using the HEC-2 computer program confirmed that the new control berm/outlet structure for water quality enhancement (i.e. 13 mm storm event) would not be submerged as a result of the two downstream undersized culverts. However, given the unsteady flow characteristics of the downstream tributary, a dynamic wave routing analysis was also required. This involved a numerical solution of the Saint Venant equations.

The Ontario Ministry of Natural Resources *Floodway Fringe Analysis Technical Guidelines* recommends the use of a dynamic wave routing model when the average reach slope is less than 0.7 m/km. Downstream of the Ninth Line SWM facility (±2,430 m) the average slope is less than 0.6 m/km.

Since the extent of the problem associated with the submergence of the concrete flood control weir (i.e. flood events greater than 2-year return period) and the downstream impact of replacing the two subject crossings was not quantified in the 1994 preliminary analysis, further detailed investigations were necessary. In 1995, the City of Mississauga initiated an MEA Class Environmental Assessment (Schedule 'C') for the reconstruction of Ninth Line. Therefore, further investigations were needed since a replacement of the Ninth Line structure was necessary, in conjunction with the road widening design, in order to achieve the original flood-reduction targets for the Ninth Line SWM facility. These additional investigations used the 1983 hydraulic design results for the SWM facility as target operational parameters. A recommendation to replace the other downstream existing culvert was also necessary to achieve the 1983 or "pre-development" approved design objectives for the SWM facility in terms of maximum outflows and storage water levels.

The City of Mississauga wanted to resolve outstanding flood susceptibility concerns as a result of the two downstream undersized culverts and the flood control operation of the Ninth Line SWM facility (CCTA, 1995). A dynamic wave routing analysis was undertaken using the BOSS-DAMBRK computer program based on the same reach length and crossing scenarios from the original FSA (1994) steady-state analysis. This study also used the same OTTHYMO and QUALHYMO hydrologic models from the Lisgar Region Water Quality Study and FSA investigations in 1993 and 1994, respectively.

24.4.2 Pine Valley Stormwater Management Pond

General

Only two end-of-pipe surface storage retrofit options were practical for the Pine Valley SWM pond. Both options included new sediment forebays at the two existing storm sewer outfalls to the pond, an extended-detention compartment for the water quality storage, a water quality control device fitted to the existing flood control outlet, and naturalization landscape plans. However, the new water quality storage was constrained by the height of the existing embankment and sideslopes, which had to be maintained. Therefore, this retrofit project could not be designed to treat a specified rainfall volume, as was the case for the Ninth Line SWM facility.

The first option included 16,000 m^3 of extended-detention excavation (corresponding to a 9 mm rainfall event) over the entire pond surface. This new cell also included the construction of a single large wet pocket area. The second option included a forebay flow-splitter structure and a bypass open channel through one-third of the pond, as well as 15,000 m^3 of additional extended-detention (corresponding to an 8 mm rainfall event) and many small wet pocket features. The purpose of the forebay flow-splitter/bypass channel structure was to direct the first 8 mm of rainfall runoff to the new water quality pond cell. The balance of the minor system flow from the storm sewer would be diverted to the bypass open channel. The primary objective of this diversion concept was to avoid re-suspension in the water quality pond cell during less frequent storms. The second option was agreed to by the regulatory agencies as the preferred retrofit solution.

Functional Design Parameters

The functional design parameters for the preferred retrofit option were:

1. capture the storm runoff from the first 8 mm of rainfall depth in a new 15,000 m^3 extended-detention storage compartment, at maximum depth of 1.5 m;
2. detain the water quality runoff volume for 12 hours (minimum) to 48 hours (maximum);
3. provide two new sediment forebays with a maximum depth of 1 m, velocity distribution chamber for the pipe outlets, and energy dissipation structure at the storm sewer outfalls;
4. provide five small wet pockets in the new water quality cell, each with a permanent water depth of from 0.25 m to 0.50 m;
5. provide a water quality control structure, at the existing flood control outlet, consisting of a hickenbottom pipe/gravel jacket structure with an orifice;
6. provide a maintenance access road to the forebay and control structures; and
7. designate an area on the sideslope of the original dry pond, away from the new cell and the tableland of the park property, to dispose of sediment that will be removed from the pond.

Detailed Modeling Procedures Leading to the Design

Similar to the Ninth Line SWM facility retrofit project, water quality enhancement for the Pine Valley SWM pond was assessed using the Chen (1975) methodology.

The STORM computer program was again used to determine the average annual runoff and overflow volume from the retrofit design. The same STORM input parameters from the Ninth Line SWM facility project were also used to analyze the retrofit features of the Pine Valley SWM pond. In addition, a sensitivity analysis was completed to assess sediment removal efficiencies for drawdown times from 12 hours to 48 hours.

24.5 Results and Discussion

24.5.1 Ninth Line Stormwater Management Facility

QUALHYMO and STORM Modeling

As illustrated in Figures 24.1 and 24.2, the QUALHYMO modeling results for the Ninth Line SWM facility retrofit design satisfied the required exceedence criteria from the Lisgar Region Water Quality Management Plan. That is, the hours and numbers of exceedence for first order pollutant (i.e. fecal coliforms) and total sediment concentrations flowing from the outlet of the retrofitted facility were typically less than the exceedence targets from the 1993 environmental assessment document.

Table 24.3 presents the results from the STORM modeling analysis. The detailed results were satisfactory relative to the pre-design target parameters.

Literature values representative of the land uses under ultimate development conditions were used to assess the impacts of sediment loading on the downstream water quality. The assumed sediment loading rates as a function of land use are shown in Table 24.4 These values were determined from recent studies in the Ottawa area by FSA.

The QUALHYMO model was used to estimate the total annual sediment loading in kg/yr, based on ultimate development conditions in the sub-watershed. The result was that approximately 860,000 kg/yr of sediment would discharge into the retrofitted facility. Assuming a conservative sediment specific gravity of 1.2, this yearly mass translates into a volume of about 715 m^3/yr. Therefore, if the active storage or extended-detention areas of the facility retained 72% of this volume, about 515 m^3/yr of sediment would accumulate in the SWM facility. The balance of sediments removed would be from the forebay, wet pockets, and wet pond compartments.

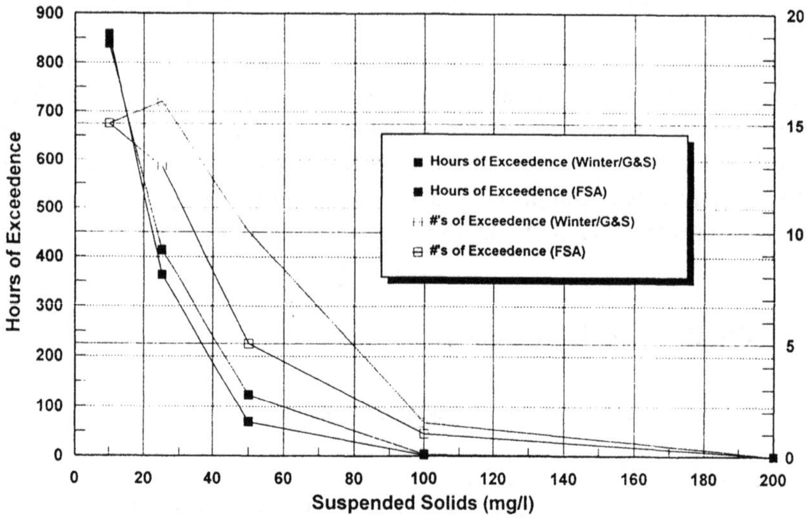

Figure 24.1 Comparison of QUALHYMO results: ultimate sub-watershed conditions - suspended solids. Sources: Winter et al (1993), and FSA (1994b).

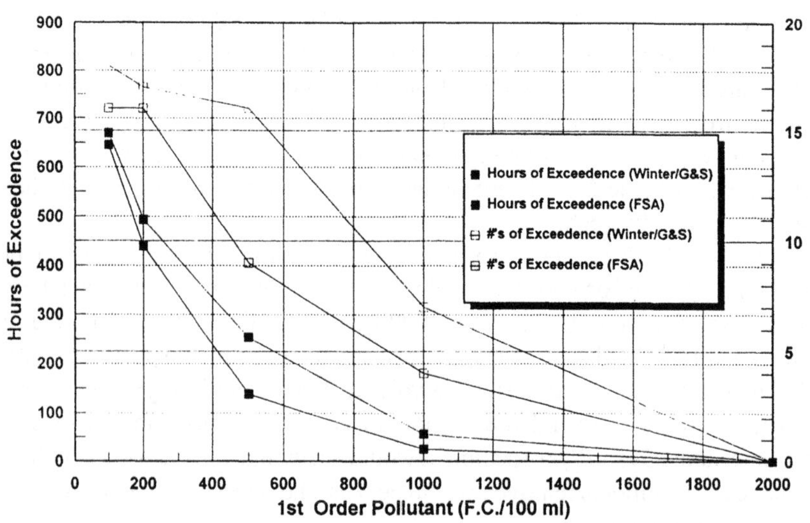

Figure 24.2 Comparison of QUALHYMO results: ultimate sub-watershed conditions - 1st order pollutant. Sources: Winter et al (1993), and FSA (1994b).

Table 24.3 Comparison of STORM modeling sediment removal efficiency results for the Ninth Line SWM facility.

Description of STORM Model	Drawdown Time (hrs)	Avg. Qr (m^3/s)	R (mm)	R_e (mm)	Sediment Removal Efficiency (F) (%)	Average Annual Trap Efficiency (E) (%)	
						[A] 20 μm Criteria	[B] 40 μm Criteria
Detail Design by Fred Schaeffer & Associates Ltd. (1994)							
1972 Rainfall	30	0.58	206	58.4	74.7	68.2	71.7
1977 - 1981 Rainfall	30	0.58	326	95.4	74.7	68.0	71.5
Environmental Study Report (ESR) Pre-Design by Fred Schaeffer & Associates Ltd. (1994)							
1977 - 1981 Rainfall	24	0.50	326	133.5	79.0	67.0	70.5

[A] Used in the 1994 FSA Environmental Study Report.
[B] Recommended in the Lisgar Region Water Quality Study.

It was calculated during the detail design that for a total detention time of 30 hours, the extended-detention storage of 61,900 m^3 (i.e. both series of design rainfall data) would provide approximately 75% removal of suspended particles. This value compared well with the FSA (1994) pre-design sediment removal efficiency of 79%. *It must be remembered that this efficiency was based on the removal of solids by gravitational settling only.* The overall sediment removal for the preferred design option was estimated in the FSA environmental study report to be approximately 7% higher than the gravitational settling efficiency.

Table 24.4 Assumed annual sediment loading rates.

Land Use	Annual Loading (kg/ha/yr)
Commercial	1100
Residential	900
Undeveloped	440

This extra efficiency was derived from Ministry Environment and Energy research performance curves used in the FSA environmental study report for other typical wet pond/wetland facilities. It is noted that these curves have been incorporated into the manual on stormwater management practices in the province of Ontario (MOEE, 1994). The total estimated efficiency of the preferred design option in the FSA report was about 79%.

Water Budget Analysis

Figure 24.3 illustrates the computed water level fluctuation from the forebay and wet pond water budget/analysis assuming both features are filled to capacity prior to the historical period investigated. It was found that minimal fluctuations of the depth of water in the forebay and wet pond compartments may be anticipated. The worst cases occurred when there was no significant rainfall for an entire month in the middle of the summer periods. That is, water levels were computed to be approximately 0.1 m lower than usual during these months.

An additional water budget simulation was completed for the same historical period assuming that both the forebay and wet pond are empty prior to the first rainfall event in May of each year examined. It was estimated that both the forebay and wet pond will be filled to capacity after the first few rainfall events of each annual period examined. These computed water budgets are therefore essentially the same as those illustrated in Figure 24.3.

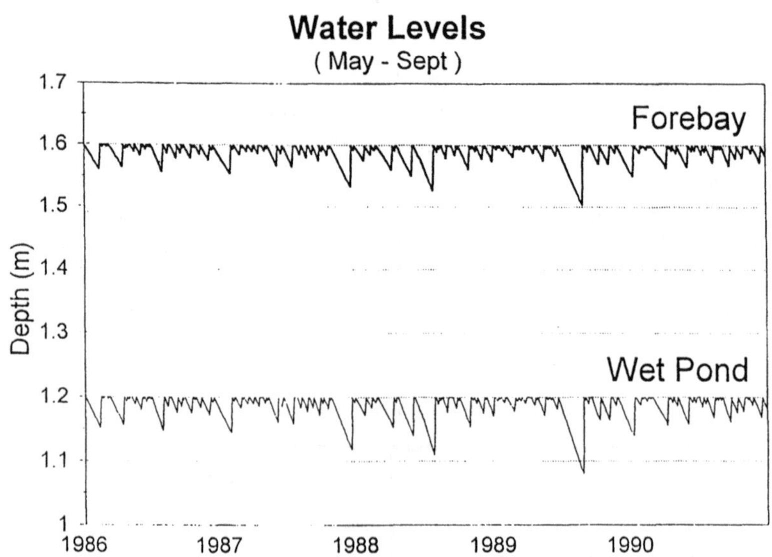

Figure 24.3 Results of water budget analysis.

BOSS-DAMBRK Modeling

Figure 24.4 presents the computed 13 mm storm event stage depletion for the SWM facility for the ultimate development conditions, if the retrofit extended-detention storage is full prior to the storm. The control berm is computed to be overtopped slightly. However, during a 13 mm storm event when the full retrofit active storage of ±61,000 m^3 is available, the SWM facility is computed to operate properly in accordance with the original (FSA, 1994) design parameters.

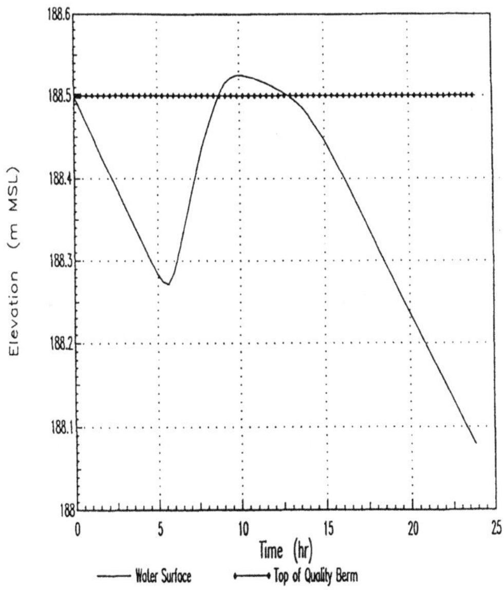

Figure 24.4 Ultimate Ninth Line SWM facility depletion stage, 13 mm storm event and initial retrofit storage at 188.5 m.

24.5.2 Pine Valley Stormwater Management Pond

Table 24.5 presents results of the STORM modeling analysis for the retrofit design of the Pine Valley stormwater management pond.

Table 24.5 Summary of STORM modeling results for the Pine Valley SWM pond.

Drawdown Time (hrs)	Qr (m^3/s)	R (mm)	R^o (mm)	F (%)	E (%)
48	0.09	426	190	85	47

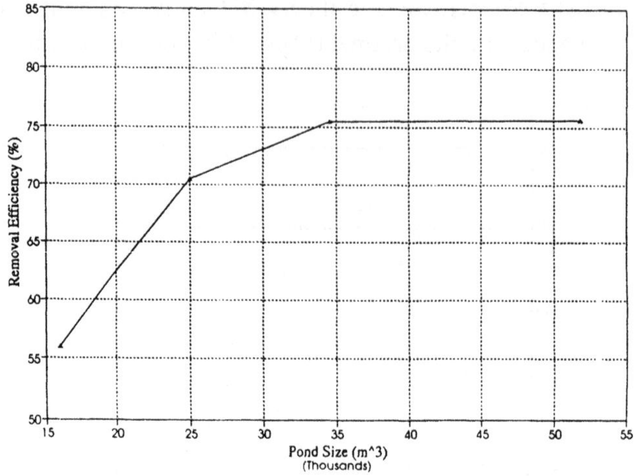

Figure 24.5 Sediment removal vs pond size. Pine Valley Pond, 48 hours detention time.

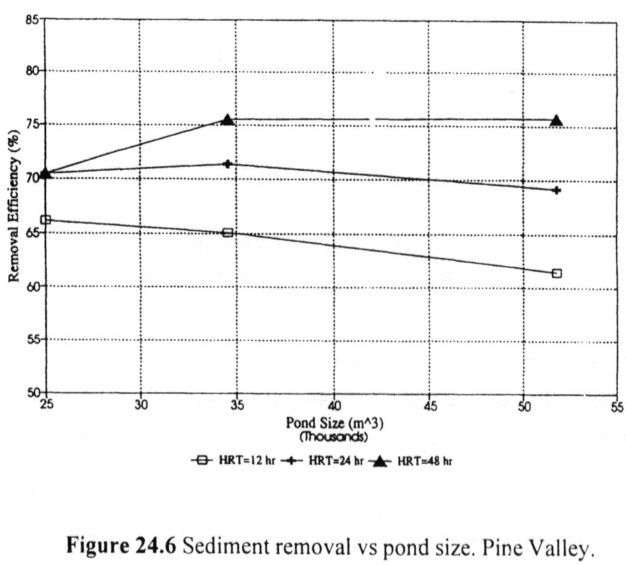

Figure 24.6 Sediment removal vs pond size. Pine Valley.

It was computed that for an average area of 14,300 m^2 and a detention time of 48 hours, the storage volume of 15,000 m^3 will provide 47% removal of suspended particles 20 μm or larger. The computed number of overflow events will be 22. Based on the same annual sediment loading rate data collected in the Ottawa area, that was also used for the Ninth Line SWM facility project, a removal efficiency of 47% is computed to lead to a yearly load of 165 tonnes - as compared to 231 tonnes computed for existing conditions and 321 tonnes computed for full development within the headwater basin.

Based on an average runoff volume coefficient for the Emery Creek watershed of 0.61, the 15,000 m^3 of extended-detention storage in the retrofitted pond is computed to correspond to a rainfall event of about 8 mm.

Using the STORM computer program, a sensitivity analysis was undertaken to investigate sediment removal efficiencies relative to various extended-detention volumes and a 48 hours drawdown time.

The computed efficiency of sediment removal versus the size of pond is illustrated on Figure 24.5. A significant improvement is computed to be achieved as the size increases from 16,000 m^3 to 25,000 m^3. Further improvements if the pond is increased are less significant.

Figure 24.6 illustrates that, for the detention time of 48 hours, the removal efficiency is computed to improve significantly compared to 24 hours.

Acknowledgements

The authors are grateful for the guidance of Allan R. Steedman of Fred Schaeffer & Associates Ltd. during these two projects. We wish to also extend our appreciation to Janice Teare and Bob Levesque with the City of Mississauga for their valuable comments during the environmental assessment study, retrofit design/construction, and culvert replacement analysis. Finally, we also acknowledge gratefully, during the Pine Valley stormwater management pond retrofit project, the assistance from Glen MacMillan with the Metropolitan Toronto and Region Conservation Authority.

References

Anacostia Restoration Team, 1992. Watershed Restoration Sourcebook: Collected Papers Presented at the Conference; "Restoring Our Home River, Water Quality and Habitat in the Anacostia", held November 6-7, 1991 in College Park M.D., Metro Washington Counc. Gov. D.C.

Andrew Brodie Associates, 1981 and 1983. Pine Valley Village Master Drainage Plan and Storm Water Management Studies; City of Vaughan.

Camp, J. R., 1946. Sedimentation and the Design of Settling Tanks, Trans. ASCE, Paper No. 2285, p.p. 895-958.

Chen, C.N. 1975. Design of Sediment Retention Basins, Trans. National Symposium on Urban Hydrology and Sediment Control, University of Kentucky, p.p. 285-290.

C. C. Tatham & Associates Ltd., 1996. Sixteen Mile Creek Culvert Improvements Study; City of Mississauga.

Fair, G.M. and J.C. Geyer, 1954. Water Supply and Waste-Water Disposal, John Wiley and Sons, New York.

Fred Schaeffer & Associates Ltd., 1993. Design Brief for Retrofitting the Pine Valley SWM Pond To Provide Run-off Quality Control for East Woodbridge Developments Ltd.; City of Vaughan.

Fred Schaeffer & Associates Ltd., 1994a. Environmental Study Report - Retrofitting of the Ninth Line Storm Water Management Facility; City of Mississauga.

Fred Schaeffer & Associates Ltd., 1994b. Design Brief - Retrofitting of the Ninth Line Storm Water Management Facility; City of Mississauga.

Herson, L., 1989. The State of the Anacostia River: 1988 Status Report; Metro. Washington Counc. Gov., D.C.

Ministry of Environment and Energy, 1994a. Stormwater Management Practices Planning and Design Manual.

Ministry of Environment and Energy, 1994b. Water Management: Policies, Guidelines, Provincial Water Quality Objectives of the Ministry of Environment and Energy.

Municipal Engineers Association, 1993. Class Environmental Assessment for Municipal Water and Wastewater Projects.

Winter Associates and Gore & Storrie, 1993. Lisgar Region Water Quality Study - Final Report; City of Mississauga.

Chapter 25

Techniques Used in an Urban Watershed Planning Study

K. R. Avery and Y. O. LaBombard

For planning level studies of urban flooding, it is important to obtain reasonable estimates of hydraulic performance and water quality improvements at a cost commensurate with the level of detail. The Buffalo District Corps of Engineers needed a planning-level study for a 1,400 acre (570 ha) flood-prone area in the City of Buffalo, New York. This planning study included: reviewing previous studies and models; and adjusting, refining and expanding an existing SWMM 3.0 model to SWMM 4.3. Improvements to the SWMM 4.3 model better reflected existing and proposed conditions. The proposed conditions included a diversion/storage plan. Hydraulic performance and water quality improvements resulting from the plan, were used to determine it's economic viability.

25.1 Introduction

This planning study for the Buffalo District Corps of Engineers evaluated the potential for Federal interest in constructing improvements to reduce flooding in a 1,400 acre (569 ha) area in the City of Buffalo, New York. The flood-prone area, located in North Buffalo, straddled twin trunk sewers on Hertel Avenue. Previous studies performed by the Buffalo Sewer Authority (BSA) identified inadequate capacity of the twin Hertel Avenue trunk sewers as the cause of

© *Advances in Modeling the Management of Stormwater Impacts - Vol. 5* W. James, Ed.
Pub. by CHI, Guelph, Canada 1997. ISBN 0-9697422-7-4. Fax: +519 767-2770

basement flooding. Of the alternatives evaluated, a diversion/storage plan called the Tunnel-Quarry Plan offered the greatest benefit at the lowest cost (Buffalo Sewer Authority (BSA), 1987). The plan would divert peak flows exceeding ten times the dry weather flow from a 1,000 acre (407 ha) area to the existing Amherst Quarry by means of a drop structure and tunnel (ECCO, Inc. and Hatch Associates Consultants, Inc., 1990).

The purpose of this planning study was to evaluate the technical adequacy of the plan. The planning study objectives were to:

- review an existing SWMM 3.0 model of the area,
- upgrade the existing model to SWMM 4.3,
- assess the adequacy of the plan for reducing basement flooding for storms having recurrence intervals up and including the BSA 5-year design storm,
- determine the impacts of the plan on flood protection of the neighborhood that currently drains to the quarry, and
- assess the water quality improvements resulting from the plan.

25.2 Tunnel-Quarry Plan Model Development

This section describes adjustment, refinement and expansion of the SWMM 4.3 model for the plan. This section also describes the potential water quality improvements provided by the plan.

25.2.1 Adjustments to SWMM 3.0 Model

In 1989, BSA developed a SWMM 3.0 model of the North Buffalo area (Calocerinos & Spina Engineers, 1987). The planning area for the SWMM 3.0 model covered approximately 5,000 acres (2,030 ha). The purpose of the SWMM 3.0 model was to determine combined sewer overflow volumes. In the model input junction ground elevations were set 100 feet (30.5 m) above ground elevations to prevent surcharge above the ground level. This planning study updated the SWMM 3.0 model to SWMM 4.3 and lowered junction ground elevations to real elevations.

25.2.2. Refinements to SWMM 4.3 Model

The first refinement made in this planning study changed the method of modeling the diversion points in the plan from the SWMM 3.0 method. When the BSA developed the plan, they established three diversion points to the tunnel from the surface sewer system. The BSA established a maximum allowable flow rate conveyed by the sewers downstream of the diversion points. This flow rate was equal to ten times the dry weather flow upstream of the diversion points. The BSA's intention was to construct diversion chambers with side overflow weirs and downstream automatic gates in the sewer. This would truncate the hydrograph peaks in the sewers downstream of the diversion points. The SWMM 3.0 model approximated the diversions as free outfalls. Improvement of the diversion concept included these changes:

- added the hydrographs, from an initial run of the SWMM 4.3 model for existing conditions, to the external dry weather flows at the junctions nearest the diversion points;
- developed an external truncated hydrograph for the sewer system, based on the maximum ten times dry weather flow criteria; and
- ran a model of only the sewer system downstream of the diversion points with the truncated hydrographs.

The second refinement added the effects of storage due to basement flooding. Basement flooding can occur whenever the HGL in a sewer main and building connection exceeds the elevation of the lowest pipe or plumbing fixture in the basement of a building. The SWMM 3.0 model did not include storage due to basement flooding. The BSA compiled a history of basement flooding complaints for use in this planning study. Development of storage due to basement flooding included:

- areal distribution of basement flooding using an ARCINFO GIS database of basement flooding complaints;
- conversion of junctions to storage junctions in the SWMM 4.3 model along the portion of the modeled sewer network corresponding to the basement flooding complaints as shown on Figure 25.1; and
- setting area of basements equal to the area of houses in the subcatchments.

Development of a storage-junction-area versus elevation relationship as shown on Figure 25.2 involved:

- set the top of the EXTRAN storage junctions to the real ground elevations of the storage junctions and set the inverts of the storage junctions to the corresponding lowest pipe inverts;
- begin basement flooding 5 feet (1.5 m) below the ground elevation of the storage junctions, (determined by assuming that basement elevations were: 8 feet (2.4 m) from floor to ceiling, 2 feet (.61 m) above ground at the building, and the ground at the building nearest the storage junctions was 1 foot (0.30 m) above ground at the downstream storage junctions);
- determine an average rise in subcatchment ground and basement elevations upstream in the subcatchment from the input length and slope data for the RUNOFF block.

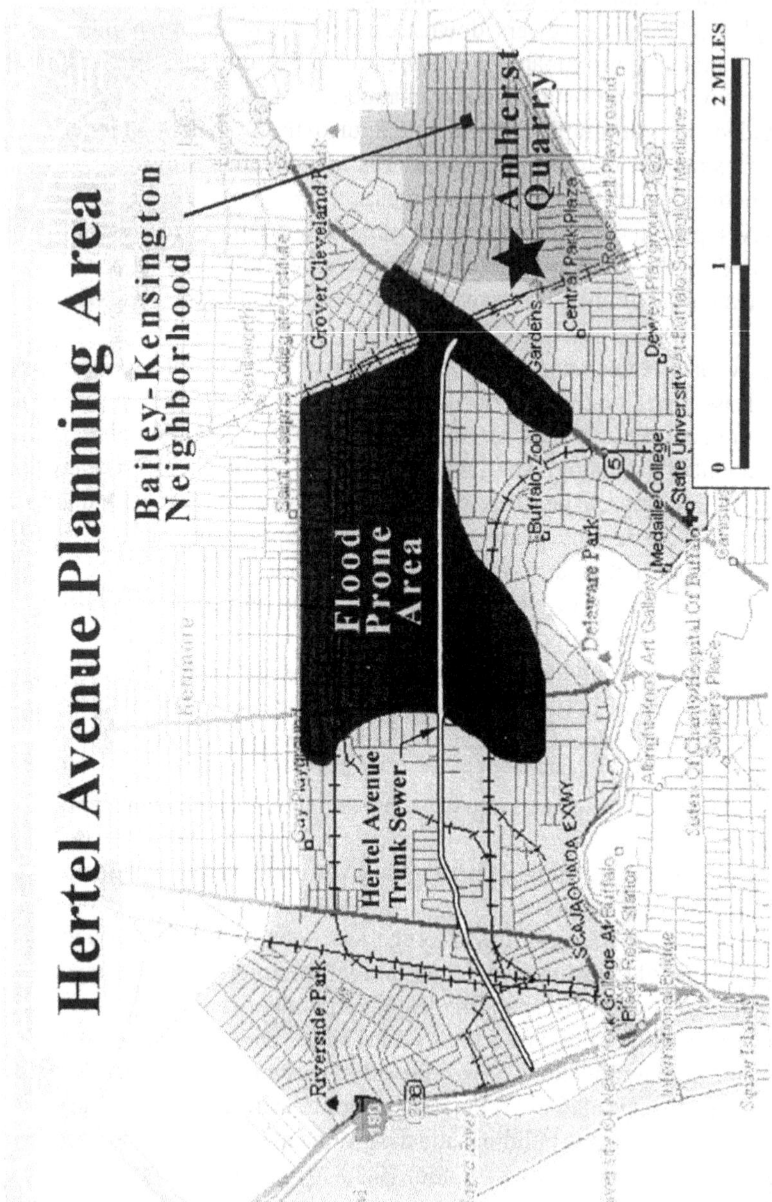

Figure 25.1 Map of basement flooding complaints.

The third refinement was the addition of connector pipes between the twin Hertel Avenue trunk sewers. Initial runs of the SWMM 4.3 model indicated differences in hydraulic grade lines (HGLs) in the twin trunk sewers up to 10 feet (3 m). Review of as-built drawings showed 24 and 36 inch (610 and 915 mm) connector pipes that joined the twin trunk sewers at six locations. The model refinement for the six connector pipes combined the junctions opposite each other into one junction. This refinement produced equal HGLs on the twin trunk sewers and results that were more consistent with the GIS database of basement flooding complaints.

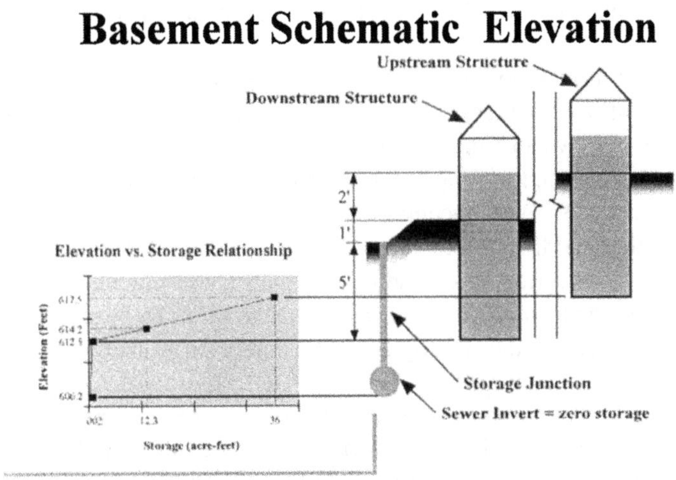

Figure 25.2 Basement storage schematic.

25.2.3 Expansion of SWMM 4.3 Model

The original SWMM 3.0 model did not include the 590 acre (240 ha) Bailey-Kensington neighborhood. This planning study expanded the model to include the Bailey-Kensington neighborhood. The neighborhood was divided into 25 subareas and run through the RUNOFF block using a five minute time step and 30 minute incremental rainfall. The SWMM 4.3 model assessed two rainfall events: the BSA 5 year, 6 hour storm and observed storm of September 17, 1976.

The SWMM 4.3 input comprised pipe diameters, inverts, and lengths obtained from sewer record maps and as-built drawings. The junctions included ground elevations obtained from 1" = 200' (1:2400) scale mapping and inverts from sewer record maps and as-built drawings. Storage junctions with cross sectional areas greater than 18 sq. ft (1.67 m^2), were input as constant cross sectional areas. The quarry was modeled from topographic mapping as a storage junction.

The model routed the hydrographs using EXTRAN, which can handle the effects of reverse flow and system storage. Reverse flow occurred when the water surface elevation in the large junction chamber on the Bailey Avenue trunk sewer exceeded elevation 645.02.

25.2.4 Water Quality Assessment

Previous studies and published documentation were used to assess the potential water quality improvements of the plan used. The six indicator constituents evaluated were: BOD5, COD, total nitrogen, total phosphorus, lead, and coliforms.

The contribution of pollutants from the watershed included components of stormwater runoff and sanitary sewage. Event mean concentrations for the six constituents in stormwater runoff were obtained from the Executive Summary of the National Urban Runoff Program (Environmental Protection Agency, Water Planning Division, 1983). Concentrations of the six constituents for domestic sewage were obtained from Metcalf & Eddy (1972). Development of the total mass loadings and diverted mass loadings for the six constituents consisted of:

- multiplying mean domestic sewage concentrations by constant dry weather flow, at the hydrograph time intervals, to calculate dry weather mass loading;
- multiplying the urban runoff event mean concentrations by the average flow in the conduit, over the time step and by the time interval, to calculate stormwater runoff mass loading;
- summing the dry weather and stormwater mass loadings to determine the total mass loading;
- subtracting ten times the dry weather flow from the total flow and dividing by the total flow to calculate a diversion ratio; and
- multiplying the total mass loading, at each time interval, by the diversion ratio to calculate the diverted mass loading.

25.3 Assessing Effects of Disconnecting Directly-Connected Impervious Areas

Previous studies and field observations confirmed that existing roof drains, with few exceptions, connected directly to the sewers. A previously-identified option to reduce basement flooding was to remove the direct roof-drain connections and install splash blocks outletting to lawn surfaces; a method to obtain a planning-level estimate of the hydraulic performance of this was developed.

25.3 Assessing Effects of Disconnecting DCIAs

Directly-connected impervious areas (DCIA) are: driveways, streets and piped connections from impervious surfaces to the sewer. Non-directly-connected impervious areas (non-DCIA) are impervious surfaces that have pervious surfaces between their boundaries and the sewers.

Evaluation of 35 representative street blocks, a total of 261 acres (106 ha), verified the percent imperviousness used in the SWMM 3.0 model. The evaluation also developed a relationship between percent impervious area and percent DCIA. Figure 25.3 is a plot of the relationship between percent impervious areas and percent DCIA used for existing and proposed conditions in the Hertel Avenue neighborhood.

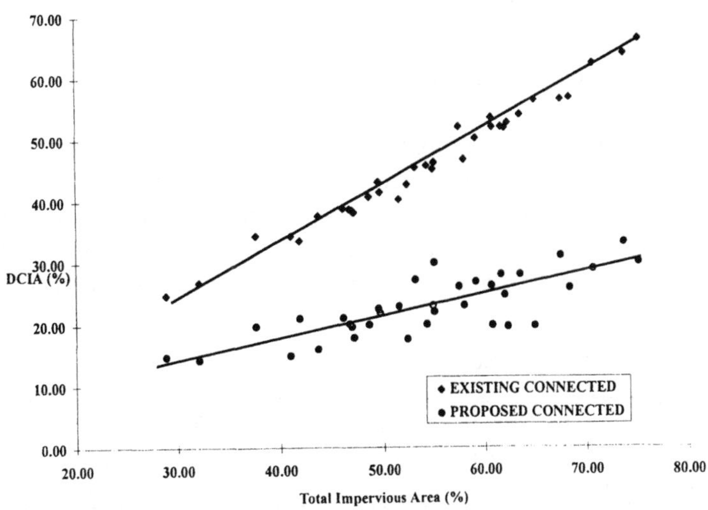

Figure 25.3 Relationship between percent impervious and percent DCIA.

Urban Hydrology for Small Watersheds describes the SCS method of modeling DCIA and non-DCIA (United States Department of Agriculture, Soil Conservation Service, Engineering Division, 1986). The SCS method provided a planning tool that could assess the potential for reducing peak runoff rates and volumes by disconnecting the roof drains. The SCS method used subcatchment RUNOFF block input data to compute: SCS curve number (CN), initial abstraction (a function of CN), and SCS lag time. Conversion of RUNOFF block input to SCS input included changing:

- RUNOFF area in acres to area in square miles,
- RUNOFF pervious and impervious area depression storage to initial abstraction,

- RUNOFF percent impervious to CN value, and
- RUNOFF area, width, slope, and roughness coefficients as input to determine the SCS lag time.

From Figure 25.3 DCIA with roof drains disconnected from the sewers can be estimated. Adjustment to CN for proposed conditions typically resulted in CN reductions of no more than 5.

SCS hydrographs for the BSA 5 year, 6 hour design storm were developed using HEC-1 Version 4.0 (United States Army Corps of Engineers, 1990). The HEC-1 hydrograph output for existing and proposed conditions was input to the EXTRAN block to assess hydraulic performance of this option.

25.4 Model Results

25.4.1 Hertel Avenue Flooding

Table 25.1 presents results for the existing conditions. The results show that basement flooding is computed to occur for events equal to and greater than the BSA 5 year design storm. The depth of basement flooding is computed to be approximately a quarter foot to a half foot (76 to 152 mm). The SWMM 4.3 model computed that basement flooding would occur only north of Hertel

Table 25.1 Hertel Avenue hydraulic gradeline results.

Street Name	Location	Ground Elevation	Minimum Basement Elevation	HGL's with Basements Modeled			
				BSA 5yr With Plan	BSA 5 yr. Existing	50% Freq. Existing	70% Freq. Existing
Parker	Trunk	615.9	610.9	597.7	615.9	602.4	601.3
	South	616.5	610.9	604.9	616.5	604.5	604.3
Stairn	Trunk	615.4	N607.8/ S610.4	596.4	603.6	600.0	598.9
	North	612.8	607.8	604.5	604.1	603.9	603.6
	North	615.5	607.8	605.3	605.1	604.8	604.6
Colvin	Trunk	604.7	N597.5 S599.7	591.4	597.8	592.5	591.6
	North	612.0	607.0	599.4	599.5	599.1	598.8
Delaware	Trunk	603.3	N594.0/ S598.3	589.1	594.5	588.9	588.0
	North	603.2	598.2	594.6	595.0	594.0	593.8
	North	605.4	600.4	594.5	598.1	594.0	593.7

Avenue; the GIS database recorded complaints south of Hertel Avenue. The GIS database of basement flooding complaints located approximately 85% in the area north of Hertel Avenue. Therefore, the results generated by the SWMM 4.3 model were generally representative of basement flooding.

Table 25.1 also presents the results computed for the plan. The results indicate that the plan removes basement flooding. This conclusion was in general agreement with previous studies that claimed that such reductions were possible. The reduction in HGL ranged between the existing conditions 50% exceedence frequency and the existing conditions 70% exceedence frequency. The 50% and 70% exceedence frequency events were taken from the depth-duration-frequency curve that includes all precipitation records, not just annual peaks.

25.4.2 Bailey-Kensington Flood Protection

The results indicate no flooding in the Bailey-Kensington neighborhood due to the plan. Increases in the water elevation in the Quarry due to the plan were approximately 8.0 and 10.5 feet (2.44 and 3.20 m) for the BSA 5 year design storm and the observed 1976 flood respectively. However, the water elevations in the Quarry with the plan were approximately 12.5 and 7.0 feet (3.81 and 2.13 m) below the zero damage point for the BSA 5 year design storm and the observed 1976 storm respectively. A check of the zero damage point, determined in previous investigations, confirmed the zero damage point to be elevation 655. Therefore, the plan would not compromise flood protection for the Bailey-Kensington neighborhood for storms equal to and less than the BSA 5 year design storm. It also was probable the plan would not compromise flood protection in Bailey-Kensington neighborhood for storms of greater recurrence interval and longer duration. The overflow to the Bailey Avenue trunk sewer, at a junction chamber, provides flood relief for these storms.

25.4.3 Water Quality

The water quality assessment was an estimate of the reduction in total mass loading at the drop structure, between existing conditions and conditions with the plan. Water diverted and stored in the Quarry would be treated later. The water bypassing the diversion structure would be conveyed in the Hertel Avenue trunk sewers to an overflow structure. The overflow structure discharges to the Niagara River. The six constituents modeled were: BOD5, COD, Total Nitrogen, Total Phosphorus, Coliforms and Lead.

Table 25.2 presents the reduction in pollutant loads along with the total loading and diverted loading for each constituent. Three of the six constituents: BOD5, Pb, and COD display greater than 50% reduction in pollutant loading. Total Nitrogen and Total Phosphorus had a greater than 40% reduction. The

Table 25.2 Reduction in pollutant loadings (BSA 5yr. design storm).

Pollutant	Total Loading (lb)	Diverted Loading (lb)	Percent Reduction
BOD5	1100	659	60
Total Phosphorus	55.7	27.1	49
Total Nitrogen	185	85.8	46
Lead	20.3	12.2	60
COD	6770	6750	55
Coliforms	2.63E+13	1.06E+13	40

reduction in pollutant loading for coliform was approximately 40%. Warm weather concentrations of coliform in urban runoff provided a worst-case scenario to determine the reduction.

25.5 Recommendations for Additional Modeling

The Hertel Avenue SWMM 4.3 model can still be improved. Recommendations for additional modeling are:

- detailed determination of the basement floor area and elevation of point of entry from the sewer lateral for the approximate 1,500 structures that experience basement flooding;
- a comparative simulation of the existing conditions and plan models for significant observed rainfall events that have occurred over a 20 year period of time; and
- further development of methodology using the SWMM 4.3 RUNOFF block to assess the impact of: disconnecting the roof drains, disconnecting downspouts and providing splash blocks to distribute the roof drainage onto lawns.

Acknowledgments

This planning study was performed under contract to the Buffalo District Corps of Engineers.

References

Buffalo Sewer Authority (1987). Discussion of Alternatives for Reducing Basement Flooding in the Hertel Avenue/North Buffalo Planning Area, Buffalo, NY.

Calocerinos & Spina Engineers, P.C. (1989). Evaluation of Anticipated Effects on Combined Sewer Overflows in the North Buffalo Area from Implementation of The Proposed North Buffalo Combined Sewage Storage Tunnel, Buffalo, NY.

ECCO, Incorporated and Hatch Associates Consultants, Incorporated (1990). Final Environmental Impact Statement for the Buffalo Sewer Authority's Proposed Hertel Avenue/North Buffalo Tunnel Project, Buffalo, NY.

United States Environmental Protection Agency, Water Planning Division (1983). National Urban Runoff Program, Volume 1 - Final Report, Washington, DC.

Metcalf & Eddy (1972). Wastewater Engineering: Collection, Treatment, and Disposal. McGraw-Hill Book Company, New York.

United States Army Corps of Engineers (1990). HEC-1 Flood Hydrograph Package User's Manual. Hydrologic Engineering Center, Davis, CA.

United States Department of Agriculture, Soil Conservation Service, Engineering Division (1986). Urban Hydrology for Small Watersheds (Technical Release 55).

Chapter 26

An Alternate Method of Finding the USDA Soil Conservation Service Runoff Curve Number for a Small Watershed

David A. Hamlet and Richard S. Huebner

Currently, one of the most widely-used methods for estimating losses for a watershed is the Soil Conservation Service (SCS) runoff curve number method (Chow et al, 1988; Viessman et al, 1989). This methods requires the identification of the hydrologic soil group, cover type, treatment, hydrologic condition, and antecedent runoff condition of a watershed (McCuen, 1982; USDA, 1986). These factors are used to select a representative curve number (CN) which, in turn, is used to estimate runoff for a given rainfall event.

Selecting a value for all of the watershed parameters listed above involves individual judgment and therefore introduces errors that might otherwise be avoided. The primary focus of this chapter was to test the hypothesis that a more accurate assessment of the curve number could be found by measuring multiple rainfall events using a recording flow meter and a rain gauge.

HEC-1 was used to analyze the hydrographs and determine an SCS runoff curve number and lag time for the watershed being studied. The work was conducted using a watershed with an area of 49 km^2 (19 mi^2) and an estimated time of concentration of 4.8 hours. Although a probable range of curve numbers (78 to 84) and lag times (3.4 to 3.8 hours) was identified, the variation in curve numbers among the five storms studied was 14. The variation in lag times was 4 hours.

© *Advances in Modeling the Management of Stormwater Impacts - Vol. 5* W. James, Ed.
Pub. by CHI, Guelph, Canada 1997. ISBN 0-9697422-7-4. Fax: +519 767-2770

26.1 Introduction

The Soil Conservation Service (SCS) runoff curve number method (USDA, 1985) was empirically developed in the 1950's from studies of small agricultural watersheds and revised several times over the last forty years. It was intended to be used to determine key hydrologic parameters like peak flow, time to peak, and volume of runoff for design events. This method requires the watershed to be characterized by the hydrologic soil group, cover type, treatment, hydrologic condition, and antecedent runoff condition. These factors are used to select a runoff curve number that is ultimately used to estimate losses and precipitation excess or runoff for a given rainfall event (USDA, 1986).

The process by which these factors are selected is a weakness of the method. For each factor listed above, the user is forced to use a discrete classification such as "Fair" or "Antecedent Moisture Condition III". In reality, watershed properties, such as cover type, tend to be variable and continuous, not discrete, values. Ultimately, the curve number is found by using a series of educated guesses rather than by direct measurement. Using the current method, errors in the value of the curve number are introduced by errors in judgment. These could be avoided by a direct measurement technique.

It is possible to derive the SCS runoff curve number and lag time for a watershed directly using the hydrograph of one or more measured rainfall events. Since it is relatively easy to accurately measure rainfall and flow, if a curve number can be found successfully in this manner, then SCS methods of calculating losses and generating a runoff hydrograph for a design storm would be improved. Measured values would replace estimation and educated guesses.

An accurate SCS runoff curve number and lag time value is important in the computation of design flows. The hydrograph developed from using the SCS runoff curve number and lag time is used to predict what the flow will be for a given rainfall event such as the 50-year storm or 100-yr storm. This data is then used for the mapping of flood plains, design of bridges and flood control structures, and for risk assessment. Using a SCS curve number that is too low will underestimate the risk of flooding and cause structures to be under-designed. Conversely, using a SCS curve number that is too high will overestimate the risk of flooding and cause structures to be over-designed.

26.2 Study Objective

The objective of this work was to investigate whether the SCS runoff curve number and lag time could be found directly by measuring the rainfall and flow at the outlet of a watershed in a reasonably short time frame rather than by using the traditional SCS runoff curve number method of finding the hydrologic soil group, cover type, treatment, hydrologic condition, and antecedent runoff condition.

26.3 Methodology

The first step was the selection of a watershed. The requirements of the watershed were that it be local, well defined, and predominantly of a single land use. These criteria were necessary in order to avoid placing multiple rain gauges and flow meters in the basin. The Fishing Creek watershed in Dauphin County, Pennsylvania met all these requirements. The drainage area is approximately 30.6 km^2 (11.8 mi^2). The watershed is located in a narrow valley between Blue Mountain and Second Mountain. Fishing Creek runs east to west the length of the valley and flows into the Susquehanna River 16.1 km (10.0 mi.) north of Harrisburg, Pennsylvania. The watershed comprises mostly forest.

Equally important to the selection of a watershed was the location of the equipment. It was desirable for the equipment to be located where flow in the stream was subcritical. In addition, the equipment had to be placed in a location that was accessible for maintenance yet secure from vandalism.

The location chosen was 3.2 km (2.0 mi.) upstream from the creek's outlet into the Susquehanna River and just downstream from a roadway structure. The stream in this location flows subcritically. The location was accessible since it was within 30.5 m (100 ft.) of the roadway and at a location where there was a parking area along the roadway. Vandalism was not a large concern since the equipment was not easily visible from the roadway, especially after the trees bloomed, and there were few houses in the area.

After a watershed and equipment location were selected, a rating curve for the stream was developed using an average Manning's n, a surveyed cross-section, and assuming uniform flow.

The next step was to collect rainfall and flow data for various rainfall events. An Isco Model 3210 Portable Recording Flow Meter and Isco Model 674 Rain Gauge were used. The system was powered by a rechargeable nickel-cadmium battery pack which was attached to the flow meter. The level of flow was measured by an ultrasonic level sensor. It was mounted approximately 1.2 m (4 ft.) above the water surface. Data was collected for nine measurable storms from April 8, 1995 through May 29, 1995 and entered into a spreadsheet. Flow data was reduced to average hourly flows in the spreadsheet.

Five of the nine storms had rainfall amounts of 6.4 mm (0.25 in.) or more. The data for these five events were used in input data files for the HEC-1 program (U.S. Army, 1987). HEC-1 is a flood hydrograph package that computes the surface rainfall-runoff response of a watershed. Of particular interest in this study was the program's ability to calibrate watershed parameters using data from observed events. In this case, the SCS curve number, SCS lag time, and initial abstraction were the optimized parameters.

Initially, two sets of HEC-1 analyses were made using data from the five rainfall events. The first set (Set 1) produced values of curve number (CN), lag time (t_p), and initial abstraction (IA) for each rainfall event using optimization algorithms found in HEC-1. The initial abstraction in the SCS Curve Number method of estimating losses is taken to be 0.2S where:

$$S = \frac{1000 - 10CN}{CN} \tag{26.1}$$

and CN is the SCS runoff curve number. As described in the following section, this set of runs resulted in values for these variables that were significantly different from storm to storm.

A second set of runs (Set 2) was made to calibrate the model and get more consistent results. HEC-1 was run for all five storms inserting selected values for SCS curve number and SCS lag time. The goal was to minimize errors and to match the peak flows.

An additional two sets of HEC-1 runs were made using values of CN, t_p, and IA optimized by HEC-1 for the 5/28/95 storm and the 4/12/95 storm (Sets 3 and 4, respectively). These storms were selected because they had the largest rainfall of the five recorded storms and, coincidentally, runoff curve numbers that were the closest to the curve number obtained when using the traditional SCS Method.

26.4 Results

26.4.1 Parameter Calibration

In general, the recession flows in the HEC-1 generated hydrographs based upon optimized parameters and the observed hydrograph matched well for each of the five recorded events. However, the peak flows did not match. In all cases, the peak flow generated by HEC-1 exceeded the observed peak flow rates. The time to peak flow rate was the same for the optimized and observed hydrographs; the optimized hydrographs peaked and receded at a faster rate than the observed hydrographs.

The optimized runoff parameters showed a large variation in the SCS curve number, SCS lag time, and initial abstraction among the five storms. A summary is displayed in Table 26.1. The SCS curve numbers ranged between 79 and 93, and the lag times ranged between 1.7 and 5.7 hours. The third optimized runoff parameter, initial abstraction (IA), varied from 4.3 mm (0.17 in.) to 10.4 mm

Table 26.1: HEC-1 optimization results summary.

Date of Storm	4/08/95	4/09/95	4/12/95	5/02/95	5/28/95
Rainfall [mm (in)]	7.6 (0.30)	9.1 (0.36)	32.0 (1.26)	13.2 (0.52)	17.3 (0.68)
SCS Curve Number	88	93	82	92	79
SCS Unitgraph Lag (hr)	1.7	5.7	3.8	5.7	3.4
Init. Abstraction [mm (in)]	4.3 (0.17)	4.3 (0.17)	10.4 (0.41)	5.8 (0.23)	7.4 (0.29)
Average Absolute Error [m^3/s (cfs)]	0.08 (3)	0.08 (3)	0.42 (15)	0.17 (6)	0.14 (5)
Average Percent Absolute Error (%)	3.6	3.5	6.9	4.6	5.0

(0.41 in.). Table 26.1 also displays an interesting relationship between curve number and rainfall amount. Higher values of the curve number are associated with smaller storm events (lower rainfall amounts).

The parameter optimization routines in HEC-1 calculate and report the average absolute error and average percent absolute error. The average absolute error is the average of the absolute value of the differences between the observed and computed hydrographs. The average percent absolute error is the average of the absolute value of percent difference between the computed and observed hydrograph ordinates.

The average absolute error ranged between 0.08 m^3/s (3 cfs) and 0.42 m^3/s (15 cfs). The average percent absolute error ranged between 3% and 7%. A second set of runs was made to try to determine a set of watershed parameters that would be consistent for all storms. This time, HEC-1 was run inserting values for SCS curve number and SCS lag time to calibrate the model, in essence, abandoning the HEC-1 optimization routines. A summary of results is listed in Table 26.2. Like the previous set of runs, the recession flows matched closely between the two hydrographs. In addition, the peak flows matched. Although the time of maximum flow rate was the same for the optimized and observed hydrographs, the optimized hydrographs peaked and receded at a faster rate than the observed hydrographs.

The second set of runs produced the same variation in SCS curve number and SCS lag time as the first set. The curve numbers ranged between 78 and 91. The lag time varied from 2.0 hours to 5.0 hours.

The average absolute error and average percentage absolute error were similar to the optimization runs. The average absolute errors ranged between 0.08 m^3/s (3 cfs) and 0.42 m^3/s (15 cfs). The average absolute percentage errors ranged between 3% and 7%.

Table 26.2: HEC-1 results summary - selected watershed parameters.

Date of Storm	4/08/95	4/09/95	4/12/95	5/02/95	5/28/95
Rainfall [mm (in)]	7.6 (0.30)	9.1 (0.36)	32.0 (1.26)	13.2 (0.52)	17.3 (0.68)
SCS Curve Number	91	91	78	88	84
SCS Unitgraph Lag (hr)	3.5	5.0	2.6	4.0	2.0
Init. Abstraction [mm (in)]	5.1 (0.20)	5.1 (0.20)	14.2 (0.56)	6.9 (0.27)	9.7 (0.38)
Average Absolute Error [m^3/s (cfs)]	0.08 (3)	0.11 (4)	0.42 (15)	0.14 (5)	0.17 (6)
Average Percent Absolute Error (%)	3.4	3.9	7.2	4.5	6.3

26.4.2. Comparing Generated Hydrograph to Other Storms

Comparison using the 5/28/94 optimized runoff parameters

A summary for the third set of HEC-1 runs is given in Table 26.3. The storm hydrographs for 5/28/95 matched closely. Since the optimized values for SCS curve number, SCS lag, and initial abstraction were used, the computed hydrograph was identical to the one produced in the first set of runs. Graphically, the 4/12/95 storm hydrographs also matched fairly well. The recession flows, total volume, and time of peak matched between the two hydrographs. As observed previously, the computed hydrograph peaked and receded at a faster rate than the observed hydrograph. The hydrographs for the other storms did not match. In all three cases, the computed hydrograph was considerably lower than the observed hydrograph.

The errors produced were higher than the previous two sets of runs. The average absolute error ranged between 0.25 m^3/s (9 cfs) and 0.42 m^3/s (15 cfs). The average absolute percent error ranged from 5% to 11%.

Comparison using the 4/12/94 optimized runoff parameters

A summary for the fourth set of HEC-1 runs is in Table 26.4. Like the third set of HEC-1 runs, the 4/12/95 storm hydrographs matched closely although this was expected since the optimized values were used. The recession flows and time of peak flow for the two hydrographs matched, however, the computed peak

Table 26.3 HEC-1 results summary using 5/28/95 storm optimized parameters for all storms.

Date of Storm	4/08/95	4/09/95	4/12/95	5/02/95	5/28/95
Rainfall [mm (in)]	7.6 (0.30)	9.1 (0.36)	32.0 (1.26)	13.2 (0.52)	17.3 (0.68)
SCS Curve Number	79	79	79	79	79
SCS Unitgraph Lag (hr.)	3.4	3.4	3.4	3.4	3.4
Init. Abstraction [mm (in)]	7.4 (0.29)	7.4 (0.29)	7.4 (0.29)	7.4 (0.29)	7.4 (0.29)
Average Absolute Error [m^3/s (cfs)]	0.25 (9)	0.28 (10)	0.42 (15)	0.28 (10)	0.14 (5)
Average Percent Absolute Error (%)	11.1	11.0	7.8	7.6	5.0

Table 26.4 HEC-1 results summary using 4/12/95 storm optimized parameters for all storms.

Date of Storm	4/08/95	4/09/95	4/12/95	5/02/95	5/28/95
Rainfall [mm (in)]	7.6 (0.30)	9.1 (0.36)	32.0 (1.26)	13.2 (0.52)	17.3 (0.68)
SCS Curve Number	82	82	82	82	82
SCS Unitgraph Lag (hr.)	3.8	3.8	3.8	3.8	3.8
Init. Abstraction [mm (in)]	10.4 (0.41)	10.4 (0.41)	10.4 (0.41)	10.4 (0.41)	10.4 (0.41)
Average Absolute Error [m^3/s (cfs)]	0.25 (9)	0.37 (13)	0.45 (16)	0.51 (18)	0.23 (8)
Average Percent Absolute Error (%)	11.1	14.6	7.1	14.8	8.8

flow exceeded the observed peak flow. Once again, the computed hydrograph peaked and receded at a faster rate than the observed hydrograph.

The hydrographs for the other storms did not match. In all four cases, the computed hydrograph was lower than the observed hydrograph. For the 4/08/95 and 4/09/95 storms, the flow receded over the entire period. This was due to the initial abstraction of 10.4 mm (0.41in) being greater than the total rainfall for those events.

The errors produced were slightly higher than for the previous set of runs. The average absolute error ranged between 0.25 m^3/s (9 cfs) and 0.51 m^3/s (18 cfs). The average absolute percent error ranged from 7% to 15%.

26.5 Discussion

From the results presented, it appears that the methodology did not yield a single SCS runoff curve number or SCS lag time but a wide range of values. In the case where the runoff parameters were optimized by HEC-1 for each storm (Set 1, Table 26.1), the SCS curve number varied by a factor of 14, and the SCS Unitgraph Lag varied by 4 hours. When HEC-1 was calibrated to match peak flows (Set 2, Table 26.2), the variation was 13 for the SCS curve number and 3 hours for the lag time. When using the HEC-1 output from the optimization run of the 5/28/95 and 4/12/95 storms in the HEC-1 input for the five storms, the computed and observed hydrographs did not match.

Despite the varied results, we cannot conclude that a SCS curve number and SCS lag time cannot be successfully found. If a factor or factors can be found that are responsible for the variation, then a correction can be applied which can improve the consistency between values.

The first factors to examine are those that are known to affect the SCS curve number using the traditional approach. These factors are hydrologic soil group, cover type, treatment, hydrologic condition, and antecedent runoff conditions. Each of these factors is briefly discussed.

Hydrologic soil group - The predominant soil types of the basin are the Calvin Series and the Dekalb Series (USDA, 1972). Both are classified as Type C soils and did not change over the course of the study.

Cover type - The cover type for the basin is classified as woods and did not change over the study period.

Treatment - Treatment pertains mainly to agricultural lands and therefore was not a factor for this study.

Hydrologic condition - Although the entire watershed is classified as being in "good" hydrologic condition, the hydrologic condition did change over the study period. At the start of the project in April 1994, the trees and shrubs had not bloomed. By the beginning of May, the trees had bloomed, and by mid to late May, the ground cover had thickened considerably.

The addition of foliage to the trees and shrubs would cause the hydrologic condition to improve and the SCS curve number to decrease. According to the SCS charts, the difference in SCS curve number between "fair" and "good" for a type C wooded soil is a decrease of 3. The optimized HEC-1 runs found that the 5/28/95 curve number was 3 less than the 4/12/95 which means that if the addition of foliage was enough to change the condition from "fair" to "good", then the difference between the 4/12/95 and 5/28/95 storm is accounted for. Although the hydrologic condition probably did not change from "fair" to "good", it definitely improved, and therefore, a difference in SCS curve numbers of 1 or 2 may be reasonable.

The hydrologic condition does not account for the variance between the other storms. The 4/08/95 and 4/09/95 storms occurred in the same week as the 4/12/95, but the curve numbers differ with the 4/12/95 storm by 6 and 11 respectively. The 5/02/95 storm occurred three weeks after the 4/12/95 storm but has a higher SCS curve number.

Antecedent runoff condition - The antecedent runoff condition can change the curve number of a watershed substantially. According to SCS charts, a watershed with a SCS curve number of 80 under normal conditions (Condition II) will have a curve number of 94 under Condition III. A watershed with a curve number of 90 under normal conditions will have a curve number of 78 under Condition I. Consequently, the antecedent runoff condition could account for the difference in curve numbers.

To verify the antecedent runoff condition, rainfall and temperature data is needed. Since the rain gauge was not activated for the entire study period, and temperature data was not recorded, data was obtained from the National Weather Service. The data although recorded 24.1 km (15.0 mi.) away from the study area should be accurate enough to indicate the soil condition for each storm. The total rainfall and average temperature for the five-day period before every storm are shown in Table 26.5.

Table 26.5 Temperature and rainfall data for Harrisburg.

Storm	Average Temperature (Previous 5 Days)	Normal Average Temperature (Previous 5 Days)	Rainfall (Previous 5 days)
4/08/95	7°C (45°F)	9°C (48°F)	0.0 mm (0.00 in)
4/09/95	8°C (47°F)	9°C (48°F)	6.1 mm (0.24 in)
4/12/95	11°C (51°F)	10°C (50°F)	23.9 mm (0.94 in)
5/02/95	12°C (53°F)	14°C (57°F)	11.9 mm (0.47 in)
5/28/95	21°C (69°F)	18°C (65°F)	2.3 mm (0.09 in)

The traditional SCS runoff curve method classifies conditions with rainfall less than 12.7 mm (0.5 in.) in the previous five days as Condition I. Rainfall of 12.7 mm (0.5 in.) to 27.9 mm (1.1 in.) is Condition II, and rainfall exceeding 27.9 mm (1.1 in.) or less than 27.9 mm (1.1 in.) with low temperatures is Condition

III. Following this criteria, the 4/08/95, 4/08/95, and 5/28/95 storms are Condition I, and the 4/12/95 and 5/02/95 storms are Condition II.

As mentioned in the introduction, the largest downfall with the SCS method is having to classify each factor into one category or another. For example, the soil at the time of the 4/12/95 storm although officially classified as Condition II was probably somewhere between Condition II and Condition III because there were two rainstorms in the previous five days. For this reason, the SCS conditions are backed away from and the following statements are made:

- the soil conditions on 4/08/95 were generally dry making the SCS curve lower than normal;
- the soil conditions on 4/09/95 were also dry, so the SCS curve number is low;
- the soil conditions on 4/12/95 were generally average bordering on wet (meaning the SCS curve number is slightly higher than normal);
- the soil conditions on 5/02/95 were average, therefore, the SCS curve number does not need to be adjusted; and
- the soil conditions on 5/28/95 were normal to dry lowering the SCS curve number.

The antecedent moisture condition does not entirely account for the large variation in SCS curve numbers. The 4/08/95 and 4/09/95 storms had a high SCS curve number to begin with, and are adjusted even higher. The 4/12/95 had a low curve number and is adjusted lower. The 5/02/95 storm had a high curve number but would not be adjusted.

Another possible explanation in the variation of SCS curve numbers is the limitations of the SCS curve number method. TR-55 (USDA, 1972) states that the SCS curve number method is less accurate for runoffs under 12.7 mm (0.5 in.). Runoff is related to the amount of rainfall and initial abstraction by the following equation:

$$Q = \frac{(P - I_a)^2}{(P - I_a) + 5I_a} \tag{26.2}$$

where:

Q = the runoff in millimeters
P = the rainfall in millimeters
I_a = the initial abstraction in millimeters

This results in runoff values for the sequence of five storms of 0.5 mm (0.02 in.), 0.8 mm (0.03 in.), 6.4 mm (0.25 in.), 1.5 mm (0.06 in.), and 2.0 mm (0.08 in.) respectively. All are below the 12.7 mm (0.5 in.) value. Since runoff decreases with rainfall, it can be said that the runoff values are less accurate for smaller storms. Therefore, the accuracy of the 4/08/95 and 4/09/95 curve numbers is much less than the others, and the curve number found from the 4/12/95 is the most accurate.

This suggests that the SCS curve number for Fishing Creek is probably in the low eighties, although the smaller storms' data suggests that it is higher. The SCS lag time is probably between 3.5 and 4 hours. As a comparison, using the traditional SCS runoff curve method the curve number is 70 and the lag time is 2.9 hours. This indicates that the results of the traditional method are low, but lacking consistent results, a corrected value cannot be recommended.

Although the main discussion centers around the variation in the SCS curve numbers for the five storms, there are other results that should be explained. The first is the failure of the HEC-1 optimization procedure to match peak flows. The optimized peak flow always exceeded the observed peak flow. This is because HEC-1's optimization procedure minimizes the difference in total flow between the computed and observed hydrographs. Since the SCS hydrograph peaked and receded at a faster rate and in a shorter time period than the observed hydrograph, to minimize the difference in total flow, the computed hydrograph had a greater maximum flow rate than the observed.

The observed hydrographs and computed hydrographs never had the same shape. The computed hydrograph always peaked and receded at a faster rate than the observed hydrograph. This is due to the fact that the method selected in the optimization runs used the SCS dimensionless unit hydrograph. The SCS dimensionless unit hydrograph is based on the equations:

$$TPEAK = 0.5t + TLAG \tag{26.3}$$

$$QPEAK = \frac{44269 \times AREA}{TPEAK} \tag{26.4}$$

where:

$TPEAK$	=	the time to the peak flow in hours
t	=	the duration of excess in hours
$TLAG$	=	0.6 x the time of concentration in hours
$QPEAK$	=	the peak flow in m^3/s
$AREA$	=	the basin area in km^2

The SCS dimensionless hydrograph may not have had the same shape as the observed hydrographs, therefore, regardless of which value for SCS curve number, SCS lag, and initial abstraction are input, the computed and observed hydrographs will not match.

The final item of discussion is the error found for the 4/08/95 and 4/09/95 storms in the third and fourth set of HEC-1 runs. The errors for the third and fourth runs match exactly for the two storms. The reason for this is that the SCS curve numbers used were so low that for a small rainfall, there was no runoff. Consequently, the hydrograph only consisted of recession flow.

Heuristically calibrating the model was as successful as using HEC-1's optimization routines in determining a single curve number and lag time for the watershed. In some cases, varying the parameters may improve the overall solution. In this study, the results did not change substantially between the two methods. Still, it is recommended that the optimization results be compared to a heuristic calibration since there is an opportunity to improve the values of the parameters with respect to some other optimization goal, such as matching peak flows.

26.6 Conclusions and Recommendations

Overall, the process of measuring rainfall and stream flow for multiple storms in the Fishing Creek basin and using the data in HEC-1 was unsuccessful in determining a single SCS runoff curve number and lag time. The SCS curve numbers and SCS lag times found by parameter calibration varied considerably from storm to storm. The hydrograph produced from running HEC-1 using the optimized values did not match the observed hydrographs. Although some of the discrepancies could be rationalized, only a broad SCS curve number range and lag time range resulted from the analysis. Nevertheless, based upon the experience of measuring flow and rainfall over a relatively short period using readily available, mobile equipment, the authors were able to identify a likely range of curve numbers for the watershed ($78 \leq CN \leq 84$) and for the basin's lag time (3.4 hr $\leq t_p \leq 3.8$ hr).

Even so, some of the factors used in the SCS runoff curve method such as the hydrologic condition and antecedent runoff condition must be accounted for even when directly measuring the rainfall and stream flow.

Since the SCS curve number method is most accurate for runoffs of 12.7 mm (0.5 in.) or more, large rainfall events are needed to assure accuracy especially in non-developed areas where the runoff is generally low. It is recommended that data for three rainfall events each exceeding 12.7 mm (0.5 in.) and occurring under antecedent runoff soil Condition II be collected to achieve a consistent SCS runoff curve number and lag time. Although the data could take weeks or months to collect, it is important to record data under these conditions. If these conditions cannot be accommodated, adjustments will have to be made which, in essence, would make this direct method no better than the traditional method of determining curve number and lag time.

References

Chow, Ven Te, Maidment, David R. and Mays, Larry W., Applied Hydrology, McGraw-Hill, New York, 1988.

McCuen, Richard H., A Guide to Hydrologic Analysis Using SCS Methods, Prentice-Hall, INC, Englewood Cliffs, New Jersey, 1982.

U.S. Army Corps of Engineers, Hydrologic Engineering Center, HEC-1, Flood Hydrograph Package User's Manual, 1981 (revised 1987).

U.S. Department of Agriculture, Soil Conservation Service, National Engineering Handbook: Section 4 - Hydrology, Washington, D.C., 1985.

U.S. Department of Agriculture, Soil Conservation Service, Urban Hydrology for Small Watersheds, Tech. Release 55, Washington, D.C., 1975 (updated 1986).

U.S. Department of Agriculture, Soil Conservation Service, Soil Survey for Dauphin County Pennsylvania, Washington, D.C., 1972.

Viessman, Warren, Lewis, Gary L., and Knapp, John W., Introduction to Hydrology, third edition, Harper and Row Publishers, Inc., New York, 1989.

Chapter 27

Role of Municipal Stormwater Management Guidelines: the Markham Experience

Paul Wisner and Les Arishenkoff

During the early 1990's, government agencies of the Province of Ontario published several guidelines dealing with best management practices (BMPs)for runoff *quality control* and other aspects of stormwater management. Prior to these publications, the Town of Markham, Ontario and, subsequently, many other municipalities had prepared specific guidelines, dealing mainly with runoff *quantity control*. There is now a need to update these older municipal guidelines and to harmonize them with recent provincial documents. This chapter gives an overview of this update by the Town of Markham, a municipality in the Greater Toronto Area with considerable experience in SWM implementation.

27.1 Needs

27.1.1 The Era of Quantity Control

In 1978, the Town of Markham was required to implement a very stringent reduction of peak flows on development in order to minimize downstream effects. Based on this experience, the Town of Markham developed for the first time, mandatory requirements for all new developments:

© *Advances in Modeling the Management of Stormwater Impacts - Vol. 5* W. James, Ed. Pub. by CHI, Guelph, Canada 1997. ISBN 0-9697422-7-4. Fax: +519 767-2770

- control of peak post-development runoff;
- utilization of storage for runoff control;
- disconnection of roof drainage; and
- use of hydrograph models.

It was only after several years of the application of these principles that an extensive municipal manual was made available by the Town of Markham to consultants undertaking work in Markham (Town of Markham, 1983). This manual described the first projects in which SWM concepts were applied in Markham. Of more importance was the introduction of the need for master drainage plans (MDPs) and recommended recreational use of SWM facilities. To facilitate project review, this manual had simplified graphs based on the OTTHYMO model, which give peak flows for small areas. Two constructed urban lakes (Mount Joy Pond, Toogood Pond) were described as examples which embody the principles of the manual.

Within two years, the Ontario Ministry of Environment and Energy (MOEE) recognized runoff quantity control principles in a manual (Brodie and Wisner, 1985), the preparation of which was co-chaired by the Town of Markham and an inter-ministerial committee. This manual was subsequently applied by other municipalities.

The Town of Markham also pursued efforts to improve local implementation of SWM techniques. In 1989, it reviewed its initial design storms based on a comparison with multi-event simulations. In the same year, the Town prepared a special on-site detention (OSD) manual which recommended:

- the use of OSDs only when their need is demonstrated in the MDP (the intent was to limit the distributed storage); and
- graphs for storage design based on modeling with OTTHYMO (Town of Markham, 1989a, 1989b);

27.1.2 The Era of Quality Control/Ecosystem Approach

In 1989, Markham sponsored an international symposium to review practical SWM experience. Over 200 participants were given the opportunity to present comments on a proposed approach for interdisciplinary MDPs including ecological considerations. Erosion control based on shear stress was also discussed at the symposium. More importantly, municipal staff and consultants had the opportunity to discuss the experiences of other Ontario municipalities, Europe, B.C. and Quebec (Town of Markham, 1989a, 1989b).

Updating the Town of Markham 1983 manual was deferred because of a succession of several provincial initiatives by the MOEE and the Ministry of Natural Resources (MNR), including:

- the MOEE/MNR (1991) interim volume runoff quality criteria based on storage of a given precipitation (similar to criteria in Washington COG);

- the 1992 MOEE SWM manual for stormwater quality best management proposing a more sophisticated ecosystem approach;
- the 1992 MOEE/MNR Subwatershed Planning Guidelines, which required runoff control studies at a larger scale than the MDPs.

Other documents were developed by the Metropolitan Toronto and Region Conservation Authority (MTRCA). A comprehensive watershed study was conducted for the Rouge River, one of the main watersheds in the Town of Markham, and the creation of the Rouge River Park was approved by the Government of Ontario.

Therefore, the Town of Markham felt the need for local guidelines to facilitate the implementation of the new provincial initiatives and account for other reports including the MNR Fish Habitat Protection Guidelines, and the MNR Natural Channel Systems Approach.

The latter phases of updating the Town of Markham manual coincided with the early preparation of MOEE's new 1994 Manual. The MOEE had numerous meetings with Markham officials to discuss the municipal position regarding the implementation of the 1994 MOEE Stormwater Management Practices Planning and Design Manual.

27.2 Organization of the Town of Markham Stormwater Management Guidelines

While a detailed description of the 1995 Markham Guidelines (Town of Markham, 1995) is beyond the scope of this chapter, its contents can be briefly summarized. An editorial effort was made so that each conceptual aspect is reviewed in a one page presentation.

- Part 1 *General Concepts* (An introduction for non-specialists is presented in eight pages): evolution of SWM concepts, ecosystem approach, subwatershed plans, types of BMP, public participation, environmental assessment, recreational aspects, safety, maintenance and need for verification of performance, and introduction to the environmental master drainage plan.
- Part 2 *Planning Aspects* (General information on SWMP/environmental BMP and selection of BMP are presented in 11 pages.) The main sections are: review of Rouge River comprehensive basin and related studies, Markham natural features, valley and corridor protection, modeling, thermal impact. A two-stage selection of BMP and economic efficiency is also included.
- Part 3 *Design Aspects* (General information on six types of municipal BMP and recommendations for SWM plans are presented in 14 pages). For SWM Plans, it reviews the dual-drainage

concept, on-site detention, bio-engineering, natural channel design, post-operational monitoring, erosion and sediment control for construction sites and content of SWM reports.

The Town of Markham Stormwater Management Guidelines also have nine appendices which deal with specific aspects, such as:

- examples of development of a short list of BMPs (in the Rouge River Watershed);
- highlights of the Rouge River Fisheries management plan;
- highlights of the Don watershed strategy; and
- fish habitat protection and stormwater management.

For consistency, some of the appendices review methods from previous Markham manuals:

- application of the rational method (RM) in Markham (the RM limitations, correction factors): the RM is still used in subdivision sewer design!
- on-site detention design graphs (needed for consistency with previous designs); and
- Markham design storms and comparison with multi-event simulations.

27.3 Review and Comparison with Provincial Guidelines

After their development, the recent Markham SWM Guidelines were submitted for review and comments to MNR, MOEE and MTRCA and several consultants, including those who prepared the provincial manuals.

The final edition accounted for comments during this review, including updates in previous agency views on some BMPs. As an example, a matrix procedure leading to an infiltration BMP was played down since MNR and MOEE expressed concerns about some of these solutions.

Some aspects of the Town of Markham Stormwater Management Guidelines are unique when compared to other manuals and include the following:

1. the synthesis of quantity, quality control and drainage design;
2. a specific Town of Markham approach to area-wide planning, called the Environmental MDP, to be used in situations where a Subwatershed Plan is not available;
3. the emphasis on economic efficiency; and
4. a careful approach to new concepts which puts greater emphasis on operational experience.

Before 1990, many new SWM concepts such as dual drainage, inlet control, etc. were mostly municipal initiatives tested by local experience and implemented gradually. The introduction of new provincial requirements between 1991 and 1993 led to concerns for municipalities such as:

- untested new recommendations;
- changes of agency emphasis (e.g. infiltration BMP) over a short period of time;
- use of "rules of thumb" instead of modeling (e.g. 25 mm storage);
- limited monitoring and maintenance experience with some BMPs;
- some attempts to propose approval of ponds without an in-depth analysis of their real need;
- proliferation of wet ponds or wetlands which are much more difficult to maintain than quantity control storage; and
- municipal review based on many documents of which some were developed by various agencies with limited municipal involvement.

The Town of Markham approach to new solutions was to combine a few experimental BMPs with the more conservative municipal SWM guidelines which reflect local practices and experience.

27.4 Discussion of Some SWM Issues

Issue 1: Phases in Planning

In the Town of Markham Stormwater Management Guidelines, the Environmental MDP (EMDP) is considered to be a compulsory step prior to the stormwater management plan (SWMP). The EMDP is an interdisciplinary, sophisticated planning study, which includes BMP selection with consideration of water quality, habitat protection, base or low flow maintenance, and other pertinent concerns. The SWMP which is the next step, uses a one-step, more straightforward conceptual design interfaced with subdivision design.

It is obvious that different types of consultants can be involved in these two steps. The SWMP does not require the same level of interdisciplinary expertise. This approach avoids duplication of environmental analysis and inventories and optimizes location and selection of BMP. Compared to other planning strategies, it also avoids review of SWMP in two phases.

The EMDP is, however, considered a temporary solution, and adequate funding should be ensured for subwatershed plans. The Markham Stormwater Management Guidelines also recommend that reviewing agencies participate in the EMDP steering committee.

Issue 2: SWM practices and regulations

SWM practices and regulations are, and will be for some time, in a state of flux. An example is the control for erosion. It is now recognized that many OSDs built for zero runoff increase, under some conditions, do not reduce downstream flooding.

Pond retrofit is an attempt to correct the situation by making better use of quantity ponds which are not considered useful anymore. We have to avoid the future retrofit of 'retrofitted' ponds.

A better understanding of the role of SWM manuals as general information is therefore necessary. Although the developers of recent provincial manuals made this aspect very clear, some consultants and reviewers interpret parts of the manuals quite literally, and therefore minimize creativity. The Markham Guidelines took a different approach.

The Markham Guidelines are a "living" document and subject to change in order to reflect the latest in stormwater management technology and it is felt this should also be valid for provincial documents.

Issue 3: Selection of BMP with Consideration of Economic Efficiency and Operation

The Markham guidelines recommend a two-stage BMP selection approach:

- *Stage I screening in sensitive projects*: the selection of a BMP from a long list is based on an ecosystem step-by-step approach, introduced in the MOEE BMP Manual (1992), and illustrated by an example;
- *Stage II selection*: a set of matrices is proposed. Such factors as capital cost, O & M cost, maintenance cost, cost per kg pollutant removed, are recognized as well as public accessibility, recreational benefits, etc.

The application of the law of diminishing returns and the economic impacts of some provincial guidelines are briefly discussed. As an example, for an increase from 60% TSS removal (Level 3) to 70% removal (Level 2) requires a wet pond volume to increase by 50% (residential areas). From 70% TSS removal to 80% TSS removal (Level 1), there is an additional increase of 75%.

More work is still necessary to refine and confirm such criteria.

Issue 4: Importance to Detailed Design Aspects

It was considered that the review of design analysis of BMPs is an essential part of municipal responsibilities. Examples are:

- Design considerations are given in concise point form for each BMP. Some additional recent information is used, compared to the MOEE 1994 manual (e.g. offline biofilters).
- A special section of the Town of Markham guidelines provides for SWM plans, detailed prescription of protection against overtopping, inlet and outlet features, including e.g. trash, back design, etc.
- The guidelines provide detailed indications of the hydraulic design of flow splitters for offline BMPs.

Issue 5: Modeling

Despite advances in modeling, there is still a trend towards simple, empirical "rules of thumb". As an example, storage of a given rainfall amount is still used. Since the first Markham manuals were issued in 1978, Markham has promoted the use of models rather than empirical design.

The 1995 Markham guidelines briefly reviewed aspects of planning and design modeling. Examples of modeling concepts presented in the Town Guidelines include:

- A caveat is presented against some uses of m^3/ha rules for quality control, including the use of MOEE manual graphs mainly beyond their 40 m^3/ha active storage criteria. These rules do not provide information on the effect of such factors as ratio of active/passive storage, rate of captured discharge, etc.;
- It is recommended that cumulative shear stress be used for erosion control;
- There is a need to consider pond monitoring results (examples are given);
- Only multi-event modeling can give an insight into quality control pond operation;
- Multi-event modeling with design storms is not encouraged;.
- Simplified graphs can be used only for OSDs.

The Town of Markham Stormwater Management Guidelines do not limit the use of any well-tested models, even if not used previously in Markham. For consistency, it recommends however, that their performance be compared to previous models. Such comparisons are available for EPA-SWMM and OTTHYMO.

The Town of Markham Guidelines also indicate, as an example, that well-documented presentation of modeling assumptions and results is essential for project approval.

Issue 6: Potential Municipal Liabilities

If the public has complaints, they are addressed to the municipality. The municipality has to consider legal aspects, and the fact that post-construction rehabilitation is expensive and sometimes impossible. Municipal liabilities, such as created bio-engineering, solutions for erosion protection, which are not *equally* flood-resistant, should be minimized.

For these reasons, not all solutions discussed in other manuals are acceptable. The Town of Markham does not recommend reduced slopes for lot grading which would be contrary to current bylaws. Neither ponding areas in backyards nor sump pumping to soak-away pits is encouraged. The Town of Markham considered that concerns for clogging and the nature of soils limit the use of perforated pipes and conveyance BMPs.

The Town of Markham agrees with the recent MNR caveats on an infiltration BMP and the guidelines discourage, in general, clogging-prone solutions, etc. While the Markham guidelines recognize the benefits for on-lot source control, use of these BMPs is only required where the municipality has access for maintenance and improvements.

Therefore, the Guidelines state that previous experience with performance and maintenance is essential in the BMP selection process. Adequate information on these aspects is also presented. Use of BMPs with short longevity is discouraged.

Finally, distributed storage (including inlet treatment) compound maintenance and liabilities. The Guidelines encourage monetary contributions to centralized facilities from small sites.

27.5 Final Remarks

It is recognized that different municipalities or agencies may have other views on some of the issues discussed, and will develop guidelines or apply the provincial manuals in other ways specific to their local conditions. The most valid part of the development of the 1995 Town of Markham Stormwater Management Guidelines was that all aspects of SWM were critically reviewed not only by the Town of Markham Engineering Department staff, but also by staff from the Planning, Parks Maintenance and Construction, and Parks & Open Space Planning Departments. This was also valid for the numerous comments received from agencies and consultants.

One of the questions raised in this process was whether these municipal guidelines duplicate provincial documents. Although the review by agencies showed that there are no contradictions between the Markham Guidelines and the provincial or MTRCA requirements, it is felt that the screening of some aspects, and the possibility to include elements otherwise dispersed in various documents, was a very useful exercise and will facilitate the future dialogue among municipalities, consultants, developers and agencies.

Acknowledgements

The active participation and support of Markham senior staff, mainly Mr. Dalo Keliar, P.Eng., Commissioner of Community Services and Mr. Alan Brown, Director of Engineering, were essential in the development of the Town of Markham Stormwater Management Guidelines. Comments from specialists at the MOEE, MNR Maple District, MTRCA and consulting firms, such as Marshall Macklin Monaghan, and Cumming Cockburn were very useful and their assistance is gratefully acknowledged.

References

Brodie, A. and P.E. Wisner, 1985. for Ontario Ministry of Environment, Urban Drainage Design Guidelines, July 1985, 200pp.

Metropolitan Toronto and Region Conservation Authority, 1990. A Comprehensive Basin Management Strategy for the Rouge River Watershed, January 1990, 102 pp.

Ministry of Natural Resources, 1994. Natural Channel Systems: An Approach to Management and Design, Development Draft, 103 pp.

Ministry of Environment and Energy and Ontario Ministry of Natural Resources, 1993. Subwatershed Planning, June 1993, 38 pp.

Ministry of Environment and Energy, 1992. Stormwater Quality Best Management Practices, June 1992, 177 pp.

Town of Markham, 1995. Storm Water Management Guidelines for the Town of Markham, (by Paul Wisner and Associates Inc.), Markham, January 1995, 61 pp.

Town of Markham, 1989a. A Design of On-Site Detention (OSD) and the Simplified Markham OSD Computation Method, (by Paul Wisner and Associates Inc.), Markham, October 1989, 74 pp.

Town of Markham, 1989b. Review of Design Storms in Markham by Comparison with Multi-event Simulation. (Report by Paul Wisner and Associates Inc.), Markham.

Town of Markham, 1983. Stormwater Management in the Town of Markham, (by P.E. Wisner, D. Mukherjee, D. Keliar and A. Brodie), Markham, June 1983.

Chapter 28

Software and Database on a CDROM for Finding Titles of 4000 Papers in the Less-accessible Literature on Urban Drainage and Related Modeling

William James, Kristi Rowe and W. Robert C. James

Provided with this book, on CD-ROM, is a very simple method to find titles and authors of 3905 papers in what has been termed "the intermittent grey literature", as well as over 100 papers related to HSPF (the Hydrological Simulation Program in Fortran). (Please note that use of the librarians' term *grey literature* refers merely to the shadowy nature of its availability in print today, and is not meant to imply anything even slightly pejorative).

For engineers, researchers and students working in stormwater management modelling, the method provides an easy way to find titles of papers in this relatively large set of papers that might otherwise not have been indexed.

28.1 Introduction

This is our fifth guide to the intermittent specialty conference literature. The database of papers is similar to the tables of papers published by the writers in the four previous guides in this present series of books. The four guides are listed in

© *Advances in Modeling the Management of Stormwater Impacts - Vol. 5* W. James, Ed.
Pub. by CHI, Guelph, Canada 1997. ISBN 0-9697422-7-4. Fax: +519 767-2770

the references at the end of this paper, and also in Table 28.6. Also included herein are the tables of conferences. New for this chapter: all papers have been placed on a CDROM, some Canadian conferences have been added, and all the past conference series have been updated.

28.2 Sources

The sources of the 3905 papers include:

- 30 European Specialty Conferences
- 26 SWMM Conferences
- 13 ASCE Specialty Conferences
- 6 International Conferences on Urban Storm Drainage
- 11 Kentucky Urban Hydrology Conferences
- 7 Canadian Conferences

Additionally, the list of over 100 papers related to HSPF, collected by Anthony Donigian from diverse sources, has been added to this database. The total is over 4000 publications, all published in the time period roughly from July 1970 to May 1995.

28.3 Installation

For this present version of the guide, it is necessary to have a microcomputer with a CDROM driver and running the Windows operating system. The CDROM at the back of this book includes a database and the program *Biblio96*. Installation of the program is intuitive: simply read the instructions on the CDROM, and activate the installation program.

Biblio96 can be run directly from the CDROM or the files can be copied to the hard drive and run from there. Instructions are given on the CDROM.

28.4 Instructions

Then, to locate titles and authors of papers of interest, follow the menu system. The software uses a character string search using AND and OR boolean operators, and parentheses. It will also search for partial words in both the author and title category. It is very fast. Once you have found a set of titles of papers of interest, you will need to determine the name of the conference in whose proceedings that paper occurred, and the edition or sequence number for that conference. This is done by noting the alphanumeric code assigned to the paper:

a typical example is I211. The letters indicate the conference set and the first digit(s) the sequence number of the proceedings in that set, thus I denotes *International* and 2 denotes the *second* conference. Meanings associated with other letters are:

A, ASCE specialty conferences;	Table 28.4
C, Canadian specialty conferences;	Table 28.2
E, European specialty conferences;	Table 28.5
I, the International conferences;	Table 28.3
K, the U of Kentucky conferences; and	see James (1994)
S, the so-called SWMM conferences.	Table 28.1
H, is reserved for the HSPF literature.	not used.

Using such a numbering system considerably reduces the size of the database and increases the speed of the search. However, clever users will note that the HSPF set does not use the numbering system, since the full source is included in each entry.

Because of the cumulative nature of this database, and a universal aversion for republishing material available elsewhere, readers must refer to the past literature guides published by the authors. Tables 28.1 through 28.5 list both the previously and the newly catalogued conferences. Table 28.6 facilitates the search for conferences by indicating the book in which the relevant guides are listed.

28.5 Retrieval of Papers

Unfortunately the database does not include the text of the papers, nor yet, the abstracts. Still worse, copies of the original papers may be difficult to obtain. Some further details are listed in the original four articles (see the references to this chapter). Papers in the SWMM conference proceedings can be obtained through NTIS, from Computational Hydraulics Int. (CHI), or from individuals as cited in Table 28.1. The Canadian conference proceedings (Table 28.2) are available from the Canadian Society for Civil Engineering, 2050 Mansfield Street, Suite 700, Montreal, PQ H3A 1Z2, Fax: 514-842-8123. Papers in the international and European conference proceedings (unfortunately mostly out of print/never in print respectively) may be located through the publisher and/or the ISBN number cited in Tables 28.3 and 28.5. Unless otherwise noted in Table 28.4, copies of the ASCE conference proceedings may be procured from the American Society Civil Engineers at 1801 Alexander Bell Drive, Reston, VA 20191-4400, Fax:703-295-6222.

Alternatively, once the title and relevant authors are located, you can with some initiative search your local technical library, using for example Science Citation Index, and/or Engineering Index (or EI Compendium Plus) against the authors' names for example, to find current papers of interest.

Table 28.1 SWMM Conferences - S23-S26 are newly added.

Date/Place	ID	Publisher/Report Number	Papers
March, 1995 Toronto, ON	S26	CHI-R191 Advances in Modeling the Management of . .. ISBN:0-9697422-5-8	18
March, 1994 Toronto, ON	S25	CHI-R183 Modern Methods for Modeling the Management of ... ISBN 0-9697422-X	26
February ,1993 Toronto, ON	S22	Lewis/CRC Press L1052 Current Practices in Modeling... ISBN 1-56670-052-3	25
February, 1992 Toronto, ON	S23	Lewis/CRC Press. L898 New Techniques in Modeling.... ISBN 0-87371-898-4	23
April 1990 Eatontown NJ		unpublished. T. Najarian Tel: 201-389-0220	
October 1988 Denver, CO	S22	EPA-600/9-89-001 NTIS: PB89 195002/AS	22
October 1987 Victoria, BC	S21	Howard Assocs. Victoria BC Tel: 604-385-0206	12
March 1987 Denver, CO	S20	EPA-600/9-87-016 NTIS: PB88-125430	17
September 1986 Toronto, ON	S19	U of Ottawa. P. Wisner, Ottawa, ON Fax: 613-744-43432	24
March 1986 Orlando, FL	S18	EPA-600/9-86-023 NTIS: PB87-117438	22
December 1985 Toronto, ON	S17	CHI-R149. Guelph, ON Fax: 519-767-2770	25
January 1985 Gainesville, FL	S16	EPA-600/9-85-016 NTIS: PB85-228302	17
September 1984 Burlington, ON	S15	CHI-R128. Guelph, ON Fax: 519-767-2770	16
April 1984 Detroit, MI	S14	EPA-600/9-85-003 NTIS: PB85-168003	17
September 1983 Montreal, PQ	S13	GREMU. 83/02 P. Beron, Ecole Polytechnique, Montreal, PQ	17
January 1983 Gainesville, FL	S12	EPA-600/9-83-015 NTIS: PB84-118454	17
October 1982 Ottawa, ON	S11	U of Ottawa. P. Wisner, Ottawa, ON Fax: 613-744-43432	15
March 1982 Washington, DC	S10	EPA-600/9-82-015 NTIS: PB83-145540	16
September 1981 Niagara Falls ON	S9	CHI-R81 Guelph, ON Fax: 519-767-2770	26
January 1981 Austin, TX	S8	CHI-R83 Fax: 519-767-2770	10
June 1980 Toronto, ON	S7	EPA-600/9-80-064 NTIS: PB81-173858	11
January 1980 Gainesville, FL	S6	EPA-600/9-80-017 NTIS: PB80-177876	15
May 1979 Montreal, PQ	S5	EPA-600/9-79-026 NTIS: PB80-105663	15
November 1978 Annapolis, MD	S4	EPA-600/9-79-003 NTIS: PB290-742/6BE	10
May 1978 Ottawa, ON	S3	EPA-600/9-78-019 NTIS: PB285-993/2BE	9
March 1977 Toronto, ON	S2	Conf. Proc. No. 5. MOEE Toronto	15
October 1976 Toronto, ON	S1	Conf. Proc. No. 4. MOEE Toronto	24

Table 28.2 Canadian conferences - these are all newly added, and don't appear elsewhere.

Date/Place	ID	Title	Papers
May 1990 Hamilton ON	C7	Engineering in Our Environment - Volume 5	122
May 1990 Hamilton ON	C6	Engineering in Our Environment - Volume 2	99
June 1989 St. John's NF	C5	9th Canadian Hydrotechnical Conference	86
May 1987 Montreal PQ	C4	8th Canadian Hydrotechnical Conference	42
May 1987 Montreal PQ	C3	Centennial Symposium on Management of Waste Contamination of Groundwater.	19
June 1985 Vancouver BC	C2	New Directions and Research in Waste Treatment and Residuals Management.	58
May 1985 Saskatoon SK	C1	7th Canadian Hydrotechnical Conference	52

Table 28.3 International conferences - none of these are newly added.

Date/Place	ID	Title	Papers
July 1993 Niagara Falls Ontario	I6	Proceedings of Sixth International Conference on Urban Storm Drainage Seapoint pub. ISBN: 1-55056-253-3	336
July 1990 Tokyo Japan	I5	Proceedings of Fifth International Conference on Urban Storm Drainage Out of print.	266
Aug. 1987 Lausanne Switzerland	I4	Proceedings of Fourth International Conference on Urban Storm Drainage Water Resources pub. Littleton, CO	186
June 1984 Goteborg Sweden	I3	Proceedings of Third International Conference on Urban Storm Drainage Chalmers Univ. ISBN: 91-7032-128-0	150
June 1981 Urbana Illinois	I2	Proceedings of Second International Conference on Urban Storm Drainage Water Resources pub. ISBN: 0918334-48-9	116
April 1978 Southampton United Kingdom	I1	Proceedings of First International Conference on Urban Storm Drainage Out of print.	57

Table 28.4 ASCE conferences - A11-A13 are newly added.

Date/Place	ID	Title	Papers
August 1991 Colorado	A13	Engineering Foundation Conference- Stormwater Runoff and Receiving Systems. Ed: E Herricks CRC Press ISBN: 1-56670-159-7	31
August 1994 Colorado	A12	Engineering Foundation Conference- Stormwater NPDES Related Monitoring Needs Ed. Harry Torno	24
April 1990 Fort Worth, TX	A11	The Water Resources Infrastructure Symposium Ed: John Scott, Reza Khanbilvardi	44
October 1989 Davos, Switzerland	A10	Urban Stormwater Enhancement - Source Control, Retrofitting and Combined Sewer Technology. Ed: Harry Torno	31
July 1986 Potosi, MO	A9	Design of Urban Runoff Quality Controls Ed: Larry Roesner, Ben Urbonas, Michael Sonnen	32
June 1986 Henniker, NH	A8	Urban Runoff Quality - Impact and Quality Enhancement Technology Ed: Ben Urbonas, Larry Roesner	33
October 1983 Niagara-on-the-Lake, ON	A7	Emerging Computer Techniques in Stormwater and Flood Management Ed: William James	29
May 1983 Baltimore MD	A6	Urban Hydrology Ed: J.W. Delleur, Harry Torno	17
August 1982 Henniker, NH	A5	Stormwater Detention Facilities: Planning, Design, Operation and Management Ed: William DeGroot	36
June 1980 Blacksburg VA	A4	Urban Stormwater Management in Coastal Area Ed: Chin Y. Kuo	45
July 1978 Henniker, NH	A3	Water Problems of Urbanizing Areas Ed: William Whipple, N. Grigg, R. Langon	35
August 1974 Rindge, NH	A2	Urban Runoff: Quantity and Quality Ed: William Whipple	37
July 1970 Deerfield, MA	A1	Urban Water Resources Management Co-chairs: D.W. Hill, K.R. Wright	14

Table 28.5 European conferences - E29-E30 are newly added. *(cont'd overleaf)*

Date/Place	ID	Title	Papers
Sept. 1994 Moscow, Russia	E30	Remote Sensing And GIS in Urban Waters.	28
May 1995 Lyon, France	E29	2nd International Conference on Innovative Technologies in Urban Storm Drainage	78
Dec 1994 St. Moritz Switzerland	E28	Closing the Gap Between Theory and Practice in Urban Rainfall Applications (Preprint edition)	27
June 1994 Goteborg Sweden	E27	Procs. of International User-Group Meeting - Computer Aided Analysis and Operation in Sewage Transport and Treatment Technology Chalmers Univ. ISSN: 0347-8165	32
June 1994 Cemise Castle-Bechyne Czech Republic	E26	Integrated Urban Storm Runoff - 7th European Junior Scientist Workshop	25
March 1994 The Netherlands	E25	Combined Sewer Overflow - a European Perspective *	7
April 1993 Vienna Austria	E24	Application of Geographic Information Systems in Hydrology and Water Resource Management	65
November 1992 Lyon France	E23	Rediscover Water. A Priority	56
June 1992 Maratea Italy	E22	Urban Drainage - Experimental Catchments in Italy	19
April 1992 St. Victor Sur Loire St Etienne France	E21	Assessment of Modelling Uncertainties and Measurement Errors in Hydrology	26
Sept. 1991 Brussels Belgium	E20	Origin, Occurence and Behaviour of Sediments in Sewer Systems	21
Sept. 1991 Terschelling The Netherlands	E19	Applications of Operations Research to Real Time Control of Water Resources Systems - 3rd European Junior Scientist Workshop	25
June 1991 Dubrovnik Yugoslavia	E18	New Technologies in Urban Drainage Elsevier ISBN: 1-85166-650-8	53
1991	E17	Highway Pollution **	5
April 1990	E16	Task Group on Source Control	13
June 1990	E15	A Symposium on Infiltration and Storage of Stormwater in New Developments	15

Table 28.5 continued European conferences.

Date/Place	ID	Title	Papers
Sept. 1989 Wageningen The Netherlands	E14	Urban Storm Water Quality and Ecological Effects Upon Receiving Waters. IAWPRC, Water Science & Technology Vol 22 (10/11) ISSN: 0273-1223	39
Sept. 1989 Munich Germany	E13	Proceedings of the 3rd International Symposium - Highway pollution **	7
June 1989 Dubrovnik Yugoslavia	E12	Computational Modelling and Experimental Methods in Hydraulics Elsevier ISBN: 1-85166-374-6	7
1989	E11	First European Junior Scientist Workshop	22
July 1988	E10	Urban Storm Drainage - Proceedings of U.S. - Italy Bilateral Seminar Water Resources pub. ISBN: 0-918334-75-6	17
July 1988 Brighton U.K	E9	Urban Discharges and Receiving Water Quality Impacts	17
April 1988 Duisburg Fed. Rep of Germany	E8	Urban Water '88 (repub: Hydrological Processes and Water Management in Urban Areas 1990). UNESCO ISBN 90-800166-1-6	108
Oct 1986 Wageningen The Netherlands	E7	Urban Storm Water Quality and Effects upon Receiving Waters	29
July 1986 London England	E6	Proc. of 2nd Int. Symposium-Highway Pollution **	17
April 1986 Dubrovnik Yugoslavia	E5	Proc. of Int. Symposium on Comparison of Urban Drainage Models with Real Catchment Data. Pergamon Press ISBN 0-08-032558-0	55
Aug. 1985 Montpelier France	E4	Urban Runoff Pollution Springer Verlag ISBN 0-387-16090-6	29
April 1985 The Netherlands	E3	Water in Urban Areas TNO, Netherlands ISBN: 90-6743-075-7	15
1984	E2	Rainfall as the Basis for Urban Run-off Design and Analysis	27
1984	E1	Proc. of 1st Int. Symposium-Highway pollution **	11

* The Smisson Foundation, Dutch Assoc. for Water Management, DHV Environment and Infrastructure, Hydro Research and Development

** Editor: R.S. Hamilton, Ctr for Pollution Research, Schl of Applied Science, Midlesex Polytechnic, Einfield, Middlesex EN3 4SF, UK.

Table 28.6 Previously collected conferences and corresponding guides.

Conference Title	Previous Publication
International Conferences	James, W. 1996. Chap. 18, Advances in Modelling the Management of Stormwater Impacts, Computational Hydraulics International, Guelph, Canada, pp. 279-338.
European Conferences	James, W. 1996. Chap. 18, Advances in Modelling the Management of Stormwater Impacts, Computational Hydraulics International, Guelph, Canada, pp. 279-338
ASCE Specialty Conferences - 309 Paper	James, W. 1995. Chap. 26, Guide to 309 papers of some ASCE Specialty Conferences on Urban Stormwater, 1970-1985. In: Modern Methods for Modeling the Management of Stormwater Impacts. Computational Hydraulics International, Guelph, Ont., Can. pp. 421-436.
Kentucky Conferences	James, W. 1994. Chap. 25, Guide to the 485 papers of Kentucky Symposia 1975-85. In: Current Practices in Modelling the Management of Stormwater Impacts. Lewis Publishers, Boca Raton, FL. pp. 389-411.
Environmental Protection Agency SWMM Conferences	James, W. 1993. Chap. 1, Introduction to Stormwater Management Model Environment. In: New Techniques for Modelling the Management of Stormwater Impacts, Lewis Publishers, Boca Raton, FL. pp. 1-28.

Acknowledgements

Kristi Rowe and Anthony Donigian created the database. Rob James wrote the access software. Bill James provided the ideas, collected the resources, and guided the work. For further information about the software and database on the CDROM, contact the present writers by any of the following methods:

Fax: (519) 767-2770 or (519) 836-0227.
E-mail: info@chi.on.ca Web: http://www.chi.on.ca

References

James, W. 1996. Chap. 18, Advances in Modeling the Management of Stormwater Impacts, Computational Hydraulics International (Tel: 519-767-0197, email info@chi.on.ca) and Ann Arbor Press (fax: 313-475-8852), Guelph, ON, Canada, pp. 279-338, ISBN 0-9697422-5-8.

James, W. 1995. Chap. 26, Guide to 309 papers of some ASCE Specialty Conferences on Urban Stormwater, 1970-1985. Modern Methods for Modeling the Management of Stormwater Impacts. Computational Hydraulics International (Tel: 519-767-0197, email info@chi.on.ca), Guelph, ON, Canada, pp. 421-436, ISBN 0-9697422-X.

James, W. 1994. Chap. 25, Guide to the 485 papers of Kentucky Symposia 1975-85. In: Current Practices in Modelling the Management of Stormwater Impacts. Lewis Publishers (Tel: 407-994-0555), Boca Raton, FL, USA, pp. 389-411, ISBN 1-56670-052-3.

James, W. 1993. Chap. 1, Introduction to the Stormwater Management Model Environment. In: New Techniques for Modelling the Management of Stormwater Impacts, Lewis Publishers (Tel: 407-994-0555), Boca Raton, FL, USA, pp. 1-28, ISBN 0-87371-898-4.

Acronyms and Abbreviations

AASHTO	American Association of State Highways and Transportation
AAT	Arc Attribute Table
ADT	Average Daily Traffic
AEE	AGRA Earth and Environmental -a consulting engineering firm
AES	Atmospheric Environment Services (Canada)
AGU	American Geophysical Union
Al	Aluminum
AM/FM	Automated Mapping/Facilities Management
AMC	Antecedent Moisture Conditions
Amer.	American
AML	Arc Macro Language
ANOVA	a two-way analysis of variance
ANSI	Area of Natural Significant Interest/American National Standards Institute
AOC	Area Of Concern
ARCS	Assessment and Remediation of Contaminated Sediment
ARS	Agricultural Research Service
AS	asphalt
As	Arsenic
ASAE	American Society of Agricultural Engineers
ASCII	American Standard Code for Information Interchange
ASCE	American Society of Civil Engineers
ASEE	American Society of Engineering Education
Assoc.	Association
ASTM	American Society for Testing and Materials
ATC	Artificial trapezoidal channel
AWMA	Air and Waste Management Association
AWRA	American Water Resources Association
AWWA	American Waterworks Association
BACT	Best Available Control Technology
BBS	Bulletin Board System
BCHD	Barnstable County Health & Environmental Department
BES	Bureau of Environmental Services
BHD	alpha-Benzene Hexachloride
BMPs	Best Management Practices
BMM	Best Management Model
BOD	Biological Oxygen Demand
BSA	Buffalo Sewer Authority
Bull.	Bulletin
BWI	Buffalo Water Intake
C $	Canadian dollars
C of A	Certificate of Approval
ca	Circa, about, approximately
CAD	Computer Aided Design/ Drafting
CADD	Computer Aided Design and Drafting
CAE	Computer Aided Engineering
CAM	Computer Aided Mapping
Can.	Canadian
CAPPI	Constant Altitude Precipitation Index maps
CBOD	Carbonaceous Biochemical Oxygen Demand
CCTA	C.C. Tatham and Associates - a consulting engineering firm
Cd	Cadmium
CDF	Centralized Detention Facility; Cumulative Density Function
CDM	Camp Dresser & McKee - a consulting engineering firm

Cdn	Canadian currency
CDROM	Compact Disk Read Only Memory
CEAM	Center for Exposure Assessment Modeling
CF	Continuous Flow
CFA	Consolidated Frequency Analysis
CFF	Confined Filtration Facility
cfs	Cubic Feet per Second 1 cfs = $4.719 \times 10^{-4} m^{3/s}$
Chap.	Chapter
CHI	Computational Hydraulics Int'l. - a consulting engineering firm
CHy	WMO Commission for Hydrology
CIESIN	Consortium for International Earth Science Information Network
Cl	Chlorine /Chloride
CMC	Central Microcomputer Controller
CN	Curve Number in SCS method
COD	Chemical Oxygen Demand
COG	Council of Governments
COWAR	Commitee on Water Research
CP	Concrete Paver
CPU	Central Processing Unit
Cr	Chromium
CRL	Communications Research Laboratory
CRA	Conestoga Rovers & Associates - consulting engineers.
CSCE	Canadian Society of Civil Engineers
CSIR	Council for Scientific and Industrial Research
CSO	Combined Sewer Overflow
CST	Central Standard Time
CSTR	Continuous Stirred Reactor
CT-DW	Common trench-Dividing Wall
CTS	Common Trench-Separate system
Cu	Copper
CWA	Clean Water Act (U.S.)
CWQG	Canadian Water Quality Guidelines
CWRA	Canadian Water Resources Association
CWWTP	Central Wastewater Treatment Plant
DBMS	Database Management System
DCIA	Directly Connected Impervious Area
DDE	Dynamic Data Exchange
DEC	Department of Environmental Conservation
DEIS	Draft Environmental Impact Statement
DEM	Digital Elevation Model
Dept.	Department
DEP	Department of Environmental Protection
DEQ	Department of Environmental Quality
DER	Department of Environmental Resources
DFO	Dept. of Fisheries and Oceans
DiCB	DiChloroBenzene
DNR	Department of Natural Resources
DNRC	Department of Natural Resources and Conservation
DNREC	Department of Natural Resources and Environemntal Control
DO	Dissolved Oxygen
DOC	Department of Conservation
DPD	n-diethyl-e-phenylenediamine
DPD method	Analytical method for determining chlorine residual utilizing the reagent DPD
DRH	Direct Runoff Hydrographs
DRI	Detroit River Interceptor
DS	Dissolved Solids
DTM	Double Trace Moments
DWF	Dry Weather Flow
DWSD	Detroit Water and Sewerage Department
DWWM	Dept. of Wastewater Management (Honolulu)
DWWTP	Detroit Wastewater Treatment Plant
DXF	Drawing Exchange File
E. coli.	Escherichia coli.
EA	Environmental Assessment
EAAC	Environmental Asessment Advisory Committee
EAs	Exposure Assessments
EBB	Electronic Bulletin Board
ECD	Electron Capture Detector
ECP	Erosion Control Plan
EDC	External drainage cell
EIA	Effective Impervious Area
EIC	Engineering Institute of Canada
EIS	Environmental Impact Statement
EMAIL	Electronic Mail
EMAP	Environmental Monitoring and Assessment Program
EMC	Event Mean Concentration

EMDP	Environmental Master Drainage Plan
Eng'rg	Engineering
EPA	Environmental Protection Agency
EPRI	Energy Production Research Institute
EPS	Environmental Protection Service
ERL	Environmental Research Laboratory
ESA	Environmentally Sensitive Area
ESR	Equivalent Solids Reservoir
ESRI	Environmental Systems Research Institute
EST	Eastern Standard Time
ET	Evapotranspiration
EV	Extreme Value
FA	Frequency Analysis
FAO	Food and Agriculture Organization (of the UN)
FC	Fecal Coliform
FDOT	Florida Department of Transportation
FDRP	Flood Damage Reduction Program
FE	Fort Erie
Fe	Iron
FEMA	Federal Emergency Management Agency
FERC	Federal Energy Regulatory Commission
FHWA	Federal Highway Administration
FLUCS	Florida Land Use Classification System
FS	Fecal Stretococci
FSA	Fred Schaeffer Associates - a consulting engineering firm
ft-NGVD	feet referenced to NGVD *(q.v.)*
FTP	File Transfer Protocol, an Internet utility
FTU	Formazin Turbidity Unit
FWS	Free Water Surface
FY	Fiscal Year
FYI	For Your Information
GC	Gas Chromatography
GDD	Growing Degree Days
GDS	Graphic Data System
GEWEX	Global Energy and Water Cycle Experiment
GI	Growth Index
GIS	Geographic Information System
GLIN	Great Lakes Information Network
gpm	Gallons Per Minute
GPS	Global Positioning System
GRU	Grouped Response Unit
GUI	Graphical User Interface
ha	Hectares
HCB	Hexachlorobenzene
HEC	Hydrologic Engineering Center (U.S. Army Corps of Engineers)
HGF	Hydraulic Gradeline Factor
HGL	Hydraulic Grade Line
HPC	Heterotrophic Plate Count
HQ	Headquarters
HRCA	Hamilton Region Conservation Authority
HRT	Hydraulic Retention Time
HRU	Homogeneous Response Unit
HSI	Habitat Suitability Indices
HSIA	Hydraulically Significant Impervious Area
HWL	High Water Level
HWY	Highway
I/O	Input/Output
IA	Initial Abstraction
IAHR	International Association for Hydraulic Research
IAHS	International Association of Hydrological Sciences
IAMAP	International Asociation of Meteorology and Atmospheric Physics
IAWPRC	International Association for Water Pollution Research and Control
IAWQ	International Association on Water Quality
IBM	International Business Machines
ICID	International Commission on Irrigation and Drainage
ID	Identification
IDF	Intensity-Duration-Frequency curves
IEEE	Inst. of Electrical and Electronic Engineering
IETD	Inter-Event Time Definition
IFIM	Instream Flow Incremental Methodology
IHD	International Hydrological Decade
IHP	International Hydrological Program
IJC	International Joint Commission
IN	Indicated Number
INAA	Instrumental Neutron Activation Analysis
Inst.	Institute
Int.	International
IRAP	Industrial Research Assistance Program (Canada)

IRC Interflow Recession Parameter
ISE Integral Square Error
IUGS International Union of Geological Sciences
IWC Industrial Waste Control
IWRA International Water Resources Association
J. Journal
JTU Jackson Turbidity Unit
LAER Lowest Achievable Emission Rate
LAN Local Area Network
LANDSAT Satellite Imagery for Geographical Information
LBIS Land Based Information System
LPD liters per day
LSZN Lower Soil Zone Nominal Storage
LWL Low Water Level
m Metres
m^3 Cubic Metres
MA Municipal Aquifer
MAB Man and the Biosphere Program
mcf Million Cubic Feet
MCI Mill Creek Interceptor
MCLs Maximum Contaminant Levels
MDA Michigan Department of Agriculture
MDP Master Drainage Plan
MDEQ Michigan Department of Environmental Quality
M.E. Master of Engineering
MEA Municipal Engineers Association
Mg Magnesium
MGD Million Gallons per Day
MH Maintenance Hatch (Manhole)
MICB Modular Interlocking Concrete Blocks
MICBEC ditto - external drainage cell type
MICBIC ditto -internal drainage cell type
MIMO Multiple Input-Multiple Output
MINTEQA2 Metal Speciation for Equilibrium for Surface- and Ground-water
MIPS Million Instruction Sets Per Second
MISA Municipal/Industrial Strategy for Abatement (Canada)
MMI Man-Machine Interface
Mn Manganese
MNR Ontario Ministry of Natural Resources
M.O. Microorganisms
MOE Ontario Ministry of the Environment (See *MOEE*)
MOEE Ontario Ministry of Environment and Energy (replaces MOE)
MPDES Montana Pollutant Discharge Elimination System
MPN Most Probable Number
MRI Mean Recurrence Interval
MS Mass Spectrometry
M.Sc. Master of Science
MSC Meteorological Service of Canada
MSL Mean Sea Level
MSD Mass Spectrometry Detector
MSMP Master Stormwater Management Plan
MSS Multi Spectral Scanner
MSU Michigan State University
MTC Ontario Ministry of Transportation and Communication
MTO now MTC (see above)
MTP Main Treatment Plant
MTRCA Metro Toronto and Region Conservation Authority
MWRA Massachusetts Water Resources Authority
Na Sodium
NAPP National Aerial Photography Program
NAQUADAT National Water Quality Data Base
NAWDEX National Water Data Exchange
NCDC National Climatic Data Center
NDMA n-nitrosodimethylamine
NEORSD North East Ohio Regional Sewer District
NGO Non-Governmental Organization
NGVD National Geodetic Vertical Datum
NH_3-N Ammonium Nitrogen
NHAP National High Altitude Photography
Ni nickel
NITR nitrate plus nitrite
NN Neural Networks
NNN Nitrate/Nitrite Nitrogen
No. Number
NO_3-N Nitrate, Nitrite
NOAA National Oceanic and Atmospheric Administration
NOD Nitrogenous Oxygen Demand
NOTL Niagara-on-the-Lake
NPD Nitrogen-Phosphorus Detector
NPDES National Pollution Discharge Elimination System (U.S.)
NPS Non Point Source
NRCC Northeast Regional Climate Center; National Research Council for Canada

NTIS	National Technical Information Service (U.S.)
NTU	Nephelometric turbidity unit
NURP	Nationwide Urban Runoff Program (U.S.)
NWF	National Wildlife federation
NWRI	National Water Research Institute (Canada)
NWIS	National Water Information System
NWS	National Weather Service (U.S.)
O&M	Operation and Maintenance
OCs	Organic Chlorinated Pesticides
ODWO	Ontario Drinking Water Objective
OLE	Object Linking & Embedding
OM	Organic Matter
OMNR	see MNR
OP	Organophosphorus
O-P	Ortho-phosphate
ORD	Office of Research and Development
ORSANCO	Ohio River Sanitation Commission
OSCTS	On-Site Containment and Treatment System
OSD	On-Site Detention
OW	Office of Water
p.	Page
P	Phosphorus
PAH	Polycyclic-Aromatic-Hydro-carbons; Polynuclear Aromatic Hydrocarbons
PAP	Porous Asphalt Pavement
PAR	Photosynthetically Active Radiation
Pb	Lead
PC	Personal Computer
PCA	Principal Component Analysis
PCB	Polychlorinated Biphenyl
PCP	Pollution Control Planning / Pollution Control Plant
PCP	Porous Concrete Pavement
PCS	Permit Compliance System
PDE	Partial Differential Equation
PDF/pdf	Probability Density Function
PDM	Probabilistic Dilution Model
PDMS	Probability Distribution Multifractal Scaling
PEL	Probable Effect Level
PET	Potential Evapotranspiration
PF	Plug Flow
PFPT	Peak Flow Presentation Table
pH	Negative Log of Hydrogen Ion Concentration
Ph.D.	Doctor of Philosophy
PHC	Petroleum Hydrocarbons
Phys.	Physics
PLC	Programmable Logic Controller
PLUARG	Pollution from Land Use Activities Reference Group
PO_4	Ortho-phosphate
pp	Pages
PP	Permanent Pools
Proc.	Proceedings
PSD	Power Spectral Density
PSWMS	Primary Stormwater Management System
pub.	Published
PVC	Poly Vinyl Chloride
PWQG	Provinicial Water Quality Guideline
PWQMN	Provincial Water Quality Monitoring Network (Canada)
PWQO	Provincial Water Quality Objectives (Canada)
QA/QC	Quality Assurance/Quality Control
QAP	Quality Assurance Plan
R&D	Research and Development
RACT	Reasonably Available Control Technology
R/LPOLY	Right/Left Polygon
RAP	Remedial Action Plan
RDI	Rainfall Derived Infiltration
RDBMS or RDMS	Relational Database Manage-ment System
RE	Relative Error
RFA	Rainfall Accumulation Map
RM	Rational Method
RMHW	Regional Municipality of Hamilton-Wentworth
RMOC	Regional Municipality of Ottawa-Carleton
ROW	Right of Way
RTC	Real Time Control
RTCDEMO	Real Time Control Demonstra-tion
RTCSIM	Real Time Control Simulation
RTU	Remote Telemetery Unit
SCC	Soil Conservation Curves
SCADA	Supervisory Control and Data Acquisition (system)
SCAS	Sewer Connection Application System
SCOWAR	Scientific Committee on Water Research
SCS	USDA Soil Conservation Service
SEE	Standard Error of Estimate

SEMCOG	South East Michigan Council of Governments
SFAS	Sewer Flow Analysis System
SFO	Stipulation and Final Order
SFS	Subsurface Flow System
SHEF	Standard Hydrometeorological Exchange Format
SI	Suitability Index
SIU	Significant Industrial Users
SJRWMD	St. Johns River Water Management District
SLRT	Sewer Level Remote Telemetry
Soc.	Society
SOD	Sediment Oxygen Demand
SO_4	Sulphate
SPAM	Spatially Averaged mean
SPDES	State Pollutant Discharge Elimination System
SPM	Storage Pumping Model
SQL	Structured Query Language
SS	Suspended Solids
SSES	Sewer System Evaluation Surveys
SSMP	Sewer System Master Plan
SSO	Sanitary Sewer Overflow
STN	Scientific and Technical Information
STORET	Storage and Retrieval of US Waterways Parametric Data
STP	Sewage Treatment Plant
STS	StormTreat Systems
SUNY	State University of New York
SWM	Stormwater Management
SWMPs	Stormwater Management Plans/ Practices
SWPPP	Storm Water Pollution Prevention Plan
SWTP	Southerly Wastewater Treatment Plant(Mill Ck, Ohio)
SYMAP	Synagraphic Mapping
TBRG	Tipping Bucket Rain/Runoff Gage
TC	Total carbon (incl. organic & inorganic)
TDN	Total Dissolved Nitrogen
TDS	Total Dissolved Solids
TEL	Threshold Effect Level
TF	Transfer Function
THOD	Theoretical Oxygen Demand
TIC	Total Inorganic Carbon
TIN	Triangulated Irregular Network
TKN	Total Kjeldahl Nitrogen
TM	Thematic Mapper
TMDL	Total Maximum Daily Load
TN	Total Nitrogen
TN-Diss	Total N dissolved
TN-Part	Total N Particulate
TOC	Total Organic Carbon
TOD	Total Oxygen Demand
TP	Total Phosphorus
TPH	Total Petroleum Hydrocarbons
Trans.	Transactions
TS	Time Series
TSI	Trophic State Indices
TSM	Time Series data Manager
TSS	Total Suspended Solids
TVS	Total Volatile Solids
U.	University
UA	Upper Aquifer
UAA	Use Attainability Analysis
UACS	Upper Aquifer Containment and treatment System
UAFS	Upper Aquifer Feasibility Study
UN	United Nations
UNCED	United Nations Commission on Environment and Development
UNDP	United Nations Development Program
UNEP	United Nations Environment Program
UNESCO	United Nations Education Scientific and Cultural Organisation
UNIDO	United Nations Industrial Development Organization
UNIX	A general purpose time-sharing operating system. Trademark of Bell Laboratories.
UPM	Universal Process Modeling
URL	Uniform Resource Locator, an Internet identification system
US $	United States dollars
USACE	United States Army Corps of Engineers
USDA	United States Department of Agriculture
USEPA	United States Environmental Protection Agency
USCS	Unified Soil Classification System
USGPM	US Gallons Per Minute
USGS	United States Geological Survey
UST	Underground Storage Tank
USZN	Upper Soil Zone Nominal Storage
UTM	Universal Transverse Mercator
UWIN	Universities Water Information Network
UWRC	Urban Water Resources Council (A.S.C.E.)
UZS	Upper Zone Storage

V	Vanadium
VDU	Visual Display Unit
VOC	Volatile Organic Compound
Vol.	Volume
VSS	Volatile Suspended Sediment
VTOC	Volatile Toxic Organic Compounds
WAN	Wide Area Network
WATSTORE	National Water Data Storage and Retrieval System
WHO	World Health Organization
WHOI	Woods Hole Oceanographic Institution
WIMP	Windows, Icons, Menus, and Pointing Devices
WIMS	Wastewater Information Management System
WMO	World Meteorological Organization
WMP	Watershed Management Plans
WPCP	Water Pollution Control Plant
WQA	Water Quality Act of 1987 (Cdn)
WQS	Water Quality Standards
WSC	Water Survey of Canada
WSEL	Water Surface Elevation Level
WSI	Western Sanitary Interceptor
WSM	Watershed Model
WTP	Water Treatment Plant
WUA	Weighted Usable Area
WWT	Wastewater Treatment
WWTP	Wastewater Treatment Plant
WWW	World Wide Web
WYSIWYG	What You See Is What You Get
Zn	Zinc
ZUM	Zones of Uniform Meteorology
30Q20	The average streamflow over a 30-day period which is equalled/exceeded on average once in every 20 yrs.
7Q20	The average streamflow over a 7-day period which is equalled/exceeded on average once in every 20 yrs.

Programs and Models

AGNPS	Agricultural Non-Point Source
ANNIE	a USGS hydrologic time-series data management system
ARCINFO	a GIS program
ASDM	Atmospheric and Sediment Deposition Model
AutoCAD®	an automated computer-aided drafting package
BASIC	a programming language
BATHTUB	a reservoir loading model
BMPPlanner	a decision support software tool
BOSS-DAMBRK	a hydrology program
CAGIS	CSO Area Geographic Information System
CASCCADE	Co-evolving Assistant Software for Changing Computational and Data Environments (links ANNIE and SWMM)
CASCADE2	Version of CASCCADE linking HECDSS and SWMM
CASS WORKS	an integrated infrastructure management software for water distribution, storm drainage etc.
CFA	Consolidated Frequency Analysis Package
CMP	Computer Mapping Program
COGO	a CAE application module
COMBINE	a time-series module of the SWMM program
CORMIX	Cornell Mixing Zone Model
DBMS	Database Management System
DEM	Digital Elevation Model
DOMECOL	a water pollution model
DOMOD7	Dissolved Oxygen Model Version 7
DOS	a computer operating system
DSPLAY	a graphical display module of HECDSS
DSS	Data Storage System, a type of file format
DSSUTL	a data storage system utility module of HECDSS
DTM	Digital Terrain Model
DUH	digital elevation models
DWOPER	Dynamic Wave Operational Model
ECOL	ecological subroutine
ECOL1	dynamic aquatic plant growth and nutrient uptake model
EPICWQ	Erosion/Productivity Impact Calculator -Water Quality
ERDAS	Earth Resource Data Analysis System - a raster-based GIS
EXAMSII	Exposure Analysis Modeling Systems II
EXSUDS	extended SUDS program
EXTRAN	extended transport, a computational module of SWMM
FDAM	flood damage analysis model
FORTRAN	a high-level programming language
GAMES	Guelph evaluation effects of Agricultural Management on Erosion and Sedimentation

GAWSER	Guelph All-Weather Sequential-Events Runoff model
Geo/SQL®	a geographical program
GISFPM	GIS floodplain management
GPS-X	a dynamic model for the design, operation and control of wastewater treatment plants
GRASS	Geographical Resources Analysis Support System
GRSM	Grand River simulation model
GUI	Graphical User Interface
HABTAV	habitat simulation program
HEC1	hydraulic modeling program to determine discharges
HEC2	hydraulic modeling program to determine water surface elevations
HECDSS	HEC hydrologic time-series data management system
HECRAS	HEC river analysis program
HSPF	Hydrologic Simulation Program-Fortran
HWY DSS	Highway Decision Support System
HYDRA	Hydrologic Data Retrieval and Alarm system
HYPERCARD	a software system developed by Apple.
HYMO	a hydrologic model
INTERHYMO	a version of HYMO
IFG4	hydraulic/velocity simulation program
IFORM	a rainfall file format defined by RAIN module of SWMM
INTERHYMO	a hydrologic model
IOWDM	a file input/output utility for ANNIE
LFA	Low Flow Frequency Analysis
MATHPK	a math and statistics module of HECDSS
MIDUSS	a stormwater design program
MODFLOW	Modular 3-Dimensional Finite-Difference Ground-water Flow Model
MTOPOND	Ontario Ministry of Trans-portation quality control performance model for stormwater ponds
MTV	Model Turbo View
MTVE	Model Turbo View - EXTRAN

OASIS	On-line Access and Service Information System
OTTHYMO	Version of the USDA runoff program (HYMO)
PCSWMM	Windows-based SWMM shell
PDF	Probability Density Function
PDM	Probabilistic Dilution Model
PHABSIM	physical habitat simulation modeling program which calculates the relationship between stream flow and physical habitat for various life stages of an aquatic organism
PLAN	a CAE applications module
PLUS	a software system developed by Spinnaker
PSRM	Pen State Runoff Model
QUALHYMO	a version of HYMO
QQS	Quantity Quality Simulation
Q'URM	Queens University runoff model
QUAL2E	Enhanced Stream Water Quality Model
QUALHYMO	a hydrologic program
QuattroPRO	a spreadsheet program
RAIN	a time-series module of the SWMM program
RAP	Rainfall Analysis Program
REPGEN	a report presentation module of HECDSS
RUNOFF	a computational module of the SWMM program
RUNSTDY	Dynamic wave routing model for analysis of complex sewer systems
SEWHYMO	a hydrological model
SIMPLE	a geo-referenced hydrologic model for remotely sensed data
SIMPTM	Simplified Particulate Transport Model
SMIFF	Spatial Mapping for Integrated Flood Forecasting
SPIDA	An unsteady routing model of hydrologic/hydraulic processes for highly looped storm drainage networks
SPINNAKER	a hypercard program
S/T	Storage/Treatment module of SWMM
STATS	a time-series module of the SWMM program
STORAGE	a computational module of the SWMM program

STORET	Storage and Retrieval System
STORM	Storage Treatment Overflow Runoff Model
SUDS	Statistical Urban Drainage Simulator
SWMenu	a SWMM shell
SWMM	Stormwater Management Model (USEPA)
SWRRB-WQ	Simulator Water Resources In Rural Basins - Water Quality
SWSTAT	a statistics module of ANNIE
SYNOP	statistical rainfall analysis program
TEMP	a time-series module of the SWMM program
TIGER	Topographically Integrated Geographic Encoding and Referencing
TOXIWASP	See WASP4
TR-20	A watershed hydrology model to route a design storm hydrograph through a pond
TRANSPORT	a computational module of the SWMM program
WASP	Water (Quality) Analysis Simulation Program
WASP5E	Developed Wetland Simulation Procedure (USEPA model)
WATFLOOD	a hydrologic data base management system for real-time flood forecasting
WATSTORE	Water data storage and retrieval system
WDM	Watershed Data Management, a type of file format
WEPP	water erosion prediction
WINSTDY	Dynamic wave routing model for analysis of complex sewer systems
WSWMM	a Windows shell for SWMM
XP-EXTRAN	An expert systems version of EXTRAN
XP-SWMM	An expert systems version of SWMM

Converting SI units to U.S. Customary Units

To convert from (SI)	Conversion factor	To get U.S. Customary
Length		
Meters (m)	Multiply by 3.28	Feet (ft)
Meters	Multiply by 1.094	Yards
Meters per second (m/s)	Multiply by 2.237	Miles per hour (mph)
Centimeters (cm)	Multiply by 0.39	Inches (in.)
Milimeters (mm)	Divide by 25.4	Inches
Kilometers (km)	Divide by 1.608	Miles (mi)
$m^3/(m^2\text{-day})$ = m/day	Multiply by 24.6	gpd/ft^2
Area		
cm^2	Divide by 6.45	Square inch
m^2	Multiply by 10.76	Square foot (ft^2)
Hectare (ha = 10 000 m^2)	Multiply by 2.46	Acre
km^2 (= 100 ha = 106 m^2)	Multiply by 0.387	Square mile
Volume		
Liters (= 1 dm^3)	Divide by 28.3	Cubic foot (ft^3)
m^3	Multiply by 35.4	ft^3
Liters	Divide by 3.78	U.S. gal
Liters	Divide by 4.54	Imp. gal
Liters/(s-ha)	Multiply by 0.014	Inches/hour
m^3	Divide by 3 780	Million U.S. gal (mg)
m^3	Multiply by 8.54	Barrel
m^3	Divide by 1 230	Acre-ft
Liters/s	Divide by 43.75	Million gal/day (mgd)
Liters/s	Multiply by 15.87	gal/min (gpm)
Liters/m^2	Divide by 40.76	gal/ft^2
Mass		
Gram (g)	Divide by 454	Pounds (lb)
Gram	Multiply by 15.43	Grain
Kilogram[a](kg = 1 000 g)	Multiply by 2.2	Pounds
Newton (= 0.1 kg[b])	Multiply by 0.225	Pounds
Metric ton (= 1 000 kg)	Multiply by 1.1	U.S. ton
Metric ton	Multiply by 0.98	English ton

Concentration		
Milligram per liter (mg/liter = g/m^3)	Multiply by 1.0	Parts per million (ppm)
mg/liter	Divide by 2.29	Grain/ft^3
Microgram per liter (ug/liter = 10^{-3} g/m^3)	Multiply by 1.0	Parts per billion (ppb)
Density		
kg/m^3	Divide by 16	lb/ft^3
g/m^3	Multiply by 6.24 x 10^{-5}	lb/ft^3
m^3/kg	Multiply by 16.03	ft^3/lb
Pressure		
g/m^2	Divide by 4 885	lb/ft^2
Bar (= 10^5 N/m^2)	Multiply by 14.2	psi (= lb/in.^{2})
kg/m^2	Divide by 4.89	lb/ft^2
kg/cm^2	Multiply by 14.49	psi
Energy etc.		
Watt (W = N x m/s)	Multiply by 3.41	Btu/hr
Kilowatts (kW = 1 000 W)	Multiply by 1.34	Horsepower (hp)
Kilowatt-hours (kW-hr)	Multiply by 3 409.5	Btu
W/m^3	Multiply by 5	hp/mg
kW-hr/(m^{2}x^{o}C)	Multiply by 176	Btu/ft^2/^{o}F
kW-hr/(m^{3}x^{o}C)	Multiply by 53.6	Btu/ft^3/^{o}F
Calories (gram)	Divide by 252	Btu
1 calorie = 1.16 x 10^{-6} kW-hr		
Degree celsius (^{o}C)	Multiply ^{o}C by 1.8 and add 32	Degrees Fahrenheit (^{o}F)

Some constants

1 m^3 of water weighs 1 000 kg.
1 ft^3 of water weighs 62.4 lb.
1 U.S. gal of water weighs 8.34 lb.
1 Imp. (English) gal of water weighs 10 lb.
1 day has 1 440 minutes and 86 400 seconds.

[a]Metric kilograms in this table are weight kilograms which equal 9.81 (m/s^2) x kg (mass) = 9.81 Newtons.

About the Editor

William James received his degrees in Civil Engineering from the University of Natal, South Africa; Delft Technological University, Holland; and Aberdeen University, Scotland. He started his professional career as a Provincial Water Engineer in Natal. With time out for graduate studies, he has also worked with city engineers, as a consulting engineer, and professor, at the University of Natal, McMaster University, the University of Alabama, Wayne State University, and the University of Guelph.

Reviewers

Robert Ambrose
US EPA
Ctr Exposure Assessmt Modeling
Athens, Georgia

Tom Barnwell
US EPA
Ctr Water Quality Monitoring
Athens, Georgia

Lars Bengtsson
Lund University
Water Resources Engineering
S-221 00 Lund, Sweden

Ana Deletic
University of Aberdeen
Dept of Engineering
Aberdeen, Scotland, UK

David Hansen
Tech Univ of Nova Scotia
Dept Civil Engineering
Halifax, Nova Scotia

James P. Heaney
University of Colorado
Civ Env & Arch Engineering
Boulder, Colorado

Isobel Heathcote
University of Guelph
School of Engineering
Guelph, Ontario

Wayne Huber
Oregon State University
Dept Civil Engineering
Corvallis, Oregon

Reviewers (continued)

R. Scott Huebner
U. Pennsylvania at Harrisburg
Dept Civil Engineering
Middletown, Pennsylvania

Kim Irvine
State Univ College at Buffalo
Dept Geography & Planning
Buffalo, New York

Robert Johanson
Univ of the Pacific
Dept of Engineering
Stockton, California

Douglas Joy
University of Guelph
School of Engineering
Guelph, Ontario

Jim Kells
Univ of Saskatchewan
Dept Civil Engineering
Saskatoon, Saskatchewan

Ralph Kummler
Wayne State University
Dept Chemical Engineering
Detroit, Michigan

James Li
Ryerson Polytechnic University
School of Civ & Survey Eng
Toronto, Ontario

Ivan Muzik
University of Calgary
Dept of Civil Engineering
Calgary, Alberta

Reviewers (continued)

Stephan Nix
University of Alabama
Dept of Civil Engineering
Tuscaloosa, Alabama

Robert Pitt
Univ of Alabama at Birmingham
Dept. of Civil Engineering
Birmingham, Alabama

Uzair Shamsi
State Univ College at Buffalo
Dept Geography & Planning
Buffalo, New York

Hugh Whiteley
University of Guelph
School of Engineering
Guelph, Ontario

Steve Wright
University of Michigan
Dept of Civil Engineering
Ann Arbor, Michigan

Not Pictured

John Bryan Ellis
Natl Env Research Ctr
Polaris House, N. Star Ave
Swindon, Wilts, UK

Roland Price
Hydraulics Research Wallingford
Wallingford, Oxon, UK

Authors and Affiliations

Affiliations of all the authors are given below. Their photographs follow at the end of this section - except in the few cases where a photograph was not submitted, or an author has also acted as reviewer. In the latter case, the photograph appears in the list of reviewers, immediately preceding this.

Mary Abrahms
Bureau of Environmental Services
City of Portland
1120 S.W. 5th Avenue, Room 400
Portland, Oregon 97204

Barry J. Adams, P.Eng.
Department of Civil Engineering
University of Toronto
Toronto, Ontario M5S 1A4

Matahel Ansar
Iowa Institute of Hydraulic Research
University of Iowa,
Iowa City, Iowa 52242

Les Arishenkoff, P.Eng.
Town of Markham
101 Centre Boulevard
Markham, Ontario L3R 9W3

Ken Avery, P.E.
Donald J. Bergmann & Associates, P.C.
One South Washington Street
Rochester, New York 14614

William R. Blackport
Terraqua Investigations
12 Dupont Street West,
Waterloo, Ontario N2L 2X6

Graham J. Bryant, P.Eng.
Stormceptor Canada Inc.
Westmetro Corporate Centre
195 The West Mall, Suite 405
Etobicoke, Ontario M9C 5K1

David Crawford, P.E.
CH2M HILL
825 N.E. Multnomah, Suite 1300
Portland, Oregon 97232
now: Crawford Engineering
3120 N.E. U.S. Grant Place
Portland Oregon 97212

Brett A. Cunningham, P.E.
Camp Dresser Mckee Inc.
6650 Southpoint Parkway
Jacksonville, Florida 32216

David Elrick
Department of Land Resource Science
University of Guelph
Guelph, Ontario N1G 2W1

John Fitzgibbon
School of Rural Planning and Development
University of Guelph
Guelph, Ontario N1G 2W1

Edward I. Graham, P.Eng.
Greenland Engineering Group
64 Jardin Drive
Concord, Ontario L4K 3P3

Philip Gray
XCG Consultants
1 Port Street East, Suite 201
Mississauga, Ontario L5G 4N1

Wayne Green, P.Eng.
City of Toronto, Engineering Dept.
100 Queen St. W., East Tower, 14th Flr
Toronto, Ontario M5H 2N2

Raymond T. Guther, P.Eng.
Philips Planning and Engineering Limited
P. O. Box 220, 3215 North Service Road
Burlington, Ontario L7R 3Y2

David A. Hamlet
Gannett Fleming, INC
P.O. Box 67100
Harrisburg, Pennsylvania 17106-7100

James P. Heaney *P.E.*
Dept. of Civil, Environmental, and Architectural Engineering
Center for Advanced Decision Support for Water and Environmental Systems
Campus Box 421
University of Colorado
Boulder, Colorado 80309-0421

Isobel Heathcote
School of Engineering
University of Guelph
Guelph Ontario N1G 2W1

Peter Hicks, P.Eng.
Conestoga-Rovers & Associates
651 Colby Drive
Waterloo, Ontario N2V 1C2
Tel: 519-884-0510 Fax: 519-884-0525

Christine Hill, P.Eng.
XCG Consultants
1 Port Street East, Suite 201
Mississauga, Ontario L4G 4N1

Scott W. Horsley
STORMTREAT™ Systems Inc.
110 Breed's Hill Road, #9
Hyannis Massachusetts 02601
Tel: 508-778-4449 Fax: 508-778-4596

Yinlun Huang
Department of Chemical Engineering
Wayne State University
Detroit, Michigan 48202

Richard Scott Huebner, P.E.
The Pennsylvania State University
W-209, Olmsted, 777 W. Harrisburg Pike
Middletown Pennsylvania 17057-4898

Mike Hulley, P.Eng.
XCG Consultants
1 Port Street East, Suite 201
Mississauga, Ontario L4G 4N1

Subhash C. Jain, P.E.
Iowa Institute of Hydraulic Research
University of Iowa,
Iowa City, Iowa 52242

W. Robert C. James
CHI
36 Stuart St.
Guelph Ontario N1E4S5
Tel: 519-767-0197; Fax: 519-767-2770
Email: jamesr@chi.on.ca.

William James, P.Eng.
University of Guelph
School of Engineering
Guelph Ontario N1G 2W1
Email: james@net2.eos.uoguelph.ca
Web: http://www.eos.uoguelph.ca/~james
Fax: + 519-836-2770

Seth L. Jelen, P.E.
Kurahashi & Associates, Inc.
12600 SW 72nd Avenue, Suite 100
Tigard Oregon 97223

Christopher Kresin
University of Guelph
School of Engineering
Guelph, Ontario N1G 2W1

Ralph H. Kummler
Wayne State University
Department of Chemical Engineering
Detroit, Michigan 48202

Yvette LaBombard, P.E.
Donald J. Bergmann & Associates, P.C.
One South Washington Street
Rochester, New York 14614
now: Stearns and Wheler
415 N. French Road, Suite 100
Amherst New York 14228
Tel: 716-691-8503; Fax: 716-691-8506

Alan S. Lam, P.Eng.
Greenland Engineering Group
64 Jardin Drive,
Concord Ontario L4K 3P3
Tel: 905-738-1818 Fax: 905-738-6875
e-mail: alam@grnland.com

Felix Limtiaco, P.E.
City and County of Honolulu.
Department of Wastewater Management
650 S. King St.
Honolulu, Hawaii 96813

Karina Lopez
School of Engineering
University of Guelph
Guelph, Ontario N1G 2W1

John Lyons
Ohio River Sanitation Commission
5735 Kellogg Avenue
Cincinnati, Ohio 45228

Brian W. Mack, P.E.
Camp Dresser Mckee Inc.
6650 Southpoint Parkway
Jacksonville, Florida 32216

Ed McBean, P.Eng.
Conestoga-Rovers & Associates
651 Colby Drive
Waterloo, Ontario N2V 1C2
Tel: 519-725-3313. Fax: 519-725-1394
email: emcbean@rovers.com

R. Mark Palmer, P.Eng.
C.C. Tatham and Associates Ltd.
115 Hurontario St. Suite 201,
Collingwood, Ontario L9Y 2L9
Tel: 705-444-2565 Fax 705-444-2327

Fabian Papa
University of Toronto
Department of Civil Engineering
35 St. George Street
Toronto, Ontario M5S 1A4

Bruce Polan
Conestoga-Rovers & Associates
651 Colby Drive
Waterloo, Ontario N2V 1C2
Tel: 519-725-3313 Fax: 519-725-1394

Steve Quigley, P.Eng.
Conestoga-Rovers & Associates
651 Colby Drive
Waterloo, Ontario N2V 1C2
Tel: 519-725-3313 Fax: 519-725-1394
email: squigley@rovers.com

Young-Yun Rhee
Wayne State University
Department of Chemical Engineering
Detroit, Mchigan 48202

Kristi Rowe
University of Guelph
School of Engineering
Guelph, Ontario N1G 2W1

Ronald B. Scheckenberger, P.Eng.
Philips Planning and Engineering Limited
P. O. Box 220, 3215 North Service Road
Burlington, Ontario L7R 3Y2

Michael F. Schmidt, P.E.
Camp Dresser Mckee Inc.
6650 Southpoint Parkway
Jacksonville, Florida 32216

Uzair M. Shamsi, P.E.
Chester Engineers
600 Clubhouse Drive
Pittsburgh, Pennsylvania 15108

Roger C. Sutherland, P.E.
Kurahashi & Associates, Inc.
12600 SW 72nd Avenue, Suite 100
Tigard, Oregon 97223

Michael K. Thompson, P.Eng.
Ministry of Environment & Energy
125 Resources Road, Rm E252
Etobicoke, Ontario M9P 3V6

Neil R. Thomson, P.Eng.
University of Waterloo
Department of Civil Engineering
Waterloo, Ontario N2L 3G1

Brian Verspagen
University of Guelph
School of Engineering
Guelph Ontario N1G 2W1
now: Wilson Miller Barton Peak Inc.
#200,, 3200 Bailey Lane @ Airport Rd.
Naples Florida 33942

Yiwen (Jenny) Wang
University of Guelph
School of Engineering
Guelph Ontario N1G 2W1

Hugh R. Whiteley, P.Eng.
University of Guelph
School of Engineering
Guelph, Ontario N1G 2W1

Paul Wisner, P.Eng.
Wisner Hydrology Consulting
61 St. Clair Avenue West, Ste 407
Toronto, Ontario M4V 2Y8

Leonard T. Wright, P.E.
Dept. of Civil, Environmental, and Architectural Engineering
Center for Advanced Decision Support for Water and Environmental Systems
Campus Box 421
University of Colorado
Boulder, Colorado 80309-0421

Steven J. Wright, P.E.
The University of Michigan
Dept of Civil and Environmental Engineering
113 EWRE
Ann Arbor, MI 48109-2125

Jennifer D. Xie
W2O Inc.
1 Port Street East
Mississauga Ontario L5G 1J9

Betsy Yingling
North East Ohio Regional Sewer District
3826 Euclid Avenue
Cleveland, Ohio 44115-2504

Dante Zettler
Montgomery Watson
1300 East 9th Street #2000
Cleveland, Ohio 44114

Chapter Authors

Barry Adams

Matahel Ansar

Ken Avery

William Blackport

Graham Bryant

David Elrick

John Fitzgibbon

Wayne Green

Raymond Guther

David Hamlet

Peter Hicks

Yinlun Huang

Subhash Jain

Robert James

Seth Jelen

Christopher Kresin

Chapter Authors (continued)

Yvette LaBombard

Karina Lopez

Ed McBean

R. Mark Palmer

Fabian Papa

Bruce Polan

Steve Quigley

Young-Yun Rhee

Kristi Rowe

Ron Scheckenberger

Roger Sutherland

Michael Thompson

Brian Verspagen

Jenny Wang

Paul Wisner

Leonard Wright

Index

Editor's Note: This index is basically a keyword-in-context list. It is, unfortunately, not comprehensive, nor sufficiently cross-referenced. Readers are therefore urged to be energetic and imaginative in pursuit of their topic of interest. The grouping of pages against an entry does not indicate that the subject is continuously covered throughout those pages; merely that it is mentioned on them.

Individual authors are not cited; nor are companies, except as they appear directly in the text. Authors and affiliations are listed on pages 495-498.

Symbols

A

C

D

H

N

Q

R

X

Z

Milton Keynes UK
Ingram Content Group UK Ltd.
UKHW021923071024
449327UK00022B/1700

9 780367 448356